RADIATION TRANSPORT IN SPECTRAL LINES

GEOPHYSICS AND ASTROPHYSICS MONOGRAPHS

AN INTERNATIONAL SERIES OF FUNDAMENTAL TEXTBOOKS

Editor

B. M. McCormac, *Lockheed Palo Alto Research Laboratory, Palo Alto, Calif., U.S.A.*

Editorial Board

R. Grant Athay, *High Altitude Observatory, Boulder, Colo., U.S.A.*

P. J. Coleman, Jr., *University of California, Los Angeles, Calif., U.S.A.*

D. M. Hunten, *Kitt Peak National Observatory, Tucson, Ariz., U.S.A.*

J. Kleczek, *Czechoslovak Academy of Sciences, Ondřejov, Czechoslovakia*

R. Lüst, *Institut für Extraterrestrische Physik, Garching-München, F.R.G.*

R. E. Munn, *Meteorological Service of Canada, Toronto, Ont., Canada*

Z. Švestka, *Fraunhofer Institute, Freiburg im Breisgau, F.R.G.*

G. Weill, *Institut d'Astrophysique, Paris, France*

VOLUME 1

RADIATION TRANSPORT IN SPECTRAL LINES

by

R. GRANT ATHAY

High Altitude Observatory, National Center for Atmospheric Research
(sponsored by the National Science Foundation)

D. REIDEL PUBLISHING COMPANY

DORDRECHT-HOLLAND

Library of Congress Catalog Card Number 72-188002

ISBN 90 277 0241 1

All Rights Reserved
Copyright © 1972 by D. Reidel Publishing Company, Dordrecht, Holland
No part of this book may be reproduced in any form, by print, photoprint, microfilm
or any other means, without written permission from the publisher

Printed in The Netherlands by D. Reidel, Dordrecht

To Twila

ACKNOWLEDGMENTS

The material presented in this book has resulted from an interaction with several of my colleagues and students. To each of them I express my gratitude and respect for their contributions. I am particularly indebted to A. Skumanich, who was a collaborator in much of the developmental work, and who has contributed greatly to my conceptual understanding of the transport problem. The programming and extensive computational work on which this book is based were skillfully done by W. B. Frye. Without his help the book would not have materialized. I am indebted to our CDC 6600 computer for its willingness to manipulate and invert thousands of large matrices and to Mrs Ruby Fulk for her prodigious typing ability.

PREFACE

The usual book on the theory of spectral line formation begins with an in-depth discussion of radiation transfer, including the elegant methods of obtaining analytical solutions for special cases, and of the physics of line broadening. Neither of those features will be found in this book. It is assumed that the reader is already familiar with the essentials of transport theory and of line broadening and is ready to investigate some of the particular applications of the theory to the flow of line photons through the outer layers of a star, or other tenuous media.

The main thrust of this book is toward the compilation and presentation of a vast quantity of computational material available to the author in the form of computer output. The material presented represents a highly filtered sample of the published work in this subject plus an extensive set of previously unpublished results. To present large quantities of computer output in an intelligible and efficient way is a difficult task, for which I have found no really satisfactory solution. Chapters III and IV, in particular, contain almost exclusively this type of presentation. The reader may find these chapters somewhat tedious because of the level of condensation of the material. I have tried to reach a reasonable balance between over condensation and excessive detail, which in the long run may be irrelevant.

Although this book is based largely upon numerical computations, I have tried to introduce enough discussion of the basic physics of the problems to allow the reader to relate the numerical results to the physical processes involved and to the mathematical form of the transport equations. This attempt has been only partially successful, mainly because the proper physical concepts are only now beginning to emerge. In this sense, this book must be regarded as a state-of-the-art progress report; hopefully, to be superceded in a few years by a treatise in which the ratio of physical insight to numerical results has reached a more acceptable level.

The end result of any good program of numerical solution of physical problems should be to guide the development of the basic physics of the problems to the point where the numerical solutions are no longer necessary. The problems associated with radiation transport in spectral lines have not yet reached that stage. Nevertheless, the need for a review of the current status of the problem is clearly evident at this time and it is in this context that I offer this book.

May 1971   R. Grant Athay

TABLE OF CONTENTS

ACKNOWLEDGMENTS   VII

PREFACE   IX

CHAPTER I. INTRODUCTION

1. Historical Summary   1
2. Radiative Transfer Equations   3
3. Local Thermodynamic Equilibrium   8
4. Pure Scattering   10

CHAPTER II. THE LINE SOURCE FUNCTION

1. The General Form of S_L 12
2. Steady State Equations   13
3. Specific Form of S_L 16
4. An Integral Form for S_L 18
5. The Form of Φ_ν and the Escape Probability   22
6. The Destruction Probability   24
7. Photon Random Walk, Degradation Length and Thermalization Length   27
8. Effectively Thick and Effectively Thin   31
9. The Escape Coefficient   32
10. Rate Coefficients   34
11. Free-Bound Continua   35

CHAPTER III. THE TWO-LEVEL CASE: ONE SPECTRAL LINE

1. ε, r_0, ϕ_ν, and B Constant   40
2. The Influence of a Temperature Gradient   43
3. The Influence of a Gradient in ε 48
4. The Influence of a Gradient in r_0 49
5. The Influence of a Gradient in ϕ_ν 51
6. The Influence of a Moving Atmosphere   53
7. The Influence of Frequency Redistribution   55

8. The Influence of Anisotropic Scattering — 63
9. The Influence of a Finite Atmosphere — 63
10. The Influence of Horizontal Structure — 65
   a. Periodic Structure — 67
   b. The Isolated Cylinder — 69
   c. The Imbedded Cylinder — 70

CHAPTER IV. THE MULTILEVEL CASE: TWO OR MORE LINES

1. General Comments — 79
2. Consistency Checks — 83
3. Two Levels Plus Continuum — 86
4. Upper Level Multiplets — 92
5. Lower Level Multiplets — 98
6. Metastable Levels — 100
7. Two Lines in Series — 105
8. Three Line Loops — 106
9. Four-Line Metastable Loops — 110
10. Four-Line Closed Loop — 114
11. Interlocking Effects on a Strong Line of Fixed ε, r_0 and λ — 117
12. Discussion — 119

CHAPTER V. LINE PROFILES

1. The Eddington-Barbier Relation — 121
2. Depth Dependence of S_ν Near Line Center — 124
3. Frequency Dependence of S_ν — 125
4. Mapping of S_ν into I_ν — 126
5. Microturbulence, Macroturbulence, and Differential Motion — 130
6. Line Cores — 136
7. Line Wings — 137
8. Center-Limb Effects — 141
9. Profile Synthesis — 144
10. A Standard Set of Data — 147
11. Evaluation of $\Delta\lambda_D$ from Cores of Strong Lines – One μ Position — 152
12. Evaluation of $\Delta\lambda_D$ from Cores of Strong Lines – Center-Limb Data — 156
13. Analysis of Line Wings — 164
14. Analysis of Line Shoulders — 172
15. Comments on Weak and Moderately Strong (Photospheric) Lines — 175
16. A Test for LTE Using Equivalent Widths — 176
17. Comparisons of Empirically Derived Line Source Functions to the Continuum Source Function — 179
18. Analysis and Restoration of Line Profile Data — 183

CHAPTER VI. TOTAL INTENSITIES OF LINES

1. Introduction	186
2. Curve-of-Growth for Absorption Lines	186
3. Emission Line Fluxes	190
4. The Two Level Atom	193
5. Upper Level Doublets	195
6. Lower Level Doublet and Metastable Level	195
7. Three Line Loop	196
8. Added Comments	200

CHAPTER VII. THE LINE BLANKETING EFFECT

1. Definition of Terms	202
2. Historical Summary	203
3. Mathematical Derivation of Blanketing Terms	204
4. The c Term	206
5. The t Term	209
6. Integrated Quantities	214
7. Influence on Temperature Structure	222
8. Multilevel Effects	224
9. Early Stellar Types	227
10. The Solar Case	227
11. The Cayrel Mechanism	234

CHAPTER VIII. NUMERICAL METHODS

1. Introduction	237
2. The Integral Flux-Divergence Equations	237
3. Required Frequency Bandwidth	244
4. Frequency Mapping	246
5. Free-Bound Continua	247
6. Simultaneous Solution of the Integral Flux-Divergence Equations	249
7. A Differential Equation Method	253
8. Extension to the Multilevel Case and Linearization	258
9. Core Saturation Method	258

INDEX OF SUBJECTS — 262

CHAPTER I

INTRODUCTION

1. Historical Summary

In a very real sense, our understanding of the universe around us is dependent upon our understanding of the generation and transport of radiation in gaseous media. The physics of radiation and the phenomena of radiative transport of energy have been extensively studied and many excellent books have been written on the subject. The field has by no means been exhausted, however. Major new techniques for handling transport problems are being developed, and, with them, new concepts of radiation processes are emerging.

We are here particularly concerned with the transport processes and physical phenomena giving rise to spectral lines. Spectral lines carry a wealth of information concerning the medium in which they form. The line shape and intensity are influenced by the abundance of the element producing the line, by the temperature and density structure of the atmosphere, and by the state of motion and magnetic field strength in the atmosphere. If properly interpreted, the line spectrum can be used to obtain accurate empirical models of stellar atmospheres, complete with such quantities as temperature, gas pressure, chemical composition, gravitational acceleration, state of motion, etc. Conversely, if improperly treated the line spectrum can lead to a bewildering array of incorrect information and mistaken notions about stellar atmospheres.

Spectral line studies in astronomy and astrophysics may be divided, broadly speaking and somewhat arbitrarily, into three phases, which we label: spectroscopic, algebraic, and numerical. Fraunhofer's introduction of the spectroscope ushered in a rich spectroscopic era in astrophysics during which the wavelengths, intensities, and identifications of many thousands of Fraunhofer lines were established. This basic work served as a pivotal point in astronomy for many years.

The development of quantum mechanics and statistical mechanics provided the basis for a physically meaningful methodology devoted to the analysis of line intensities and shapes. There followed a second and still richer period in astrophysics marked by the extraction of a great wealth of information about the thermodynamic and hydrodynamic structure of stellar atmospheres. This rich harvest of information was made possible by the introduction of certain physical assumptions concerning the nature of line formation, assumptions that made possible algebraic solution of the equations of radiative transfer. Interpretation of the line spectrum was then relatively straightforward and the desired information was readily extracted.

This second phase of spectral line studies in astrophysics was, and is, essentially a cream-skimming operation. It is devoted to rapid, algebraic analysis. To achieve this end, it has been necessary to suppress much of the physics of the problem in the interest of simplicity. The particular simplifying assumptions that are used, viz., pure scattering and local thermodynamic equilibrium (pure absorption), have no real justification other than the resulting algebraic simplicity. They represent, with only slight modification, idealized extremes that are rarely, if ever, completely achieved in nature.

This does not imply that the algebraic approach, even if overly simplified, does not have its purpose. Indeed, it is still a fruitful approach for certain problems and will likely continue to be so for some time to come. On the other hand, we must recognize that in this approach we have set aside most of the physics of line formation. Among other things, we have either deliberately suppressed or deliberately misrepresented the effects of collisional processes. We have, in other words, partially suppressed or misrepresented the true effects of temperature and density on the emergent radiation in the lines. It follows, that any information derived from the lines using this simplified approach that depends upon or relates to temperature and density may be quite erroneous.

When the effects of collisional processes are taken explicitly into account in line formation theory, the equation of radiative transfer becomes generally too complicated for algebraic solution. One resorts, therefore, to numerical techniques and to electronic computers. It is natural that this phase of the problem relying upon numerical solutions to the radiative transfer equation developed after the algebraic phase, and natural, also, that the numerical phase did not really develop until electronic computers became a common part of the astrophysicist's research equipment.

Because one no longer insists on the nicety of algebraic solutions, it is no longer necessary to suppress the physical processes of line formation to the same extent as characterized the algebraic phase. More of the physics can be retained without sacrificing answers. Correspondingly, the information that is derived from the lines becomes more meaningful, if not more plentiful.

This book is devoted to the physical understanding and to some of the results emerging from the numerical approach to line formation. The numerical phase of the problem is still relatively young, and much remains to be done both in applications and in the development of a suitable methodology. Much of what has been done is still viewed, perhaps, in poor perspective. New concepts are still emerging and old concepts are still evolving. Thus, in most respects, this text is a progress report.

We shall comment only briefly on the algebraic phase of line formation theory. Several excellent texts on this subject are available to the interested reader. Also, we shall not treat the topic of line broadening. This latter decision represents a strong departure from the usual approach to line formation. It is made with considerable reluctance but in conformity with a desire to avoid too lengthy a book and to concentrate on the radiative transfer aspects of line formation.

Modern terminology among astrophysicists has placed unfortunate labels on different approaches to line formation theory. The pure absorption, or LTE, approach is amenable to algebraic treatment. It prescribes the nature of the collisional effects (thermodynamic equilibrium) and excludes all other possibilities. Perhaps largely because of its relative simplicity, the LTE theory has attracted a large devoted following. For this reason there is a tendency to classify opposing theories of line formation as either LTE or non-LTE. The non-LTE nomenclature is misleading because it implies an exclusion of LTE as its basis, which is not the case. The numerical approach accounting explicitly for collisions is the least exclusive approach that has been used. It neither favors nor rejects LTE over non-LTE. It simply treats the problem in a less restricted way. To label this more general approach as non-LTE is completely misleading, therefore, and has been the cause of much misunderstanding.

The case of pure scattering, regardless of whether the scattering is coherent or non-coherent, is an extreme case of what is now popularly called non-LTE. It is the case of no collisions and radiative detailed balance. From one point of view, the assumption of LTE is equivalent to the assumption of pure collisions, i.e., to collisional detailed balance. The radiative detailed balance is still preserved in the LTE case. Hence, both pure scattering and LTE preserve detailed balance in the radiative processes. In the intermediate case where collisional processes are included, but do not dominate, it is not necessary that the radiative processes remain in detailed balance. They will, in general, be out of balance in the external layers of the line forming region. For this reason it is inappropriate to think of the general problem of line formation as being intermediate to pure scattering and LTE.

In LTE all processes are in equilibrium with their reverse process, as in strict thermodynamic equilibrium. In pure scattering, radiative equilibrium is imposed in the absence of collisions. The more general case where collisional and radiative processes combine to establish a steady state population of energy levels without presupposing that either the collisional or radiative processes are in detailed balance has not been given a satisfactory name in astrophysics. Throughout the remainder of this text we shall adopt the name *kinetic equilibrium* to denote this situation. Thus, we use kinetic equilibrium (KE) synonymously with the more common, but misleading, label non-LTE.

2. Radiative Transfer Equations

We define the following quantities to be used in this chapter:
- I_v = specific intensity of radiation
- $J_v = \int I_v \, d\omega/4\pi$ = mean intensity of radiation
- $H_v = \int \mu \, I_v \, d\omega/4\pi$ = net flux of radiation
- $\mu = \cos\theta$ (see Figure I-1)
- B_v = the Planck function
- j_v = volume emissivity
- κ_v = volume absorbtivity
- κ_0 = volume absorptivity at line center due to the line alone

$S_\nu = j_\nu/\kappa_\nu =$ the source function
$d\tau_\nu = -\kappa_\nu\, dh =$ opacity (see Figure I-1)
$\tau_0 =$ optical depth at line center due to the line alone
$\tau_c =$ continuum optical depth at the wavelength of the line
$\phi_\nu =$ line absorption profile normalized to unity at line center
$\Phi_\nu =$ line absorption profile normalized to unit area $\int \Phi_\nu\, d\nu = 1$
$r_0 = (d\tau_C/d\tau_0)$

The differential equation of radiative transfer for a plane parallel atmosphere (see

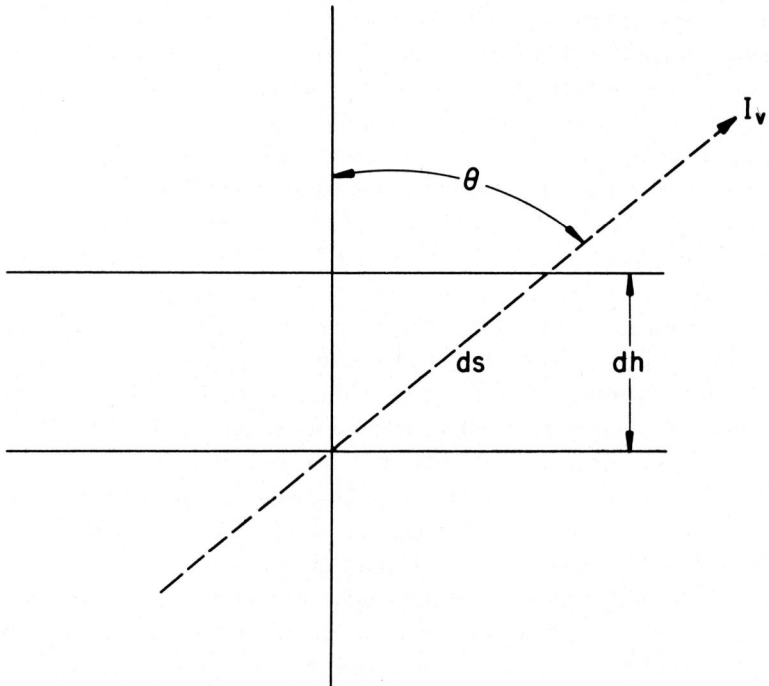

Fig. I-1. Illustration of beam of intensity I_ν in direction θ through an infinitesimal, plane-parallel slab of thickness dh.

Figure I-1) may be written

$$dI_\nu/ds = j_\nu - \kappa_\nu I_\nu. \tag{I-1}$$

Introducing $d\tau_\nu$ and S_ν, we find

$$\mu(dI_\nu/d\tau_\nu) = I_\nu - S_\nu. \tag{I-2}$$

The formal solution of Equation (I-2) is

$$I_\nu = -e^{-\tau_\nu/\mu} \int_c^{\tau_\nu} S_\nu e^{-t/\mu} \frac{dt}{\mu}, \tag{I-3}$$

where the constant c is determined by the sign of μ. For radiation traveling inward in the star,

$$I_\nu(-) = - e^{-\tau_\nu/\mu} \int_0^{\tau_\nu} S_\nu \, e^{-t/\mu} \frac{dt}{\mu}, \qquad (I-4)$$

and for radiation traveling outward

$$I_\nu(+) = e^{-\tau_\nu/\mu} \int_{\tau_\nu}^{\infty} S_\nu \, e^{-t/\mu} \frac{dt}{\mu}. \qquad (I-5)$$

Equation (I-4) is subject to the boundary condition $I_\nu(-)=0$ at $\tau_\nu=0$, and Equation (I-5) is subject to the boundary condition $S_\nu \, e^{-\tau_\nu} \to 0$ as $\tau_\nu \to \infty$ (cf., Chandrasekhar, 1960; Kourganoff, 1963).

Equation (I-5) gives for the intensity of radiation flowing out of a star at $\tau_\nu=0$

$$I_\nu(\mu, 0) = \int_0^{\infty} S_\nu \, e^{-\tau_\nu/\mu} \frac{dt_\nu}{\mu}. \qquad (I-6)$$

Thus, it is sufficient to specify S_ν as a function of τ_ν in order to compute $I_\nu(\mu, 0)$. Conversely, if enough is known about $I_\nu(\mu, 0)$ it is possible, in principle, to determine S_ν as well as the frequency dependence of τ_ν and relative values of τ_ν for different lines.

In the presence of a spectral line both j_ν and κ_ν are made up of continuum and line components. Let the continuum components be j_c and κ_c and the line components be j_L and κ_L. Equation (I-1) then becomes

$$(dI_\nu/ds) = j_c + j_L - (\kappa_c + \kappa_L) I_\nu, \qquad (I-7)$$

and, when divided by κ_0, the value of κ_L at line center, Equation (I-7) becomes

$$\mu \frac{dI_\nu}{d\tau_0} = \left(\frac{\kappa_c}{\kappa_0} + \frac{\kappa_L}{\kappa_0} \right) I_\nu - \frac{j_c}{\kappa_c} \frac{\kappa_c}{\kappa_0} - \frac{j_L}{\kappa_L} \frac{\kappa_L}{\kappa_0}$$
$$= (r_0 + \phi_\nu) I_\nu - r_0 B - \phi_\nu S_L. \qquad (I-8)$$

We have assumed in Equation (I-8) that the continuum source function, j_c/κ_c, is equal to the Planck function and we have used S_L to denote the line source function. Equation (I-8) may be rewritten in the form

$$\frac{\mu}{\phi_\nu + r_0} \frac{dI_\nu}{d\tau_0} = I_\nu - S_\nu, \qquad (I-9)$$

where

$$S_\nu = \frac{\phi_\nu}{\phi_\nu + r_0} S_L + \frac{r_0}{\phi_\nu + r_0} B. \qquad (I-10)$$

Note that even when S_L and B are independent of frequency over the width of a spectral line S_v may be frequency dependent because of the quantity ϕ_v.

The purpose of studying spectral lines of stars is to learn as much as possible about the atmosphere of the stars. The information carried by the lines is contained in the quantities S_L, ϕ_v and r_0; each are important. There are two basic ways of using spectral line data: One method, which we call spectrum synthesis, is to specify a series of model atmospheres sufficiently to compute a set of spectra $I_v(\mu, 0)$ and then to select that model atmosphere that most closely reproduces the observed spectrum. One then assumes that the model atmosphere selected is somewhat unique and that it bears a close relationship to the actual stellar atmosphere. The second way of proceeding is to use the observations themselves to determine the quantities S_L, ϕ_v, and r_0, which, in turn, are used to determine such quantities as temperature, density, velocity, etc.

Either the spectrum synthesis method or the empirical method has its own advantages and limitations. Both methods are productive and they can be used in complimentary ways. Our purpose in this text is to emphasize the spectrum synthesis method. We shall, however, comment on the empirical method, both now and in a later chapter.

To illustrate the use of the empirical method, consider a simple application to two lines of a multiplet with the following restrictions on S_L and τ_v:

$S_L \neq S_L(v)$ near line center
$S_L(1) = S_L(2)$ at a common geometrical depth
$\tau_v(1) = m\tau_v(2)$

$$S_L(1) = a + b_v(1)\, \tau_v(1) \tag{I-11}$$

$$S_L(2) = a + b_v(2)\, \tau_v(2),$$

$$= a + b_v(2)\, \frac{\tau_v(1)}{m}. \tag{I-12}$$

$\phi_v \neq \phi_v(\mu)$
$\phi_v(1) = \phi_v(2)$
$r_0(1) \ll \phi_v$ and $r_0(2) \ll \phi_v$, $\Delta v < \Delta v_1$.

With the latter restriction Equation (I-9) reduces to

$$\mu(dI_v/d\tau_v) = I_v - S_L, \quad \Delta v < \Delta v_1 \tag{I-13}$$

for both lines. It follows from Equations (I-6), (I-11), and (I-12) and the restrictions on ϕ_v that

$$I_v(1, \mu, 0) = a + b_v(1)\,\mu, \tag{I-14}$$

and

$$I_v(2, \mu, 0) = a + b_v(2)\,\mu$$
$$= a + mb_v(1)\,\mu. \tag{I-15}$$

The condition $b_\nu(2) = mb_\nu(1)$ follows from Equations (I-11) and (I-12) and the equality of $S_L(1)$ and $S_L(2)$. The requirement that $S_L \neq S_L(\nu)$ requires, in turn, that near line center $b_\nu = b\phi_\nu^{-1}$ since $\tau_\nu = \tau_0 \phi_\nu$.

Equations (I-14) and (I-15) together with observed values of $I_\nu(1, \mu, 0)$ and $I_\nu(2, \mu, 0)$ can be used in several ways. Firstly, observations of either line at two values of μ can be used to determine both a and b, hence $S_L(\tau_\nu)$. Secondly, the equations predict that $I_\nu(1, \mu, 0) = I_\nu(2, \mu, 0)$ at $\mu(1) = m\mu(2)$. Thus, by finding the values of μ that give congruent line profiles (near line center) we can determine m, or if m is known we can test the validity of the assumption $S_L(1) = S_L(2)$ at a common geometrical depth. Waddell (1962; 1963) has applied this latter test to the solar Na D lines and Mg b lines with positive results. Thirdly, Equations (I-14) and (I-15) predict that for a fixed value of μ, $I_\nu(1, \mu, 0) = I_\nu(2, \mu, 0)$ when $b_\nu(\nu_1) = mb_\nu(\nu_2)$. This latter condition gives $\phi_\nu(\nu_2) = m\phi_\nu(\nu_1)$. Thus, if m is known the frequency dependence of ϕ_ν can be derived. In particular, if the observations are made in the Doppler core of the lines $\phi_\nu = \exp - (\Delta \nu / \Delta \nu_D)^2$, where $\Delta \nu_D$ is the Doppler half width, then the two lines will have equal intensity at $\Delta \nu_1^2 - \Delta \nu_2^2 = \Delta \nu_D^2 \ln m$. If $\Delta \nu_1$, $\Delta \nu_2$, and m are known, $\Delta \nu_D$ can be determined. This technique for determining $\Delta \nu_D$ was introduced by Goldberg (1958) and has been used extensively.

Note that the use of Equations (I-14) and (I-15) as functions of μ carries the explicit restriction that $\phi_\nu \neq \phi_\nu(\mu)$, a fact that is often overlooked in empirical studies of solar Fraunhofer lines. If, for example, ϕ_ν is broadened by macroscopic motions and if these motions are anisotropic, ϕ_ν does depend upon μ and Equations (I-14) and (I-15) are invalid.

The preceding illustrations are given in order to establish that the spectral lines can be used to evaluate empirically such quantities as $S_L(\tau_\nu)$, m and $\phi_\nu(\nu)$. Each of these quantities contains valuable information on the properties of the stellar atmosphere, but the derivation of each requires explicit assumptions of the nature of those listed at the beginning of this section.

In the spectrum synthesis method of analysis one seeks a model atmosphere that predicts the proper values of $S_L(\tau_\nu)$, m and $\phi_\nu(\nu)$ (r_0 values are needed also). This method, in principle, is not limited by restrictive assumptions. In current practice, however, restrictive assumptions are still necessary. The severity of these restrictions is largely unknown, but they could perhaps be as restrictive as those used in empirical studies.

An additional difficulty of both the spectrum synthesis method and the empirical method arises when one asks questions about uniqueness. Is there one and only one model atmosphere that will produce a given spectrum or, conversely, a given set of S_ν? The answer to these questions obviously depends upon how much of the spectrum is specified and with what accuracy. Certainly if we restrict ourselves to one or two spectral lines we have no guarantee of uniqueness, particularly if the exact shapes of those lines are not known with high accuracy and if the line broadening mechanisms contained in ϕ_ν are not known accurately. Hopefully, if enough of the spectrum is specified with reasonable accuracy, say, a few percent, there will be a close correspondence of the model atmosphere to the actual atmosphere. The question of how much of the spectrum is needed to insure uniqueness to a given level of accuracy has hardly been studied. It is an important problem and deserves careful attention.

3. Local Thermodynamic Equilibrium

The purpose of determining such quantities as S_L is to learn something about the regions of the stellar atmosphere where the lines arise. The purely formal evaluation of S_L as illustrated by Equations (I-11) and (I-14) is a relatively useless exercise unless we can use S_L to provide some form of thermodynamic information.

In thermodynamic equilibrium S_L is given by the Planck function and $I_\nu(\mu, 0)$ is directly related to the temperature of the radiating medium. There are, of course, no spectral lines and no gradients in S_L in strict thermodynamic equilibrium. Thus, the integration of Equation (I-6) is trivial. It was recognized early in astrophysics that the spectral lines in the Sun could not be explained in terms of thermodynamic equilibrium models. However, much of the simplicity of thermodynamic equilibrium is retained if one assumes that 'locally', at each point, in a stellar atmosphere all processes are in thermodynamic equilibrium at the local temperature. All microscopic processes are in detailed balance. Collisional excitations between two energy levels are exactly balanced by collisional de-excitations and photon excitations (absorptions) are exactly balanced by photon de-excitations (spontaneous emissions and stimulated emissions). The conditions of strict thermodynamic equilibrium are relaxed only to the extent that temperature gradients are now permitted. The temperature gradients give rise to the spectral lines. Schwarzschild (1914) was the first to introduce the basic concept of LTE, though he referred to it as 'pure absorption.' In LTE, S_L is still given by B, the Planck function. Equations (I-11) and (I-14) give $I_\nu(\mu, 0) = S_\nu(\mu, 0)$ at $\tau_\nu = \mu$. Thus, when we look at the center of the Sun ($\mu = 1$) at any given frequency we see a radiation intensity equal to the Planck function at a temperature, T, corresponding to that at $\tau_\nu = 1$. If the temperature decreases with decreasing τ_c, $I_\nu(\mu, 0)$ will decrease toward the center of the line (increasing ϕ_ν) and toward the edge of the solar disk (decreasing μ). In both cases the intensity decrease results from the fact that the surface $\tau_\nu = \mu$ moves outward in the atmosphere, i.e., moves to a lower temperature. An absorption line, therefore, means uniquivocally that T decreases outward in the atmosphere and absorption lines are consistent with limb darkening. At the extreme solar limb ($\mu = 0$) all lines should theoretically disappear in LTE (Schwarzschild, 1914) since $\tau_\nu = \mu$ is now at $\tau_\nu = 0$ where there is no longer a temperature gradient. The Fraunhofer lines do not appear to disappear at the limb, and this fact was, for a time, used as an argument against LTE. The argument was later dropped, however, when it was realized that observations of the extreme limb were invariably contaminated by light from the disk.

The LTE assumption has other advantages in addition to the simplicity of having $S_L = B$. All energy states are populated just as they are in thermodynamic equilibrium. In particular, the relative populations, n_1 and n_2, of two energy levels are given by

$$\frac{n_1}{n_2} = \frac{\tilde{\omega}_1}{\tilde{\omega}_2} e^{\chi_{12}/kT}, \qquad (I\text{-}16)$$

where $\tilde{\omega}_1$ and $\tilde{\omega}_2$ are the statistical weights of the levels, χ_{12} is the energy difference between the levels, measured positively when level 1 has lower energy than level 2, and k is the Boltzmann constant. If level 2 is a continuum, the Saha equation gives

$$\frac{n_1}{n_c} = \left(\frac{h^2}{2\pi mkT}\right)^{3/2} n_e \frac{\omega_1}{2U} e^{\chi_{1c}/kT}, \qquad (I\text{-}17)$$

where h is Planck's constant, n_e is the electron density, U is the partition function of the ion, n_c is the ion density and χ_{1c} is the ionization potential.

In LTE the source functions of any closely spaced lines are equal at a given geometrical depth. Furthermore, the relative optical depths in different lines are related by simple formulas depending only upon temperature or upon temperature and electron density. Most of the spectrum, therefore, is redundant in that it provides no new information. For example, if T, n_e, ϕ_ν, and the abundance of an element are known as functions of τ_ν at one wavelength, the entire spectrum of the element may be predicted (assuming that atomic oscillator strengths are known).

The redundancy of data in LTE permits self-consistency checks. Most such checks are hampered by uncertainties in atomic f-values and inaccurate observations. Nevertheless, a considerable amount of self-consistency exists, and many have used evidences of self-consistency in arguments justifying the assumption of LTE. This is a highly misleading approach, however. An inconsistency between two spectral lines of order Δ does not mean that Equations (I-16) and (I-17) are accurate to order Δ. It means only that whatever inaccuracies are present are related. As a numerical illustration, consider two energy levels, 1 and 2, with ionization potentials of 6 and 3 V, respectively. Spectral lines are observed with each of these levels as ground states. Assume that the optical depths derived from these lines give relative populations, at a particular depth, corresponding to a temperature in Equation (I-16) of 4000°. It is also discovered that at this same depth the line source functions, including the 1-2 transition, correspond to Planck functions at temperatures near 4000°. Similarly, the line opacities themselves give absolute values of n_1 and n_2 that, when used with Equation (I-17), give temperatures near 4000°. Question. Is the kinetic temperature of the free particles near 4000° or is the 4000° simply a parametric description of the radiation field? The question cannot be answered without additional data. If, for example, we multiply the population of level 1, (as predicted by Equation (I-17)) by a factor of 10^2 and the population of level 2 by a factor of 35 and also change the value of T in the exponential terms in Equations (I-16) and (I-17) to 6650°, the equations are still consistent with the observations. All the consistency proves, therefore, is that the populations n_1 and n_2 inferred from observations can be described to accuracy of order Δ, with a single parameter characterized by $T \approx 4000°$ in the exponential term. It does not prove that this parameter is even approximately the same as the temperature characterizing the velocity distribution of free particles.

Literature on the LTE formulation of spectral lines is extensive and quite complete in a variety of texts and will not be reviewed here. We shall discuss in the following chapters in some detail the conditions under which LTE is valid and the conditions under which we expect it to fail. Suffice it to say here that, although it is a powerful method by virtue of its simplicity, and although it has been useful for the rapid, coarse analysis of much data, LTE is generally a poor approximation to reality.

The concepts of LTE were originally introduced in the context of the continuum spectrum, where there is much more reason to suppose that they are valid and where, in the case of the H^- continuum, it seems to be a very good approximation.

4. Pure Scattering

The first notably successful attempts to interpret the solar Fraunhofer lines were made by Schuster (1905) and Schwarzschild (1914). Schuster discussed the transmission of radiation through a 'foggy atmosphere.' He defined foggy to be synonymous with scattering and argued that all bodies are therefore foggy. Schwarzschild refined the mathematical treatment of the scattering problem and introduced the pure absorption case (LTE) as a comparison. His results showed that in the case of scattering the relative appearance of the spectrum should be unchanged from center to limb on the solar disk but that all intensities should decrease uniformly by a factor of 3 from center to limb. On this basis, he concluded that scattering is a better approximation for the solar Fraunhofer lines than is pure absorption (LTE).

Although the scattering hypothesis was initially favored over LTE, the arguments used to arrive at this conclusion were not sufficiently convincing. Eventually, the proponents of LTE dominated.

Both pure absorption and pure scattering are extreme conditions unlikely to be encountered in stellar atmospheres. In the case of pure scattering, the atmosphere is essentially passive. Photons are redirected by the scattering process but not destroyed. Collisional processes of excitation and de-excitation occur so seldom that they are of no importance.

In theory, it is possible to have an atmosphere that acts as a pure scatterer for spectral lines but as a pure absorber for the continuum. Then, since a line photon has a finite chance of being absorbed by a continuum transition it can be destroyed after undergoing a number of scatterings. This aspect of the scattering problem will be discussed in subsequent chapters. As soon as continuum absorption is permitted the problem is no longer one of pure scattering, of course.

Scattering theory was developed around the concept that the scattering is coherent in frequency, i.e., that the emitted and absorbed photons have identical frequencies. The line source function is then frequency dependent and is given by

$$S_L(v) = J_v. \qquad (I\text{-}18)$$

Much of the simplicity of LTE theory lies in the fact that S_L depends only on temperature and is independent of I_ν. In scattering almost the opposite is true, S_L depends only upon J_ν (or I_ν) and has little to do with temperature when τ_c is small.

In the scattering case, solution of the transfer equation is complicated but still tractable. Excellent discussions of the scattering problem are given by Chandrasekhar (1960) and Kourganoff (1963).

It was pointed out by Thomas (1957), and earlier by Henyey (1940) and Unno (1952a, b) that scattering in stellar atmospheres is not coherent in frequency. An atom may absorb and re-emit a photon at the same frequency in its own frame of reference. However, if the atom is moving and if the photon is emitted in a different direction than that traveled by the incident photon the emitted photon will be Doppler shifted with respect to the incident photon. In stellar atmospheres, Doppler broadening greatly predominates over natural broadening and the Doppler redistribution of emitted photons renders the scattering almost completely non-coherent in those parts of the line dominated by Doppler broadening (Thomas, 1957).

Several studies (Jefferies and White, 1960; Hummer 1965) have shown that the true scattering function is strongly non-coherent in the Doppler core of a line and becomes quite strongly coherent in the damping wings. Nevertheless, these same authors have shown that when scattering is treated as though it were completely non-coherent throughout the line serious errors are not necessarily introduced.

For non-coherent scattering,

$$S_L = \int J_\nu \Phi_\nu \, d\nu, \tag{I-19}$$

which further complicates the problem of solving the transfer equation. In both LTE and coherent scattering one frequency is independent of another so far as the transfer problem is concerned. This is not true in non-coherent scattering. All frequencies in the line are coupled and the entire ensemble of photons within the line act as a whole.

References

Chandrasekhar, S.: 1960, *Radiative Transfer*, Dover, New York.
Goldberg, L.: 1958, *Astrophys. J.* **127**, 308.
Henyey, L. G.: 1940, *Proc. Nat. Acad. Sci.* **26**, 50.
Hummer, D. G.: 1965, *Proc. Second Harvard Smithsonian Conf. on Stellar Atm.*, p. 13.
Jefferies, J. T. and White, O. R.: 1960, *Astrophys J.* **132**, 767.
Kourganoff, V.: 1963, *Basic Methods in Transfer Problems*, Dover, New York.
Schuster, A.: 1905, *Astrophys. J.* **1**, 1.
Schwarzschild, K.: 1914, in D. H. Menzel (ed.), *Selected Papers on the Transfer of Radiation*, Dover, New York, 1966.
Thomas, R. N.: 1957, *Astrophys. J.* **125**, 260.
Unno, W.: 1952a, *Publ. Astron. Soc. Japan* **3**, 158.
Unno, W.: 1952b, *Publ. Astron. Soc. Japan* **4**, 100.
Waddell, J. H. III.: 1962, *Astrophys. J.* **136**, 231.
Waddell, J. H. III.: 1963, *Astrophys. J.* **137**, 1210.

CHAPTER II

THE LINE SOURCE FUNCTION

1. The General Form of S_L

In this chapter we shall have need of the following quantities and relationships in addition to those defined in Chapter I:

$j > i$
A_{ji} = spontaneous transition probability
B_{ji} = Induced transition coefficient

$$B_{ij} = \frac{\tilde{\omega}_j}{\tilde{\omega}_i} B_{ji} \tag{II-1}$$

$$A_{ji} = \frac{2h\nu^3}{c^2} B_{ji} \tag{II-2}$$

B_{ij} = absorption coefficient
C_{ij} = collisional transition probability from i to j
R_{ij} = combined transition probability from i to j

$$P_{ij} = \tilde{\omega}_i \, e^{-\chi_i/kT} R_{ij} \tag{II-3}$$
$$\phantom{P_{ij}} = W_i R_{ij}$$
$$P_{ii} = \sum_j P_{ij}$$

χ_i = excitation energy
$C_{ij} W_i = C_{ji} W_j$
ψ_ν = profile of A_{ji} normalized to unit area.

We have defined the source function as the ratio of emissivity to absorptivity. Stimulated emissions are treated as negative absorptions, as usual, in order to retain standard forms for the source functions. The emissivity in a spectral line is

$$j_\nu = h\nu A_{ji} n_j \psi_\nu, \tag{II-4}$$

and the absorptivity is

$$\kappa_\nu = h\nu \left(B_{ij} n_i - B_{ji} n_j \right) \Phi_\nu. \tag{II-5}$$

We assume here that the profile for stimulated emission is the same as that for absorptions, viz. Φ_ν. (This point is discussed briefly in Chapter III.) Hence,

$$S_L = \frac{A_{ji}n_j\psi_\nu}{(B_{ij}n_i - B_{ji}n_j)\Phi_\nu}$$

$$= \frac{A_{ji}}{B_{ji}} \frac{1}{(\tilde{\omega}_j n_i/\tilde{\omega}_i n_j) - 1} \frac{\psi_\nu}{\Phi_\nu} \qquad \text{(II-6)}$$

$$= \frac{2h\nu^3}{c^2} \frac{1}{(\tilde{\omega}_j n_i/\tilde{\omega}_i n_j) - 1} \frac{\psi_\nu}{\Phi_\nu}.$$

If we replace n_i/n_j with its Boltzman ratio (Equationn I-16) and set $\psi_\nu \equiv \Phi_\nu$, we obtain

$$S_L = \frac{2h\nu^3}{c^2} \frac{1}{e^{h\nu/kT} - 1}, \qquad \text{(II-7)}$$

which is the Planck function. The condition $\psi_\nu \neq \Phi_\nu$ will be retained temporarily but will soon be replaced with the condition $\psi_\nu \equiv \Phi_\nu$. In adopting this latter condition, we are adopting a frequency independent S_L, except for the slow frequency dependence of the Planckian form. For a discussion of this point the reader is referred to Jefferies (1968) and to Chapter III, Section 8.

In general, n_i/n_j is not given by the Boltzmann equation, and is not a function of temperature only. Instead it will depend upon J_ν and the equation for S_L will be coupled to the transfer Equation (I-9).

2. Steady State Equations

All that we really know about the populations of atomic energy levels in stellar atmospheres is that they are relatively constant in time. Thus, for level i we may write

$$\sum_j R_{ij} n_i = \sum_j R_{ji} n_j. \qquad \text{(II-8)}$$

A similar equation may be written for each atomic energy level in a given model atom. One equation will be redundant. Upward transition rates are of the form

$$R_{ij} = C_{ij} + B_{ij} \int J_\nu \Phi_\nu \, d\nu \qquad \text{(II-9)}$$

and downward rates are of the form

$$R_{ji} = C_{ji} + B_{ji} \int J_\nu \Phi_\nu \, d\nu + A_{ji}. \qquad \text{(II-10)}$$

The set of equations, of which (II-8) is one member, can be solved for the ratio n_i/n_j, as a function of the various transition rates. For a multilevel atom, such a procedure involves tedious algebra, none of which is very instructive. As an alternative, we may solve the single Equation (II-8) for n_i/n_j. This exhibits the nature of the problem with considerably simpler algebra.

It is convenient to introduce two dimensionless parameters in Equation II-8 before proceeding further. We multiply Equation (I-17) by a factor b_j to obtain

$$n_j = \left(\frac{h^2}{2\pi mkT}\right)^{3/2} n_e n_c \frac{b_j \tilde{\omega}_j}{2U} e^{\chi_{jc}/kT}, \qquad \text{(II-11)}$$

which is the modified Saha equation. The factor b_j measures the departure of level j from its LTE population at the same values of n_e and n_c. Since n_e and n_c may both differ from their LTE value at any given temperature, b_j is not a true departure coefficient. In terms of the b_j coefficients, Equation (II-8) becomes

$$\sum_j P_{ij} b_i = \sum_j P_{ji} b_j, \qquad \text{(II-12)}$$

where P_{ij} is as defined at the beginning of this section. Also, Equation (II-6) becomes

$$S_L = \frac{2h\nu^3}{c^2} \frac{1}{(b_i/b_j) e^{h\nu/kT} - 1} \frac{\psi_\nu}{\Phi_\nu}. \qquad \text{(II-13)}$$

The second dimensionless parameter to be introduced is ϱ_{ij} defined by

$$\varrho_{ji} = \frac{A_{ji} n_j + B_{ji} n_j \int J_\nu \Phi_\nu \, d\nu - B_{ij} n_i \int J_\nu \Phi_\nu \, d\nu}{A_{ji} n_j}. \qquad \text{(II-14)}$$

This Equation reduces to

$$\varrho_{ji} = 1 - \frac{(B_{ij} n_i - B_{ji} n_j)}{A_{ji} n_j} \int J_\nu \Phi_\nu \, d\nu \qquad \text{(II-15)}$$

and we see from Equation (II-6) that for $\psi_\nu = \Phi_\nu$ it further reduces to

$$\varrho_{ji} = 1 - \frac{\int J_\nu \Phi_\nu \, d\nu}{S_L}. \qquad \text{(II-16)}$$

It is understood in Equation (II-16) that S_L and Φ_ν have the implicit indices ji.

Much will be said in this chapter and in later chapters about the meaning and use of ϱ_{ji}. For the moment we simply note that a set of the quantities ϱ_{ji} for each transition in an atom is sufficient to eliminate the quantities $\int J_\nu \Phi_\nu \, d\nu$ from Equations (II-8) and (II-12) and thus to solve these equations explicitly for the quantities n_i/n_j or b_i/b_j for given values of n_e, n_c and T. If n_i/n_j is known, S_L is known. Hence, the ϱ_{ji} quantities represent parametric solutions to the coupled radiative transfer and source function equations. Now ϱ_{ji} cannot be determined of course, without solving these coupled equations, so we cannot use the ϱ_{ji} to obtain real solutions in any way other than as convenient parameters.

Note that when the ϱ_{ji} are introduced we may write $R_{ij} = C_{ij}$ and $R_{ji} = C_{ji} + A_{ji} \varrho_{ji}$. Thus, Equation (II-12), expanded for the particular transition ji, becomes

$$C_{ij} W_i b_i + \sum_{m \neq i,j} C_{im} W_i b_i = (C_{ji} + A_{ji} \varrho_{ji}) W_j b_j + \sum_{m \neq i,j} P_{mi} b_m. \qquad \text{(II-17)}$$

The symbol $\sum_{m \neq i, j}$ denotes a sum over all m except $m=j$ and $m=i$. Equation (II-17) becomes, after it is divided by $A_{ji}W_j b_i$, and has terms rearranged,

$$\left(\frac{C_{ji}}{A_{ji}} + \varrho_{ji}\right)\frac{b_j}{b_i} = \frac{C_{ij}W_i}{A_{ji}W_j} + \frac{1}{A_{ji}W_j}\sum_{m \neq i,j}\left(C_{im}W_i - \frac{b_m}{b_i}P_{mi}\right). \quad \text{(II-18)}$$

We now make use of the equality $C_{ji}W_j = C_{ij}W_i$ and define

$$\varepsilon_{ji} = \frac{C_{ji}}{A_{ji}}, \quad \text{(II-19)}$$

and

$$\theta_{ji} = \frac{1}{A_{ji}W_j}\sum_{m \neq i,j}\left(C_{im}W_i - \frac{b_m}{b_i}P_{mi}\right). \quad \text{(II-20)}$$

With this notation, Equation (II-18) becomes

$$(\varepsilon_{ji} + \varrho_{ji})\frac{b_j}{b_i} = \varepsilon_{ji} + \theta_{ji}. \quad \text{(II-21)}$$

This equation will be of use later on.

Let us now repeat the above manipulation of Equation (II-12) where we introduce ϱ_{mi} for all transitions other than $m=j$. We then have in place of Equation (II-17)

$$\left(C_{ij} + B_{ij}\int J_\nu \Phi_\nu \, d\nu\right)W_i b_i + \sum_{m \neq i,j} C_{im}W_i b_i$$
$$= \left(C_{ji} + A_{ji} + B_{ji}\int J_\nu \Phi_\nu \, d\nu\right)W_j b_j + \sum_{m \neq i,j} P_{mi} b_m, \quad \text{(II-22)}$$

and in place of Equation (II-21)

$$\left(1 + \varepsilon_{ji} + \frac{B_{ji}}{A_{ji}}\int J_\nu \Phi_\nu \, d\nu\right)\frac{b_j}{b_i} = \varepsilon_{ji} + \frac{W_i B_{ij}}{W_j A_{ji}}\int J_\nu \Phi_\nu \, d\nu + \theta_{ji}. \quad \text{(II-23)}$$

Note that $(W_i/W_j)B_{ij} = B_{ji} e^{-h\nu/kT}$. Thus, if we define

$$\beta = e^{-h\nu/kT} \quad \text{(II-24)}$$

and

$$\eta_{ji} = (B_{ji}/A_{ji})\int J_\nu \Phi_\nu \, d\nu, \quad \text{(II-25)}$$

we obtain

$$\frac{b_j}{b_i} = \frac{\beta^{-1}\frac{B_{ji}}{A_{ji}}\int J_\nu \Phi_\nu \, d\nu + \varepsilon_{ji} + \theta_{ji}}{1 + \varepsilon_{ji} + \eta_{ji}}. \quad \text{(II-26)}$$

3. Specific Form of S_L

Equation (II-13) may be rewritten

$$S_L = \frac{2h\nu^3}{c^2} \frac{b_j}{b_i} e^{-h\nu/kT} \left(\frac{1}{1 - (b_j/b_i) e^{-h\nu/kT}} \right) \frac{\psi_\nu}{\Phi_\nu} \qquad \text{(II-27)}$$

or

$$S_L = \frac{2h\nu^3}{c^2} \frac{b_j}{b_i} \beta \left(\frac{1}{1 - (b_j/b_i)\beta} \right) \frac{\psi_\nu}{\Phi_\nu}. \qquad \text{(II-28)}$$

Solutions for b_j/b_i obtained in the previous section can be substituted into Equation (II-28) to obtain the explicit dependence of S_L upon J_ν. In order to keep S_L linear in J_ν, we follow a formulation suggested by E. H. Avrett and use Equation (II-26) for b_j/b_i in the numerator of Equation (II-28) and Equation (II-21) for the stimulated emission term in the denominator of Equation (II-28).

Note that Equation (II-28) may be written in terms of the Planck function as

$$S_L = \frac{b_j}{b_i} \frac{(1-\beta)B}{(1 - (b_j/b_i)\beta)} \frac{\psi_\nu}{\Phi_\nu}. \qquad \text{(II-29)}$$

Note, also, that

$$\frac{2h\nu^3}{c^2} \frac{B_{ji}}{A_{ji}} = 1.$$

Thus, substituting Equation (II-26) for b_j/b_i in the numerator of Equation (II-28), we obtain

$$S_L = \frac{\int J_\nu \Phi_\nu \, d\nu + (\varepsilon + \theta)(1-\beta)B}{(1+\varepsilon+\eta)(1 - (b_j/b_i)\beta)} \frac{\psi_\nu}{\Phi_\nu}. \qquad \text{(II-30)}$$

The subscripts ji are dropped for convenience, but they are implicit on all quantities.

Consider the quantity $(1+\varepsilon+\eta)(b_j/b_i)\beta$ in the denominator of Equation (II-30). We note from Equation (II-26) that this quantity is given by

$$(1+\varepsilon+\eta)(b_j/b_i)\beta = \eta + (\varepsilon+\theta)\beta,$$

and from Equation (II-21) that it is given further by

$$(1+\varepsilon+\eta)(b_j/b_i)\beta = \eta + \beta(\varepsilon+\varrho)(b_j/b_i).$$

Thus the denominator of Equation (II-30) reduces to $1+\varepsilon-\beta(\varepsilon+\varrho)(b_j/b_i)$.

We replace $\varepsilon+\theta$ in the numerator of Equation (II-30) with Equation (II-21) to obtain

$$S_L = \frac{\int J_\nu \Phi_\nu \, d\nu + \varepsilon^* B}{1 + \varepsilon^\dagger} \frac{\psi_\nu}{\Phi_\nu} \qquad \text{(II-31)}$$

where
$$\varepsilon^* = (1-\beta)(\varepsilon+\varrho)(b_j/b_i) \qquad \text{(II-32)}$$
and
$$\varepsilon^\dagger = \varepsilon - \beta(\varepsilon+\varrho)(b_j/b_i). \qquad \text{(II-33)}$$

The purpose of writing S_L in this form is to express it entirely in terms of quantities related to that specific transition. Since we have used only one equation (the lower level equation) to solve for b_j/b_i, we still have ϱ in the numerator. Had we used the full set of equations ε^* and $\varepsilon^\dagger$ would be defined differently and ε^* would not contain ϱ. If only the upper level equation is used, we obtain by a similar set of algebraic steps
$$\varepsilon^* = (1-\beta)(\chi+\varrho)(b_j/b_i) \qquad \text{(II-34)}$$
and
$$\varepsilon^\dagger = \chi - \beta(\chi+\varrho)(b_j/b_i), \qquad \text{(II-35)}$$
where
$$\chi = \varepsilon + \sum_{m \neq i, j} R_{jm}/A_{ji}. \qquad \text{(II-36)}$$

When the full set of equations are used it is difficult to write general expressions for ε^* and $\varepsilon^\dagger$ except in matrix form. However, the equation for ε^* is of the form

$$\varepsilon^*_{ji} = \left(\varepsilon_{ji} + \frac{\mathscr{P}_{ij}}{W_j A_{ji} \mathscr{P}^{ij}}\right)(1-\beta) \qquad \text{(II-37)}$$

and Equation (II-35) is still valid for $\varepsilon^\dagger$ with

$$\chi_{ji} = \varepsilon_{ji} + (\mathscr{P}_{ji}/W_j A_{ji} \mathscr{P}^{ij}). \qquad \text{(II-38)}$$

The quantities $\mathscr{P}_{ij}$ and $\mathscr{P}_{ji}$ represent all of the alternative routes from i to j and j to i excluding routes through levels i and j and excluding all redundant loops. For example, consider a five level atom, with levels numbered serially to 5, and consider the form of $\mathscr{P}_{12}$. We have

$$\begin{aligned}\mathscr{P}_{12} =\ & P_{13}P_{32}(P_{44}P_{55} - P_{45}P_{54}) \\ & + P_{13}(P_{34}P_{42}P_{55} + P_{34}P_{45}P_{52} + P_{35}P_{52}P_{44} + P_{35}P_{54}P_{42}) \\ & + P_{14}P_{42}(P_{33}P_{55} - P_{35}P_{53}) \\ & + P_{14}(P_{43}P_{32}P_{55} + P_{43}P_{35}P_{52} + P_{45}P_{52}P_{33} + P_{45}P_{53}P_{32}) \\ & + P_{15}P_{52}(P_{33}P_{44} - P_{34}P_{43}) \\ & + P_{15}(P_{53}P_{32}P_{44} + P_{53}P_{34}P_{42} + P_{54}P_{42}P_{33} + P_{54}P_{43}P_{32}). \quad \text{(II-39)}\end{aligned}$$

$\mathscr{P}_{21}$ is obtained by permuting the indices 1 and 2. Equation (II-39) can be reduced to four levels by setting $P_{55}=1$ and $P_{5j}=0$ for $j \neq 5$.

The quantity $\mathscr{P}^{ij}$ in Equations (II-37) and (II-38) represents the product of the total rates $\prod^m P_{mm}$ for $m \neq i$ or j minus any closed transition loops within the levels $m \neq i$ or j. Thus, for the five level case

$$\mathcal{P}^{12} = P_{33}P_{44}P_{55} - P_{34}P_{43}P_{55} - P_{35}P_{53}P_{44}$$
$$- P_{45}P_{54}P_{33} - P_{34}P_{45}P_{53} - P_{35}P_{54}P_{43}. \qquad \text{(II-40)}$$

Again, the four level case is obtained by setting $P_{55}=1$ and $P_{5m}=0$ for $m \neq 5$.

It may be seen from Equation (II-37) that the terms entering ε^* represent different means of getting electrons from level i to level j. Hence, these are source terms for the production of ij photons. Similarly, it may be seen from Equation (II-38) that the terms in $\varepsilon^\dagger$ are electron de-excitation terms, or, equivalently, sink terms for ij photons. When only the lower level equation is used all photon sinks other than the direct collisional sink C_{ji} are treated as negative source terms and are included in ε^*. The upper level equation formulation counts some sources as negative sinks, and the formulation using the full set of equations treats all sources and sinks as positive quantities. All three formulations are useful and will be discussed further in later sections.

In using the upper level and lower level equations, we find that ε^* depends upon ϱ, hence, upon S_L. Furthermore, b_j/b_i depends upon the full set of ϱ's for all transitions, hence, upon the full set of S_L's. Even when the full set of equations are used each of the quantities P_{mn} depends upon ϱ_{mn}. Thus, no matter how the problem is formulated, the various S_L's for a given atom are each dependent upon the remaining members of the set and, in principle, the full set must be solved for simultaneously. In practice, of course, only a few transitions interlock significantly and it is usually not necessary to solve for S_L for all possible transitions.

In the particular case of a two-level atom, i.e., when all interlocking processes and multiplet structure are ignored, there is only one line to treat, one photon source, C_{12}, and one photon sink, C_{21}. Equation (II-31) reduces, in this case, to

$$S_L = \frac{\int J_\nu \Phi_\nu \, d\nu + \varepsilon B}{1 + \varepsilon} \frac{\psi_\nu}{\Phi_\nu}. \qquad \text{(II-41)}$$

This is a much simpler problem to treat than any of the problems with interlocking transitions. It is not surprising, therefore, that the two-level problem has been much more thoroughly discussed than multilevel problems.

4. An Integral Form for S_L

Equation (II-31) for S_L is to be solved simultaneously with the radiative transfer equation, which we repeat here in the form

$$\mu \frac{1}{\phi_\nu + r_0} \frac{dI_\nu}{d\tau_0} = I_\nu - \frac{r_0}{\phi_\nu + r_0} B - \frac{\phi_\nu}{\phi_\nu + r_0} S_L. \qquad \text{(II-42)}$$

Many workers prefer to work with this differential form of the transfer equation. Techniques for solving the differential form of the transfer equation have been developed and discussed by a number of authors following an initial method developed by Feutrier (1964). For a discussion of these methods the reader is referred to recent papers by Auer and Mihalas (1968), Rybicki (1970, 1971), and Auer (1971) and to Chapter VIII.

Various integral equation techniques for solving for S_L have been developed. For a review of these methods the reader is referred to Avrett (1971) and Skumanich and Domenico (1971). The integral equation form for S_L to be discussed here follows that developed by Athay and Skumanich (1967). Our choice of this method for illustration here is purely a matter of familiarity on the part of the author. Other methods, including differential forms, are equally versatile. The particular method used by a given worker is largely a matter of taste and past experience. Those who are experienced in solving differential equations numerically will naturally prefer the differential forms and others may prefer the integral forms. Our purpose here is to concentrate on the nature of the solutions rather than the particular manner of obtaining them. Thus, we illustrate only one particular method.

In the following, we derive the integral form for S_L upon which all subsequent illustrations of S_L are based. Numerical methods for solving this form of the equation are discussed in Chapter VIII.

We wish to combine Equations (II-31) and (II-42) into a single equation for S_L. To do this we first operate on Equation (II-42) with $\int d\omega/4\pi$. This gives

$$\frac{1}{\phi_v + r_0} \frac{dH_v}{d\tau_0} = J_v - \frac{r_0}{\phi_v + r_0} B - \frac{\phi_v}{\phi_v + r_0} S_L, \tag{II-43}$$

since ϕ_v, r_0, S_L and B are all assumed to be independent of angle. Next we operate on Equation (II-43) with $\int \Phi_v \, dv$.

Note that we may write

$$S_L = S_{L0}(\psi_v/\Phi_v)$$

$$\frac{\phi_v}{\phi_v + r_0} = 1 - \frac{r_0}{\phi_v + r_0}$$

and

$$\int \Phi_v \, dv = \int \psi_v \, dv = 1.$$

Thus, the operation $\int \Phi_v \, dv$ on Equation (II-43) yields

$$\int \frac{\Phi_v}{\phi_v + r_0} \frac{dH_v}{d\tau_0} = \int J_v \Phi_v \, dv - \delta B - (1 - \delta') S_{L0}, \tag{II-44}$$

where

$$\delta = \int \frac{r_0 \Phi_v}{\phi_v + r_0} \, dv, \tag{II-45}$$

and

$$\delta' = \int \frac{r_0 \psi_\nu}{\phi_\nu + r_0} d\nu. \tag{II-45a}$$

Combining Equation (II-44) with Equation (II-31) to eliminate $\int J_\nu \Phi_\nu d\nu$ we obtain

$$S_{L0} = \frac{\varepsilon^* + \delta}{\varepsilon^\dagger + \delta'} B + \frac{1}{\varepsilon^\dagger + \delta'} \int \frac{\Phi_\nu}{\phi_\nu + r_0} \frac{dH_\nu}{d\tau_0} d\nu. \tag{II-46}$$

In the two-level case $\varepsilon^* = \varepsilon^\dagger = \varepsilon$, and we find

$$S_{L0} = \frac{\varepsilon + \delta}{\varepsilon + \delta'} B + \frac{1}{\varepsilon + \delta'} \int \frac{\Phi_\nu}{\phi_\nu + r_0} \frac{dH_\nu}{d\tau_0} d\nu. \tag{II-47}$$

The integral term in Equation (II-47) consists of a bandwidth factor $\Phi_\nu/(\phi_\nu + r_0)$ and a flux divergence factor $dH_\nu/d\tau_0$. The bandwidth factor may be written as

$$\frac{\Phi_\nu}{\phi_\nu + r_0} = M \frac{\phi_\nu}{\phi_\nu + r_0}, \tag{II-48}$$

where

$$M^{-1} = \int \phi_\nu d\nu = \sqrt{\pi} \Delta\nu_D, \quad \text{for a Doppler profile}. \tag{II-49}$$

For a Voigt profile with $a \ll 1$, the normalizing factor M^{-1} is still given to good approximation by Equation (II-49). When $r_0 \ll 1$ and ϕ_ν is Doppler, the factor $\phi_\nu/(\phi_\nu + r_0)$ acts as a unit step function of approximate width given by $\phi_\nu = r_0^{-1}$. Thus, the bandwidth is $\Delta\nu = \Delta\nu_D (\ln r_0^{-1})^{1/2}$, and even for $r_0 = 10^{-6}$, $\Delta\nu = 3.7 \Delta\nu_D$ only. On the other hand, for a Voigt profile with $a = 10^{-3}$, ϕ_ν is of the form $(a/\sqrt{\pi}(\Delta\nu/\Delta\nu_D)^2)$ in the line wings, and ϕ_ν reaches 10^{-6} at $\Delta\nu \approx 30 \Delta\nu_D$. The effective bandwidth of $\phi_\nu/(\phi_\nu + r_0)$ therefore depends critically upon both r_0 and ϕ_ν.

Equation (II-46) demonstrates that the basic physical cause of a difference between S_L and B is a non-zero flux divergence within the bandwidth of $\phi_\nu/(\phi_\nu + r_0)$. In general, the flux divergence for monochromatic radiation is not zero in a stellar atmosphere even though the total flux divergence may be zero. Thus, we expect the result $S_L \neq B$ to be a common condition rather than a rarity in stellar atmospheres. Within a particular spectral line there may be a significant flux divergence in any layer of the atmosphere from which line photons escape to external space in any part of the line. By definition such layers extend to depths near $\tau_c = 1$. However, will find that for very strong lines the flux divergence in the deeper layers often becomes of negligible importance. The same is not true for weak lines, and we will demonstrate shortly that differences between S_L and B may exist throughout the photospheres of stars.

The flux divergence term in Equation (II-46) is multiplied by factor $(\varepsilon^\dagger + \delta')^{-1}$, which may either cut-off or enhance the effect of this term. One of the distinguishing characteristics of stellar atmospheres is that ε tends to be small. A crude estimate of ε in a stellar atmosphere is obtained by adopting a collisional de-excitation cross-section of πa_0^2, a mean electron velocity of 10^8 cm s^{-1}, an electron density of 10^{12} cm^{-3} and a spontaneous transition probability of 10^8 s^{-1}. We then have

$$\varepsilon \approx \frac{\pi a_0^2 v_e n_e}{A} = \frac{10^{-16} \times 10^8 \times 10^{12}}{10^8} = 10^{-4}.$$

For strong lines, δ' is small also, and $(\varepsilon+\delta')^{-1}$ tends to be large. Thus, for strong lines we expect relatively large differences between S_L and B.

In the case of very weak lines, it is possible to solve Equation (II-47) explicitly. We note that for $r_0 \gg 1$

$$\delta = \int \frac{r_0 \Phi_\nu}{\phi_\nu + r_0} \, d\nu = \frac{r_0}{r_0} \int \Phi_\nu \, d\nu = 1. \tag{II-50}$$

By the same token $\delta' = 1$. Thus, if $\varepsilon \ll 1$, we have from Equation (II-47) and from the definition of r_0,

$$S_{L0} = B + \int \frac{\Phi_\nu}{r_0} \frac{dH_\nu}{d\tau_0},$$

$$= B + \int \Phi_\nu \frac{dH_\nu}{d\tau_c}.$$

Since the line is weak, we may set $(dH_\nu/d\tau_c) = J_c - B$, which gives

$$S_{L0} = B + \int \Phi_\nu (J_c - B) \, d\nu = J_c. \tag{II-51}$$

Hence, for weak lines with small values of ε, S_{L0} is given by the continuum mean intensity.

To the approximation that B_ν can be represented by

$$B_\nu = a_\nu + b_\nu \tau_\nu, \tag{II-52}$$

we have (cf. Kourganoff, 1963)

$$J_\nu = a_\nu [1 - \tfrac{1}{2} E_2(\tau_\nu)] + b_\nu [\tau_\nu + \tfrac{1}{2} E_3(\tau_\nu)] =$$
$$= B_\nu - \frac{a_\nu}{2} E_2(\tau_\nu) + \frac{b_\nu}{2} E_3(\tau_\nu), \tag{II-53}$$

where the E_n are exponential integrals defined by

$$E_n(\tau_\nu) = \int_1^\infty \frac{e^{-\tau_\nu \chi}}{\chi^n} \, d\chi. \tag{II-54}$$

At $\tau_\nu = 0$, $E_2 = 1$ and $E_3 = \tfrac{1}{2}$ and at $\tau_\nu = 1$, $E_2 = 0.15$ and $E_3 = 0.11$. Thus, Equations

(II-53) and (II-54) give for $\tau_v = 0$

$$S_{L0} = B_v \left[1 - \frac{1}{2} E_2(\tau_v) + \frac{1}{2} \frac{b_v}{a_v} E_3(\tau_v) \right]$$

$$= B_v \left[\frac{1}{2} + \frac{1}{4} \frac{b_v}{a_v} \right] \tag{II-55}$$

and for $\tau_v = 1$

$$S_{L0} = B_v \left[1 - \frac{1}{2} \frac{a_v E_2(\tau_v)}{a_v + b_v} + \frac{1}{2} \frac{b_v E_3(\tau_v)}{a_v + b_v} \right]$$

$$= B_v \left[1 - \frac{0.075}{1 + (b_v/a_v)} + \frac{0.055 \, b_v/a_v}{1 + b_v/a_v} \right]. \tag{II-56}$$

These equations give $S_{L0} = B_v$ at $\tau_v = 0$ for $b_v/a_v = 2$ and at $\tau_v = 1$ for $b_v/a_v = 0.075/0.055 = 1.4$. In the Sun b_v/a_v varies from about 5 at $\lambda 3500$ to about 3.8 at $\lambda 4500$, 1.5 at $\lambda 6000$ and about 1 at $\lambda 7500$. Over this same wavelength range, S_{L0}, as given by Equations (II-55) and (II-56), varies at $\tau_v = 0$ from about 1.75 B_v to about 0.75 B_v, and S_{L0} varies at $\tau_v = 1$ from about 1.034 B_v to about 1.01 B_v. Although these do not represent large departures of S_{L0} from B_v, they are nevertheless significant for some problems.

Before considering other aspects of the solutions of Equation (II-47) it is instructive to consider further the meaning of some of the parameters and to establish some of the concepts that are useful in discussing the solutions.

5. The Form of Φ_v and the Escape Probability

The function Φ_v plays an especially important role in determining the behavior of S_L. One of the primary means by which a line photon at large τ_0 is able to sense the boundary of the atmosphere is by diffusion* in frequency to a point where $\tau_v \leqslant 1$. The diffusion in frequency is controlled by Φ_v.

At the beginning of this chapter we defined the emission profile ψ_v such that the probability of a photon being emitted between y and $y + dy$ is given by $\psi_v \, dy$, where $y = \Delta v / \Delta v_D$. Thus, the probability of an emission occurring at a distance from line center greater than y_1 is given by $2 \int_{y_1}^{\infty} \psi_y \, dy$. Since we have set $\psi_y \equiv \Phi_y$, the same probability is given by $2 \int_{y_1}^{\infty} \Phi_y \, dy$. We now assume that all photons for which $\tau_y \leqslant 1$ escape the atmosphere whereas no photons for which $\tau_y > 1$ escape. Then, if we define y_1, such that (following Osterbrock, 1962)

$$\tau_0 \phi_{y_1} = 1, \tag{II-57}$$

---

* Strictly speaking the change in frequency of a line photon as a result of scattering is a Markovian process rather than a diffusion process. We shall continue to speak of 'diffusion' in frequency, however, as a matter of convenience.

the probability of photon escape, $P(e)$, is

$$P(e) = 2 \int_{y_1}^{\infty} \Phi_y \, dy. \qquad \text{(II-58)}$$

It is clear that $P(e)$ is closely related to the flux divergence $dH_y/d\tau_y$. Since the flux divergence is directly related to the difference between S_L and B, as shown by Equation (II-46), it is equally clear that the magnitude of S_L will depend upon the form of Φ_y. This will be particularly true at large values of τ_0 where $P(e)$ is small and, hence, where $P(e)$ depends critically upon the shape of Φ_y.

It is customary in texts dealing with spectral line formation to devote a substantial portion of the text to the discussion of line broadening mechanisms, which give the form of Φ_y. We omit such a discussion from this text for two reasons: firstly because the discussion would add nothing new to the literature and secondly because the discussion would substantially increase the length and expense of the text.

For a discussion of line broadening mechanisms the interested reader is referred to Cooper (1969), Jefferies (1968), Bohm (1961) and Mihalas (1970). Here, we shall simply adopt the common astrophysical form for ϕ_y. Thus, we write

$$\phi_y = \frac{a}{\pi} \int_{-\infty}^{\infty} \frac{\exp(-x^2)}{a^2 + (y-x)^2} \, dx = H_0(y) + a H_1(y) + a^2 H_2(y) + \cdots, \qquad \text{(II-59)}$$

where

$$a = (\Gamma/4\pi \, \Delta v_D). \qquad \text{(II-60)}$$

The functions $H_0, H_1, H_2 \ldots$ are tabulated in a number of places in the literature (cf., Harris, 1948; Hummer, 1965). In stellar atmospheres collisional damping is often of importance and may exceed radiation damping. Thus, Γ is the total collisional and radiation damping constant.

To a rather good approximation when $a \ll 1$, which is typical of the astrophysical applications, we may write

$$\phi_y = e^{-y^2} + \frac{a}{\sqrt{\pi} \, y^2}. \qquad \text{(II-61)}$$

It is understood that the term in a is included only for $y > 1$.

In many astrophysical problems a is of the order of 10^{-2} to 10^{-3}. Because the term e^{-y^2} varies much more rapidly with y than does the term $a/\sqrt{\pi} y^2$, there are two characteristic regions in the profile ϕ_y: a Doppler region where $e^{-y^2} \gg (a/\sqrt{\pi} y^2)$ and a wing region where the reverse inequality holds. The transition from core to wing occurs quickly since $e^{-y^2} \approx 1.8 \times 10^{-2}$ at $y=2$ and $e^{-y^2} \approx 1.2 \times 10^{-4}$ at $y=3$.

It is convenient for purposes of comparison to consider also the Lorentz profile given by

$$\phi_x = \frac{1}{\sqrt{\pi}} \frac{1}{1 + x^2}, \qquad \text{(II-62)}$$

where x is measured in units of the collision or natural width rather than the Doppler width.

The escape probabilities are readily evaluated from Equations (II-58), (II-61) and (II-62). When Φ_y is of a Doppler form $P(e)$ is given directly by the probability integral, which is tabulated. From the asymptotic conditions we find the approximate equality

$$P(e) = \frac{1}{\sqrt{\pi}} \frac{e^{-y_1^2}}{y_1}, \quad \text{Doppler},$$

and from Equation (II-57) we find

$$P(e) = \frac{1}{\sqrt{\pi}} \frac{1}{\tau_0 (\ln \tau_0)^{1/2}}, \quad \text{Doppler}. \tag{II-63}$$

For the Voigt profile we consider the case where $(a/\sqrt{\pi}y^2) \gg e^{-y^2}$. Equations (II-61) and (II-58) then give

$$P(e) = \frac{2}{\pi} \frac{a}{y_1}, \quad \text{Voigt}$$

and Equation (II-57) gives

$$P(e) = \frac{2a^{1/2}}{\pi^{3/4} \tau_0^{1/2}}, \quad \text{Voigt}. \tag{II-64}$$

Similarly, Equation (II-62) yields

$$P(e) = \frac{2}{\pi^{3/4} \tau_0^{1/2}}, \quad \text{Lorentz}, \tag{II-65}$$

for $x \gg 1$. Note that for large τ_0, $P(e)$ Doppler $\ll P(e)$ Voigt $(\tau_0 \gg a^{-1}) \ll P(e)$ Lorentz. Thus, we expect differences between S_L and B to extend to larger values of τ_0 as the profile of Φ_y becomes more nearly of the Lorentz form. For $a = 10^{-2}$ and $\tau_0 = 10^4$, for example, the three forms for $P(e)$ give approximately 10^{-5}, 10^{-3} and 10^{-2}, respectively. Note, also, that none of the expressions for $P(e)$ is valid at small τ_0 since they each yield $P(e) \to \infty$ as $\tau_0 \to 0$ rather than the correct answer $P(e) \to 1$ as $\tau_0 \to 0$.

6. The Destruction Probability

The significance of the parameters $\varepsilon^\dagger$ and δ is most readily understood in terms of the probability that a line photon will be converted back to thermal energy or will degrade to other spectral lines before it escapes the atmosphere. Consider first the parameters δ and δ'.

It is clear from the definition of δ and δ' (Equations (II-45) and (II-45a)) that the two parameters are very similar. In the particular case of detailed balance, in fact, $\psi_v \equiv \Phi_v$ and $\delta \equiv \delta'$. When a spectral line is present, however, and when collisional effects are not important the forms of ψ_ν and Φ_ν may differ. These effects are discussed in Chapter III, Section 8. For the moment we will simply assume that to a sufficiently good approximation $\psi_\nu \equiv \Phi_\nu$ and $\delta' = \delta$.

The physical significance of the parameter δ can be illustrated in the following way: A photon emitted in a line transition at frequency ν has a probability, δ_ν, of being absorbed by a continuum process. This probability is given by

$$\delta_\nu = \frac{\kappa_c}{\kappa_0 \phi_\nu + \kappa_c} = \frac{r_0}{\phi_\nu + r_0},$$

and the average of δ_ν over the profile of Φ_ν is given by

$$\frac{\int \delta_\nu \Phi_\nu \, d\nu}{\int \Phi_\nu \, d\nu} = \int \delta_\nu \Phi_\nu \, d\nu = \int \frac{\Phi_\nu r_0}{\phi_\nu + r_0} \, d\nu,$$

the latter form of which is just the definition of δ. We may interpret δ, therefore, as the average probability that a line photon will be absorbed in a continuum transition. Since continuum absorption and re-emission are not coherent, δ is a sink for line photons entirely analogous to the sinks in $\varepsilon^\dagger$. It arises through the coupling of S_L and B via the transfer equation. An entirely analogous interpretation may be given δ', of course.

The limits on δ are unity for weak lines and zero for infinitely strong lines. An approximate relationship between δ and W, the equivalent width of a line, can also be established. Note that δ can be written in the alternative form:

$$\delta = r_0 M \int \frac{\phi_\nu}{\phi_\nu + r_0} \, d\nu = r_0 M \int \left(1 - \frac{r_0}{\phi_\nu + r_0}\right) d\nu \qquad \text{(II-66)}$$

In the particular case of $r_0 \neq r_0(\tau)$ and $\phi_\nu \neq \phi_\nu(\tau)$, we may set $r_0 = \tau_c/\tau_0$ and rewrite Equation (II-66) as

$$\delta = r_0 M \int \left(1 - \frac{\tau_c}{\tau_\nu + \tau_c}\right) d\nu,$$

which may be evaluated at $\tau_c = 1$ to give

$$\delta = r_0 M \int \left(1 - \frac{1}{1 + \tau_\nu}\right) d\nu. \qquad \text{(II-67)}$$

The integral in Equation (II-67) is an approximate expression for W, derived by Menzel (1936). Thus, to a reasonably good approximation

$$\delta \approx r_0 M W. \qquad \text{(II-68)}$$

On the linear portion of the curve of growth, i.e., when $r_0 \gg 1$, δ is near unity.

On the wing portion, we may approximate ϕ_ν by $a/\sqrt{\pi}\,(\Delta\nu/\Delta\nu_D)^2$, and we have

$$\delta = r_0 \frac{a}{\sqrt{\pi}} \int \frac{1}{a/\sqrt{\pi} + r_0(\Delta\nu/\Delta\nu_D)^2}\,d\nu.$$

This integrates to

$$\delta = r_0 \frac{a}{\sqrt{\pi}} \frac{1}{((a/\sqrt{\pi})\,r_0)^{1/2}} = \left(\frac{ar_0}{\sqrt{\pi}}\right)^{1/2}. \tag{II-69}$$

Figure II-1 contains plots of δ versus r_0 for different values of a (Avrett, 1965).

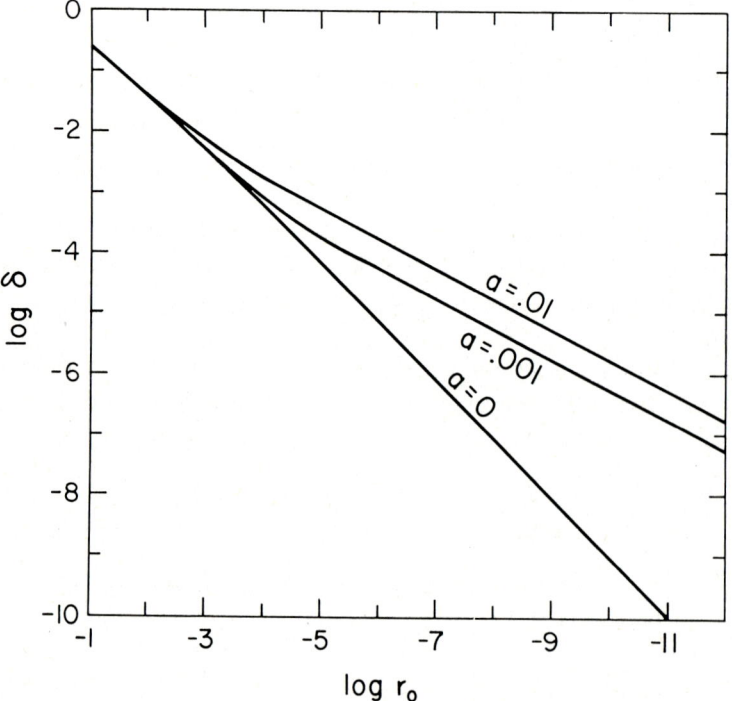

Fig. II-1. The continuum destruction probability, δ, as a function of r_0 and a (Avrett, 1965; courtesy Smithsonian Astrophysical Observatory).

To illustrate the significance of $\varepsilon^\dagger + \delta$ consider the two-level atom where $\varepsilon = \varepsilon^\dagger$. From the definition of ε, we have

$$\frac{\varepsilon_{21}}{1 + \varepsilon_{21}} = \frac{C_{21}/A_{21}}{1 + C_{21}/A_{21}} = \frac{C_{21}}{A_{21} + C_{21}}.$$

This latter form may be recognized as the probability, ignoring stimulated emission, that a photon will be destroyed by a collisional de-excitation following a line absorption event. The combined probability for line absorption at frequency v followed by collisional destruction is (dropping subscripts on ε)

$$\gamma_v = \frac{\kappa_v}{\kappa_c + \kappa_v} \frac{\varepsilon}{1 + \varepsilon},$$

and the average of this over the line is

$$\gamma = \int \frac{\Phi_v \kappa_v}{\kappa_c + \kappa_v} \frac{\varepsilon}{1 + \varepsilon} \, dv.$$

Note that

$$\frac{\kappa_v}{\kappa_c + \kappa_v} = \frac{\phi_v}{\phi_v + r_0} = 1 - \frac{r_0}{\phi_v + r_0},$$

hence that

$$\gamma = \frac{\varepsilon}{1 + \varepsilon} \int \Phi_v \left(1 - \frac{r_0}{\phi_v + r_0}\right) dv = \frac{\varepsilon}{1 + \varepsilon}(1 - \delta). \tag{II-70}$$

Since δ is the probability for destruction of line photons by continuum absorption, the total probability, $P(d)$, for a photon destruction event is

$$P(d) = \gamma + \delta = \frac{\varepsilon}{1 + \varepsilon}(1 - \delta) + \delta = \frac{\varepsilon + \delta}{1 + \varepsilon}. \tag{II-71}$$

This probability is of order unity in the case of very weak lines, $\delta = 1$, or in the case $\varepsilon \gg 1$.

Equation (II-71) provides the desired meaning of $\varepsilon + \delta$, viz., when $\varepsilon \ll 1$, $\varepsilon + \delta$ is the probability that a photon at the wavelength of the line will be destroyed following its next absorption.

The preceding interpretation based upon the two-level atom somewhat oversimplifies the problem. However, if we simply broaden the definition of ε to include interlocking processes in addition to collisions within the line the same general interpretation of $\varepsilon^\dagger + \delta$ as the probability for photon destruction is valid. It is necessary in using this interpretation of $\varepsilon^\dagger + \delta$ to require that all photon sinks within the atom are treated as true sinks.

7. Photon Random Walk, Degradation Length and Thermalization Length

Photons undergoing line scattering do not remain at a single frequency. They may, for example, scatter from frequency v into frequency v'. Since ϕ_v is frequency dependent, the opacity, and hence the mean free path to the next absorption, changes with each scattering. Nevertheless, it is convenient to define a random walk for photons.

We consider that a scattering event preserves the photon even though the photon frequency may change somewhat, and we speak of 'a photon' undergoing successive scattering, as in a random walk process. So long as a destruction event or an escape does not occur, the random walk will continue. Thus, the photon is either scattered or destroyed at the end of each step or it escapes the atmosphere.

A photon injected at optical depth τ_0, either by a creation event or by diffusion will undergo a number of scatterings, N_S, before it is either destroyed or escapes. The total average number of steps executed by a photon will be N_S+1 and will satisfy the equation

$$(N_S + 1)\left[P(e) + \{1 - P(e)\} P(d)\right] = 1. \tag{II-72}$$

For the moment we replace N_S+1 with N_S to obtain

$$N_S = P(e)^{-1}\left[1 + \frac{P(d)}{P(e)} - P(d)\right].$$

The escape probabilities given by Equations (II-63), (II-64) and (II-65) and the destruction probability for a two-level atom given by Equation (II-71) will then yield:

$$N_S = \sqrt{\pi} \tau_0 (\ln \tau_0)^{1/2} \left[1 + \frac{\varepsilon + \delta}{1 + \varepsilon} \{\sqrt{\pi} \tau_0 (\ln \tau_0)^{1/2} - 1\}\right]^{-1}, \quad \text{Doppler}, \tag{II-73}$$

$$N_S = \frac{\pi^{3/4}}{2} \frac{\tau_0^{1/2}}{a^{1/2}} \left[1 + \frac{\varepsilon + \delta}{1 + \varepsilon} \left\{\frac{\pi^{3/4}}{2} \frac{\tau_0^{1/2}}{a^{1/2}} - 1\right\}\right]^{-1}, \quad \text{Voigt}, \tag{II-74}$$

and

$$N_S = \frac{\pi^{3/4}}{2} \tau_0^{1/2} \left[1 + \frac{\varepsilon + \delta}{1 + \varepsilon} \left\{\frac{\pi^{3/4}}{2} \tau_0^{1/2} - 1\right\}\right]^{-1}, \quad \text{Lorentz}. \tag{II-75}$$

Note that Equations (II-73), (II-74) and (II-75) now have the proper asymptotic behavior at $\tau_0 = 0$, i.e., $N_S \to 0$. This is because we have compensated for the errors in $P(e)$ at small τ_0 by replacing $N_S + 1$ with N_S in Equation (II-72). Furthermore, note that Equation (II-74) applies only for $\tau_0 > a^{-1/2}$. At smaller τ_0 the Doppler form takes over.

It is evident from Equations (II-73), (II-74), and (II-75) that there is a characteristic optical depth, τ_{th}, at which $P(e)$ and $P(d)$ are equal. At greater optical depths $P(d)$ exceeds $P(e)$ and we have

$$N_S = \frac{1+\varepsilon}{\varepsilon + \delta}, \quad \tau_0 \gg \tau_{th}. \tag{II-76}$$

At smaller optical depths $P(e)$ exceeds $P(d)$ and

$$N_S = P(e)^{-1}, \quad \tau_0 \ll \tau_{th}. \tag{II-77}$$

Since the destruction probabilities ε and δ represent process which convert photon energy to thermal energy, we refer to τ_{th} as the thermalization depth. It is defined for the three profile forms by

$$\tau_{th} (\ln \tau_{th})^{1/2} = \frac{1 + \varepsilon}{\pi^{1/2}(\varepsilon + \delta)}, \quad \text{Doppler}, \tag{II-78}$$

$$\tau_{th} = \frac{4}{\pi^{3/2}} \frac{a}{(\varepsilon + \delta)^2}, \quad \text{Voigt}, \qquad (\text{II-79})$$

and

$$\tau_{th} = \frac{4}{\pi^{3/2}} \frac{1}{(\varepsilon + \delta)^2}, \quad \text{Lorentz}. \qquad (\text{II-80})$$

By definition τ_{th} is the depth below which photons are more likely to thermalize than to escape and above which they are more likely to escape. Hence, τ_{th} marks the depth at which boundary effects become important and begin to appreciably uncouple S_L and B. For $\tau_0 \ll \tau_{th}$ we cannot expect close equality between S_L and B, but for $\tau_0 \gg \tau_{th}$ we can expect relatively close equality between S_L and either J_c ($\delta \gg \varepsilon$) or B ($\varepsilon \gg \delta$).

The concept of thermalization length was introduced by Jefferies (1960) and has been of great help in understanding the behavior of S_L. Values of τ_{th} obtained for two-level atoms by computation agree well with the values given by Equations (II-78) to (II-80). More rigorous values derived by entirely different methods by Avrett and Hummer (1965) and Hummer and Stewart (1966) agree to within about 40% with the values given here. Hummer and Rybicki (1970) have considered the thermalization lengths as a function of the frequency of the original created photon.

It should be noted that in deriving the escape probabilities in Section 5 we assumed that photons escaped the atmosphere primarily by diffusion in frequency rather than by diffusion in τ_0. That this is indeed the case can be seen by comparing the results to the classical random walk results. In a classical random walk each step has the same length, which corresponds to the case of Φ_y being independent of y, i.e., where diffusion in frequency does not increase the escape probability. For the classical random walk a photon at large depth would travel a mean distance $N_S^{1/2}$. Hence, we would find

$$\tau_{th} = [(1 + \varepsilon)/(\varepsilon + \delta)]^{1/2}, \quad \text{classical}. \qquad (\text{II-81})$$

Even for the Doppler profile, therefore, τ_{th} is strongly increased over the classical value. The difference in τ_{th} is due, of course, to the effect of frequency diffusion and the results clearly support the assumption that at large τ_0 photons escape mainly by this means.

For multilevel problems where interlocking effects are important ε should be replaced by $\varepsilon^\dagger$ in the equations for N_S and τ_{th}. In this case, however, some of the interlocking processes represented in $\varepsilon^\dagger$ are nonthermal and it is no longer proper to refer to τ_{th} as the thermalization length. Instead, we replace τ_{th} with τ_{deg}, the degradation length. Because $\varepsilon^\dagger$ is always greater than ε, the degradation length is always less than the thermalization length.

The non-thermal terms in $\varepsilon^\dagger$ do not couple S_L to J_C or B directly and therefore do not necessarily tend to bring equality between S_L and B or S_L and J_C. What the non-thermal terms do represent is the coupling effects between photons in different spectral lines. The coupling effects tend to equalize the thermalization lengths in different lines and, hence, to extend the thermalization lengths in some lines and to shorten them in other lines. Large values of $\varepsilon^{\dagger}$ arising from interlocking effects usually mean that the thermalization depth in the line with large $\varepsilon^{\dagger}$ is determined by one or more of the interlocking lines and does not necessarily mean that the thermalization depth is small.

These interlocking effects can be readily visualized in terms of a close doublet with a common *upper* level. Assume, for illustration, that the two lines have equal transition probabilities and that ε is small. For both lines $\varepsilon^{\dagger}$ is then near unity. Photons degrade very quickly from one line to the other, but the degradation in no way couples the line source functions to the Planck function. The two lines in fact will thermalize together near the greater thermalization depth of the two lines considered independently. This follows from the fact that so long as $\tau_0 \ll \tau_{th}$ in either line photons in either line are more likely to escape than to thermalize, because of the strong coupling. In other words a flux divergence in either line produces a flux divergence in the companion line. These effects will be illustrated numerically in Chapter IV.

It is of interest to note from Equation (II-69) that $\delta = (a\, r_0/\sqrt{\pi})^{1/2}$ for lines with strong wings and from Equation (II-79) that for $\delta \gg \varepsilon$, $\tau_{th} = 4\pi^{-3/4}\, a\, \delta^{-2}$. Thus, for this case we find $\tau_{th} \approx 4\pi^{-1/2} r_0^{-1}$, and if we set $r_0^{-1} = \tau_{th}/\tau_c$ we find $\tau_c \approx 4\pi^{-1/2}$. It is perhaps intuitively obvious, as computations in the next chapter will verify, that for the case $\delta \gg \varepsilon$ thermalization will always occur near $\tau_c = 1$. The primary photon sink in this case is continuum absorption and this sink cannot completely dominate the transfer solution for $\tau_c < 1$, i.e., boundary effects must be present.

Alternative derivations of N_S for Doppler profiles have been given by Osterbrock (1962), Ivanov (1963), and Hummer (1964). The latter author considers the Voigt and Lorentz profiles also and obtains results in rather close agreement with those given by Equation (II-73) to (II-75). Osterbrock considers the Voigt profile but for a physical situation quite different from that assumed here. These authors define $\langle N_S \rangle$ as the ratio of the total number of emissions per unit time in a unit column between 0 and τ_0 to the total number of created photons in the same time and volume intervals. The former quantity is given by

$$\int_0^{\tau_0}\int_0^{\infty} 4\pi\kappa_\nu S_L \frac{d\nu}{h\nu}\, d\tau_0$$

and the latter is given by

$$\int_0^{\tau_0}\int_0^{\infty} 4\pi\kappa_\nu \varepsilon B \frac{d\nu}{h\nu}\, d\tau_0.$$

Since S_L, B and ε are independent of ν over the width of κ_ν, we may write

$$\langle N_s \rangle = \frac{\int_0^{\tau_0} S_L \, d\tau_0}{\int_0^{\tau_0} \varepsilon B \, d\tau_0}. \tag{II-82}$$

Note that this definition of $\langle N_s \rangle$ applies only for the case where all of the photons are generated within the layer o to τ_0 and specifically does not allow for photons diffusing into the depths o to τ_0 from greater depths. Thus, even though Equation (II-82) gives results very similar to Equations (II-73) to (II-75) the two results cannot be combined to compute $S_L(\tau_0)$.

On the other hand, it is possible to combine Equation (II-82) with Equations (II-73) to (II-75) to obtain estimates of S_L in those cases where both equations are applicable, viz., the cases where the total optical thickness of the atmosphere is $\tau_0 = 2T$ and where there is no incident radiation. In such cases, we may expect that, at $\tau_0 = T$, S_L will have a maximum and that $\int_0^T S_L \, d\tau_0$ is given approximately by $S_L T$. Then, if we take $\varepsilon B = $ constant we find from Equation (II-82) that

$$S_L/B = \varepsilon N_s. \tag{II-83}$$

Equation (II-83) combined with Equations (II-73) to (II-75) for $\tau \ll \tau_{th}$ gives

$$\frac{S_L(\max)}{B} = \sqrt{\pi \varepsilon \tau_0} (\ln \tau_0)^{1/2}, \quad \text{Doppler}, \tag{II-84}$$

$$\frac{S_L(\max)}{B} = \frac{\pi^{3/4}}{2} \varepsilon \frac{\tau_0^{1/2}}{a^{1/2}}, \quad \text{Voigt} \tag{II-85}$$

and

$$\frac{S_L(\max)}{B} = \frac{\pi^{5/4}}{2} \varepsilon \tau_0^{1/2}, \quad \text{Lorentz}. \tag{II-86}$$

Scaling laws very similar to these have been found by several authors from numerical solutions for S_L. These results will be discussed further in Chapter III.

8. Effectively Thick and Effectively Thin

If an atmosphere, or a particular layer of an atmosphere has an optical thickness less than τ_{deg}, most of the line photons generated within that region will escape without degradation from the line. Thus, if one defines effective thinness in terms of the probability of photon escapes the criterion for thinness is $\tau_0 \ll \tau_{deg}$. Conversely, if $\tau_0 \gg \tau_{deg}$ the atmosphere is effectively thick.

An atmosphere whose optical thickness is less than τ_{th} will never reach the LTE condition of $S_L = B$. This provides a second possible definition of effective thinness, viz., the condition $\tau_0 \ll \tau_{\text{th}}$. The meaning of effectively thin in this latter case is in the sense that S_L does not approach B, and this is an entirely different definition than the preceding one. In the two-level case τ_{deg} and τ_{th} are synonymous, of course, so it is not necessary to distinguish between the two definitions of effective thinness.

To avoid confusion, we adopt as the condition for effectively thin $\tau_0 \ll \tau_{\text{th}}$, i.e., we define thinness in terms of the coupling between S_L and B, which is the definition most commonly used in the literature.

The concept of degradation length has been used much less frequently than the concept of thermalization length and it is not necessary perhaps to coin a special name to describe the condition $\tau_0 \ll \tau_{\text{deg}}$. However, the reader should note that the term effectively thin does not imply that all photons generated in the effectively thin layer escape from the atmosphere in the line in which they originally appeared. They may appear in any transition with which they are interlocked.

In particular cases τ_{deg} may be orders of magnitude smaller than τ_{th}. Consider, for example, the Lyman-α and Lyman-β lines of hydrogen. As we shall later show (Chapter IV) the source functions for the two lines thermalize at the same depth. Relatively few of the Lyman-α photons are converted to Lyman-β photons because the electron excitations from level 2 to level 3 in Balmer-α proceed much more slowly than the spontaneous de-excitation transitions from levels 2 to level 1 in Lyman-α. For this reason the degradation length for Lyman-α photons is close to the thermalization length, i.e., the two-level approximation is reasonably good.

Lyman-β photons, on the other hand, decay readily into Balmer-α and Lyman-α pairs. Because the atmosphere has much lower opacity in Balmer-α than in Lyman-β the Balmer-α photons will likely escape so the decay of Lyman-β photons to Balmer-α and Lyman-α is essentially irreversible in the outer atmospheric layers where the Balmer-α line is effectively thin.

An electron in the $n = 3$ level of hydrogen has transition probabilities of $A_{31} = 5.5 \times 10^7$ and $A_{32} = 4.4 \times 10^7$. Thus, for every 2.2 absorption in Lyman-β one Lyman-β photon will decay to Balmer-α and Lyman-α and the degradation length for Lyman-β is therefore 2.2. It will have this value at all densities such that $A_{32} \gg C_{32}$, i.e., for electron densities less than about 10^{15} cm^{-3}. By contrast the thermalization length in Lyman-β may be very large because of the large thermalization length in Lyman-α. In the Sun, $\tau_{\text{th}} \approx 10^4$ for Lyman-β, and the two-level approximation is extremely poor.

The effect of the small degradation length in Lyman-β is to make the flux divergence in Lyman-β very large. This, in turn, depresses the Lyman-β source function much below that of Lyman-α, as we shall demonstrate in Chapter IV.

9. The Escape Coefficient

The quantity ϱ introduced by Equation (II-15) will be referred to extensively in the following chapters. It is useful, therefore, to give it a name and to consider its meaning in somewhat more detail. Athay and Johnson (1960) and Thomas (1960) originally called this parameter the net radiative bracket, which is descriptive of how the quantity is defined mathematically. A name conveying more of its physical significance seems desirable.

From Equations (II-16) and (II-31), we have

$$S_L = \varepsilon^* B/(\varrho + \varepsilon^\dagger), \tag{II-87}$$

hence

$$\varrho S_L = \varepsilon^* B - \varepsilon^\dagger S_L. \tag{II-88}$$

If, for the moment, we ignore continuum absorptions, Equations (II-88) and (II-46) require that

$$-\varrho S_L = \int \Phi_\nu \frac{dH_\nu}{d\tau_\nu} d\nu. \tag{II-89}$$

Thus, $-\varrho S_L$ gives the flux divergence averaged over the line. As an appropriate name, therefore, we suggest, following Giovanelli (1967) and Skumanich (1970) that ϱ be called the 'escape coefficient'.

From the defining equation for ϱ, Equation (II-15), we note that ϱ is given by one minus the ratio of the absorption rate to the emission rate in the line. This latter ratio measures the probability that the emitted photon is a scattered photon, and one minus this ratio is therefore the probability that the photon is not scattered. In other words ϱ measures the probability that a photon emitted by a spontaneous transition is a created photon resulting from collisional excitation or interlocking rather than a scattered photon.

In the case of a two-level atom,

$$\varrho = \varepsilon \left(\frac{B}{S_L} - 1 \right). \tag{II-90}$$

If we ignore stimulated emissions in B and S_L, we may write $B/S_L = b_i/b_j$. Thus,

$$\varrho = \varepsilon \left(\frac{b_i}{b_j} - 1 \right). \tag{II-91}$$

The quantity on the right side of Equation (II-91) is proportional to the net collision rate from i to j. Note that

$$C_{ij} W_i b_i - C_{ji} W_j b_j = C_{ji} W_j b_j \left(\frac{C_{ij} W_i b_i}{C_{ji} W_j b_j} - 1 \right) = C_{ji} W_j b_j \left(\frac{b_i}{b_j} - 1 \right). \tag{II-92}$$

Equation (II-91) is valid, of course, only for the two-level atom. It states simply that the net collision rate must equal the net radiative rate for a steady state condition.

Detailed balance in radiative transitions corresponds to the case $\varrho = 0$. This situation is assumed in either pure scattering or in LTE. In a kinetic equilibrium ϱ is generally not zero. It may be positive or negative, although the latter condition occurs less frequently than the former. The magnitude of ϱ is usually small. In the extreme of

$J_\nu = 0$, $\varrho = 1$, and in the opposite extreme of $S_L = \infty$, $\varrho = -\varepsilon^\dagger$. Thus, as a general rule,

$$-\varepsilon^\dagger \leq \varrho \leq 1.$$

In some strongly interlocked cases $\varepsilon^\dagger$ may be large compared to unity and ϱ may be much less than -1, i.e., $|\varrho| \gg 1$. Examples where this is true will be illustrated in Chapters IV and VII.

It is intuitively evident that the escape coefficient, ϱ, is closely related to the escape probability, $P(e)$, and hence to the mean number of scatterings. This latter relationship may be demonstrated simply by substituting in Equation (II-82) the relation

$$\varepsilon B = (\varrho + \varepsilon) S_L. \tag{II-93}$$

Thus,

$$\langle N_S \rangle = \frac{\int_0^{\tau_0} S_L \, d\tau_0}{\int_0^{\tau_0} (\varrho + \varepsilon) S_L \, d\tau_0}. \tag{II-94}$$

For $\tau_0 \ll \tau_{th}$ we expect $\varrho \gg \varepsilon$ and

$$\langle N_S \rangle = \langle \varrho \rangle^{-1}. \tag{II-95}$$

Also, for the specific case where $N_S = \langle N_S \rangle$, i.e., at the center of a finite atmosphere with $\tau \ll \tau_{th}$, we have

$$\langle N_S \rangle = P(e)^{-1}$$

and

$$P(e) = \langle \varrho \rangle.$$

When τ_0 becomes large compared to τ_{th} most of the contribution to the integral $\int (\varrho + \varepsilon) S_L \, d\tau_0$ will come from large τ_0 where $\varepsilon \gg \varrho$ and Equation (II-94) will give $\langle N_S \rangle = \langle \varepsilon \rangle^{-1}$ in conformity with our earlier conclusions. If continuum absorption is significant εB should be replaced by $(\varepsilon B + \delta J_c)$ so that $\int (\varrho + \varepsilon) S_L \, d\tau_0$ becomes $\int [(\varrho + \varepsilon) S_L + \delta J_c] \, d\tau_0$.

10. Rate Coefficients

The various rate coefficients and parameters entering the equations for S_L require a considerable knowledge of atomic cross-sections. These include collisional excitation and collisional ionization cross-sections as well as photo excitation and photo ionization cross-sections. For information on such cross-sections the reader is referred to the general literature on atomic parameters.

The largest source of information on photo excitation cross-sections (f-values) is contained in a series of monographs published by the U.S. National Bureau of Standards. Photo ionization cross-sections are relatively scarce, by comparison, as are data on collisional excitation and ionization cross-sections. Data prior to 1962 are summarized by Allen (1963).

Excitation and ionization rates for collisions with electrons are given by

$$C_{ij} = \int_{v_0}^{\infty} n_e(v) \, v Q_{ij} \, dv, \qquad (\text{II-96})$$

where Q_{ij} is the collision cross-section, usually a function of v, and $n_e(v)$ is the number of electrons having velocities between v and $v + dv$. Downward collision rates should be computed using the relation

$$C_{ij} W_i \equiv C_{ji} W_j. \qquad (\text{II-97})$$

Convenient approximate expressions for C_{ij} have been given by van Regemorter (1962).

Photoionization rates are given by

$$A_{ic} = 4\pi \int_{v_0}^{\infty} \alpha_v \frac{J_v}{h v} \, dv, \qquad (\text{II-98})$$

where α_v is the photoionization cross-section. Recombination rates may be computed from the equation

$$W_e A_{ci} = W_i A_{ic}^*, \qquad (\text{II-99})$$

where

$$A_{ic}^* = 4\pi \int_{v_0}^{\infty} \alpha_v \frac{B}{h v} \, dv, \qquad (\text{II-100})$$

with B being the Planck function at the local electron temperature.

11. Free-Bound Continua

We have said nothing at all in the preceding about the treatment of free-bound continuum radiation. In most line problems the free-bound continua are optically thin and values of $\int J_v \Phi_v \, dv$ can be computed directly from the observed radiation field of the star. There is, in this case, no problem in treating the continuum transitions since all that is needed in the steady state equations are $\int J_v \Phi_v \, dv$ and the recombination rates. The latter are readily computed if the photoionization cross-sections are known.

It sometimes happens that one needs to solve the transfer problem in the free-bound continuum. It is essential, for example, to solve the Lyman continuum transfer problem in order to obtain the degree of ionization of hydrogen in a stellar atmosphere.

The free-bound continuum transfer problem differs from the line problem only in that the frequency dependence of B cannot be neglected over the bandwidth of the continuum. Thus S_c is frequency dependent because of the frequency dependence of B.

Consider the Lyman continuum as a typical problem in free-bound continua. By a set of steps exactly analogous to those leading to Equation (II-13), we find

$$S_c = \frac{2h\nu^3}{c^2} \frac{1}{b_1 e^{h\nu/kT} - 1}. \tag{II-101}$$

In the following we neglect stimulated emissions as a matter of convenience, although it is not necessary to do so. The steady state equation for the ground state of hydrogen gives

$$b_1 W_1 \left(C_{1c} + B_{1c} \int J_\nu \Phi_\nu \, d\nu \right) = (C_{c1} + A_{c1}) W_c + \sum_{m \neq i, j} (P_{m1} b_m - b_1 P_{1m}).$$

From which, we find

$$\frac{1}{b_1} = \frac{W_1 \left(B_{1c} \int J_\nu \Phi_\nu \, d\nu + C_{1c} \right) - \sum_{m \neq i, j} (P_{m1} b_m - b_1 P_{1m})}{W_c (A_{c1} + C_{c1})},$$

or

$$\frac{1}{b_1} = \frac{(W_1 B_{1c}/W_c A_{c1}) \int J_\nu \Phi_\nu \, d\nu + \varepsilon + \theta}{1 + \varepsilon}, \tag{II-102}$$

where

$$\varepsilon = \frac{C_{c1}}{A_{c1}} = \frac{W_1 C_{1c}}{W_c A_{c1}} \tag{II-103}$$

and

$$\theta = \frac{1}{W_c A_{c1}} \sum_{m \neq i, j} \left(P_{1m} - \frac{b_m}{b_1} P_{m1} \right). \tag{II-104}$$

Note that for hydrogenic free-bound continua for which $\alpha_\nu = \alpha_0 (\nu_1/\nu)^3$

$$B_{1c} \int J_\nu \Phi_\nu \, d\nu = 4\pi \int_{\nu_1}^{\infty} \frac{\alpha_\nu J_\nu}{h\nu} d\nu$$

$$= 4\pi \frac{\alpha_0 \nu_1^3}{h} \int_{\nu_1}^{\infty} \frac{J_\nu}{\nu^4} d\nu. \tag{II-105}$$

Also, A_{C1} is given by

$$A_{C1} = \frac{W_1}{W_c} 4\pi \int_{v_1}^{\infty} \frac{\alpha_v B_v}{h\nu} dv$$

$$= \frac{W_1}{W_c} 4\pi \frac{\alpha_0 v_1^3}{h} \frac{2h}{c^2} \int_{v_1}^{\infty} \frac{e^{-h\nu/kT}}{v} dv$$

$$= \frac{W_1}{W_c} 4\pi \frac{\alpha_0 v_1^3}{h} \frac{2h}{c^2} E_1. \tag{II-106}$$

Thus, Equation (II-104) becomes

$$\frac{2h}{c^2} \frac{E_1}{b_1} = \frac{\int_{v_1}^{\infty} (J_v/v^4) \, dv + (2h/c^2) E_1 (\varepsilon + \theta)}{1 + \varepsilon}. \tag{II-107}$$

By analogy with the line problem

$$\varepsilon + \theta = \frac{1}{b_1}(\varepsilon + \varrho). \tag{II-108}$$

The transfer Equation (II-43), which we write as

$$\frac{dH_v}{d\tau_v} = J_v - S_v$$

gives

$$\int_{v_1}^{\infty} \frac{J_v}{v^4} dv = \int_{v}^{\infty} \frac{S_v}{v^4} dv + \int \frac{1}{v^4} \frac{dH_v}{d\tau_v} dv.$$

Let us ignore other sources of continuum radiation so that $S_v = B_v/b_1$. We then have

$$\int \frac{J_v}{v^4} dv = \frac{2h}{c^2} \frac{E_1}{b_1} + \int \frac{1}{v^4} \frac{dH_v}{d\tau_v} dv.$$

Upon elimination of $\int (J_v/v^4) \, dv$ from Equation (II-107), we obtain

$$\frac{2h}{c^2} \frac{1}{b_1} = \frac{\varepsilon^*}{\varepsilon^\dagger} \frac{2h}{c^2} + \frac{1}{E_1 \varepsilon^\dagger} \int \frac{1}{v^4} \frac{dH_v}{d\tau_v} dv, \tag{II-109}$$

where, for the lower level case,

$$\varepsilon^\dagger = \varepsilon \tag{II-110}$$

and

$$\varepsilon^* = \varepsilon + \theta = \frac{1}{b_1}(\varepsilon + \varrho). \tag{II-111}$$

We next multiply Equation (II-109) by $v_1^3 \, e^{-h v_1/kT}$ and define

$$S_1 = \frac{2h v_1^3}{c^2} \frac{e^{-h v_1/kT}}{b_1} \tag{II-112}$$

to obtain

$$S_1 = \frac{\varepsilon^*}{\varepsilon^\dagger} B_1 + \frac{e^{-h v_1/kT}}{E_1 \varepsilon^\dagger} v_1^3 \int \frac{1}{v^4} \frac{dH_v}{d\tau_v} dv. \tag{II-113}$$

Because of the factor v^{-4} in Equation (II-113), the frequency integration need not be carried to values of v greatly exceeding v_1. It is found in practice to be sufficient to integrate out to $v = 3v_1$. Also, because α_v is a moderate function of v it is not necessary to use a large number of frequency points. Again, it is found in practice that points at v_1, $1.25 v_1$, $1.5 v_1$ ... $3v_1$ are sufficient.

In the case of hydrogen in the solar atmosphere it is not necessary to include levels above $n = 3$. However, the Lyman lines are not in radiative detailed balance in the regions where the Lyman continuum is formed. Thus, $\varrho \neq 0$ for these lines, and it is necessary to solve the line transfer problem in conjuction with the Lyman continuum problem (Beebe and Milkey, 1972). A method for solving Equation (II-113) is outlined in Chapter VIII.

References

Allen, C. W.: 1963, *Astrophysical Quantities*, 2nd Ed., Athlone, London.
Athay, R. G. and Johnson, H. R.: 1960, *Astrophys. J.* **131**, 413.
Athay, R. G. and Skumanich, A.: 1967, *Ann. Astrophys.* **30**, 669.
Auer, L. H.: 1971, *Interdisciplinary Symposium on the Application of Transport Theory*, Oxford, Sept. 1970.
Auer, L. H. and Mihalas, D.: 1968, *Astrophys. J.* **153**, 245.
Avrett, E. H.: 1965, *Smithsonian Astrophys. Obs. Spec. Rept.* 174.
Avrett, E. H.: 1971, *Interdisciplinary Symposium on the Application of Transport Theory*, Oxford, Sept. 1970.
Avrett, E. H. and Hummer, D. G.: 1965, *Monthly Notices Roy. Astron. Soc.* **130**, 295.
Beebe, H. A. and Milkey, R. W.: 1972, *Astrophys. J. Letters* **172**, L 111.
Bohm, K. H.: 1961 in J. L. Greenstein (ed.), *Stellar Atmospheres*, Univ. of Chicago, Chicago.
Cooper, J.: 1969, in S. Geltman, K. T. Mahanthappa, and W. E. Brittin (eds.), *Lectures in Theoretical Physics*, Gordon and Breach, New York.
Feutrier, P. 1964, *Compt. Rend. Acad. Sci. Paris* **258**, 3189.
Giovanelli, R. G.: 1967, *Australian J. Phys.* **20**, 81.
Harris, D. L.: 1948, *Astrophys. J.* **108**, 112.
Hummer, D. G.: 1964, *Astrophys. J.* **140**, 276.
Hummer, D. G.: 1965, *Mem. Roy. Astron. Soc.* **70**, 1.
Hummer, D. G. and Rybicki, G. B.: 1970, *Monthly Notices Roy. Astron. Soc.* **150**, 419.
Hummer, D. G. and Stewart, J. C.: 1966, *Astrophys. J.* **146**, 290.
Ivanov, V. V.: 1963, *Soviet Astron-AJ* **7**, 199.
Jefferies, J. T.: 1960, *Astrophys. J.* **132**, 775.
Jefferies, J. T.: 1968, *Spectral Line Formation*, Blaisdell, Waltham.
Kourganoff, V.: 1963, *Basic Methods in Transfer Problems*, Dover, New York.
Menzel, D. H.: 1936, *Astrophys. J.* **84**, 462.
Mihalas, D.: 1970, *Stellar Atmospheres*, Freeman, San Francisco.
Osterbrock, D. E.: 1962, *Astrophys. J.* **135**, 195.

Rybicki, G. B.: 1970, *Extended Atmosphere Stars*, NBS Special Publication 332.
Rybicki, G. B.: 1971, *Interdisciplinary Symposium on the Applications of Transport Theory, Oxford. Sept. 1970*.
Skumanich, A.: 1970, private communication.
Skumanich, A. and Domenico, B.: 1971, *Interdisciplinary Symposium on the Applications of Transport Theory, Oxford, Sept. 1970*.
Thomas, R. N.: 1960, *Astrophys. J.* **131**, 429.
van Regemorter, H.: 1962, *Astrophys. J.* **136**, 906.

CHAPTER III

THE TWO-LEVEL CASE: ONE SPECTRAL LINE

In this chapter we illustrate solutions to Equation (II-47) for the case $\psi_\nu = \phi_\nu$ and $\delta = \delta'$. The solutions discussed are achieved using the numerical techniques outlined in Chapter VIII. We start with the simplest case where each of the parameters entering Equation (II-47) are constant with depth and where r_0 is specified. We then proceed to consider the effect of depth variations in the different parameters.

The approach followed in this chapter is largely of academic interest rather than of practical interest. It is a necessary precursor to the more practical cases considered in Chapter IV, however, if we are to understand why the source function behaves as it does.

In any real stellar atmosphere each of the parameters r_0, ε, ϕ_ν, and B vary with depth, sometimes in complex ways. Furthermore r_0 is itself dependent upon S_L since both r_0 and S_L depend explicitly on the population of the lower level of the line transition. Thus, in prescribing r_0 and in giving simple forms to $\varepsilon(\tau_c)$, $\phi_\nu(\tau_c)$, and $B(\tau_c)$ we are oversimplifying both the physics and the mathematics. On the other hand, this process helps develop better insight as to the relative importance of each of the parameters.

1. ε, r_0, ϕ_ν, and B Constant

The simplest problem to treat in line transfer problems by most techniques is the case where all parameters are constant with depth and where τ_0 is fixed relative to τ_c. It is not necessary in this case to iterate the solutions since r_0 is independent of S_L.

This problem was first discussed in its current context by Jefferies and Thomas (1958) who used three frequency points and were able to obtain analytical solutions. Unfortunately, the three point frequency quadrature does not properly allow for effects of frequency diffusion within the line and therefore does not give the proper thermalization length. It did, however, succeed in giving a proper value for S_L at $\tau_0 = 0$.

The two-level problem has been extensively discussed by Avrett (1965), Avrett and Hummer (1965), and Hummer (1968), and by others. We give here a review of some of the more important aspects of the problem.

Figure III-1a exhibits a plot of S_L/B vs. $\varepsilon\tau_0$ for the case $\delta \ll \varepsilon$ and for a Doppler profile for ϕ_ν. When plotted in this way the results for $\varepsilon \ll 1$ tend to define a more or less common curve whose slope is given by

$$\log \frac{S_L}{B} \approx \tfrac{1}{2}(\log \varepsilon\tau_0) + \text{const}.$$

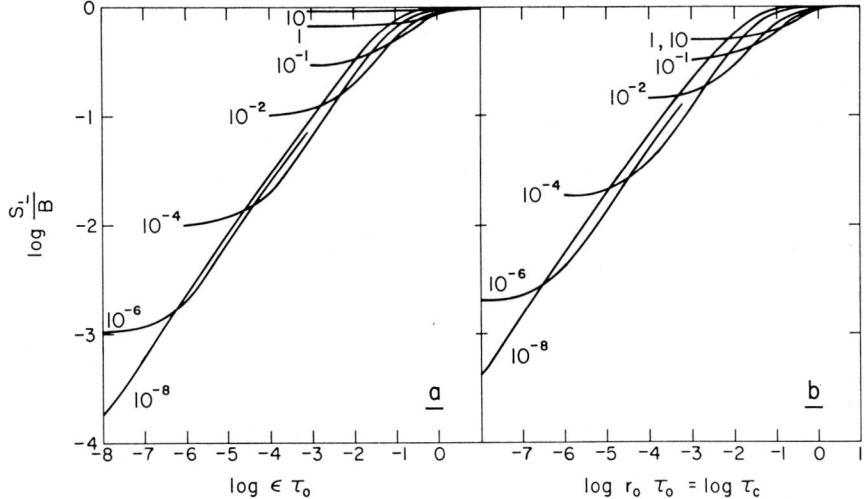

Fig. III-1. Solutions for S_L/B as a function of: (a) $-\varepsilon$ ($\varepsilon \gg r_0$) and (b) $-r_0$ ($\varepsilon \ll r_0$). All parameters are constant with depth and ϕ_ν is Gaussian.

or

$$\frac{S_L}{B} \approx \varepsilon^{1/2}(\tau_0 + 1)^{1/2}. \qquad (\text{III-1})$$

Note that this latter result is approximately $S_L/B = (\varepsilon/P(e))^{1/2}$. This approximation holds, roughly, between the limits $\tau_0 = 0$ and $\tau_0 = (10\varepsilon)^{-1}$. The surface value of S_L/B is given by $\varepsilon^{1/2}$.

Figure III-1b exhibits a similar plot of S_L/B vs. $r_0\tau_0 = \tau_c$ for the case $\delta \gg \varepsilon$ and for a Doppler profile. These results give the approximate relationship

$$\log \frac{S_L}{B} \approx \tfrac{1}{2} \log \delta_0 \tau_0 + \text{const}.$$

or

$$\frac{S_L}{B} \approx \delta^{1/2}(\tau_0 + 1)^{1/2} \approx (\delta/P(e))^{1/2}, \qquad (\text{III-2})$$

between $\tau_0 = 0$ and $\tau_c = 0.1$. Surface values of S_L/B are given by $\delta^{1/2}$. Hence, there is a complete equivalence between ε and δ for the cases considered, as is clearly indicated by Equation (II-47).

Note from Figures III-1a and III-1b that S_L approaches B for $\varepsilon \gg 1$ but that S_L does not approach B for $r_0 \gg 1$. This latter result arises from the fact that δ approaches unity as r_0 becomes large. Thus, the continuum sink can never be sufficiently strong to force $S_L = B$ in the layers where $\tau_c < 1$. In the limit $\delta = 1$, S_L approaches J_c, as shown in the preceding chapter.

The dependence of S_L/B on $\tau_0^{1/2}$ is strictly a result of the Doppler form of ϕ_ν. We note from Figures III-1a and III-1b that S_L approaches B at a depth given by

$\tau_0 \approx (\varepsilon + \delta)^{-1}$. Thus, the thermalization length is $\tau_{th} \approx (\varepsilon + \delta)^{-1}$. Note the approximate agreement between this result and the result predicted by Equation (II-78).

The thermalization depth, i.e., the depth at which $\tau_0 = \tau_{th}$, is reached for the Doppler case and for $\varepsilon > \delta$ at $\tau_0 \approx \varepsilon^{-1}$ independently of the value of τ_c. Thus, if $\varepsilon = 10^{-4}$ and $\delta = 10^{-8}$ thermalization sets in at $\tau_0 = 10^4$, which corresponds to $\tau_c = 10^{-4}$. For very strong lines, then, we expect the approximation $S_L = B$ to hold even to quite small values of τ_c, hence, through a considerable portion of the line wings.

In the case $\delta > \varepsilon$, the thermalization depth is near $\tau_c = 1$ (Figure III-1b) independently of τ_0, as was suggested in the preceding chapter for the case of lines with strong wings. In the Doppler case δ and r_0 do not differ greatly, thus $\tau_{th} \approx r_0^{-1}$ as well as δ^{-1}.

In Figures III-2a and III-2b we show the effect on S_L/B of a Voigt profile for ϕ_v. For all of the cases shown in Figure III-2a $\varepsilon \gtrsim \delta$. Thus, Figure III-2a is to be compared to Figure III-1a. For Figure III-2b, $\delta \gg \varepsilon$ and this figure is to be compared to Figure III-1b.

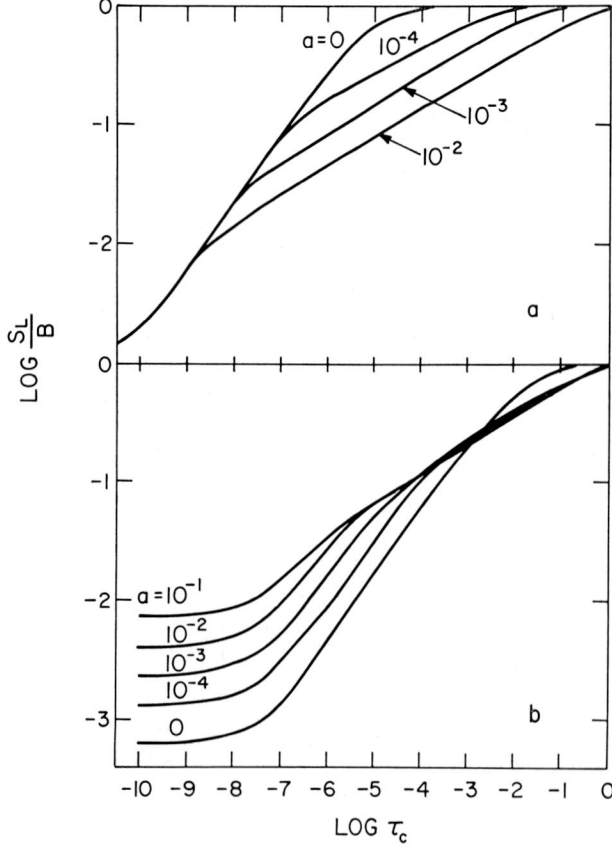

Fig. III-2. Solutions for S_L/B as a function of the Voigt parameter a for: (a) — $\varepsilon = 10^{-6}$, $r_0 = 10^{-10}$, and (b) — $\varepsilon = 10^{-8}$, $r_0 = 10^{-7}$.

We recall from Equation (II-69) that for the Voigt profile

$$\delta \approx (ar_0/\pi^{1/2})^{1/2}.$$

Thus, for a given r_0, δ increases in proportion to $a^{1/2}$. Since the surface value of S_L/B is given by $\delta^{1/2}$ when $\delta \gg \varepsilon$, the surface value of S_L/B changes in proportion to $a^{1/4}$. When $\delta \ll \varepsilon$, a has only a slight effect on the surface value of S_L/B.

At values of $\tau_c < r_0/a$, i.e., $\tau_0 < a^{-1}$, the curves in Figures III-2a and III-2b show the same general characters as the curves in Figures III-1b and III-1a except that the curves in Figure III-2b are displaced upward from each other by $S_L(1)/S_L(2) = [\delta(1)/\delta(2)]^{1/2}$.

Near $\tau_c = r_0/a$, $(\tau_0 = a^{-1})$, the slopes of the curves change. This reflects the change from radiative transfer in the Doppler core $(\tau_0 < a^{-1})$ to radiative transfer in the line wings. The average slopes of the curves in Figures III-2a and III-2b for $\tau_0 > a^{-1}$ lie between the slopes found for the pure Doppler case ($\frac{1}{2}$) and the pure dispersion case ($\frac{1}{4}$).

We note also from Figures III-2a and III-2b that the thermalization depth for the Voigt case is given by

$$\tau_{th} \approx a/(\varepsilon + \delta)^2, \qquad \text{(III-3)}$$

in approximate agreement with Equation (II-79). Since $a/\delta^2 \approx r_0^{-1}$, Equation (III-3) again gives $\tau_c \approx 1$ at $\tau_0 = \tau_{th}$ for $\delta > \varepsilon$, as was noted in the preceding discussion.

In realistic applications to stellar atmospheres line wings become strong when $r_0 < 10^{-4}$. Also, we expect ε to be of the order of 10^{-4} and a to be of the order of 10^{-3}. Hence, ε and δ may be of the same order in lines with well developed wings. Those lines with weak or moderate wings will tend to favor $\delta > \varepsilon$ and those with very strong wings will favor $\varepsilon > \delta$. Interlocking effects increase the effective ε and tend to favor $\varepsilon > \delta$.

2. The Influence of a Temperature Gradient

In the photospheric layers of stars where most of the spectral lines form the temperature gradient is such that B_ν is approximately of the form

$$B_\nu = B_\nu(0)(1 + \beta_\nu \tau_c). \qquad \text{(III-4)}$$

Since this expression gives (see Chapter V)

$$I_\nu = B_\nu(0)(1 + \beta_\nu \mu) \qquad \text{(III-5)}$$

when integrated in the transfer equation, β_ν is called the limb-darkening coefficient. In the Sun, β_ν varies from about 4.5 at the violet limit of the visual spectrum to about 1 at the red limit.

It follows from Equation (III-4) and the fact that $\beta_\nu \leqslant 4.5$ through the visual spectrum that there is relatively little variation in B for $\tau_c \lesssim 10^{-1}$. For this reason we do not expect temperature gradients of the form given by Equation (II-4) to have much effect on $S_L/B(0)$ at small values of τ_c when ε and r_0 are constant. Numerical calculations bear this out. An exception occurs when ε or r_0 is varying rapidly in the layers where B is changing. This will be illustrated in Section 4 where we discuss the effects of gradients in r_0.

There is, however, an interesting consequence near $\tau_c = 1$ of a temperature gradient of the photospheric form. For B_ν of the form (Equation III-4), $J_c(0)$ is given by

$$J_c(0) = \tfrac{1}{2} B_\nu(0) \left(1 + \frac{\beta_\nu}{2}\right). \tag{III-6}$$

Thus $J_c(0)$ varies from about $0.75 \, B_\nu(0)$ in the red to about $1.7 \, B_\nu(0)$ in the violet. Since weak lines tend to the limit $S_L = J_c$, they approach the condition

$$\frac{S_L(0)}{B_\nu(0)} = \frac{1}{2}\left(1 + \frac{\beta_\nu}{2}\right). \tag{III-7}$$

Thus $S_L(0)/B_\nu(0)$ for the weak lines exceeds unity in the violet by a factor of about 1.7 and falls below unity in the red by a factor 0.75. At optical depths just below $\tau_c = 1$, S_L is closer to B_ν but departs from B_ν in the same direction. These effects are illustrated in Figure III-3b where we show the effects on lines of different r_0 for $\beta = 4.5$, $\varepsilon = 10^{-4}$, and $a = 10^{-3}$. Note that in this plot $\delta > \varepsilon$ for $r_0 \leqslant 10^{-4}$ and $\delta < \varepsilon$ for $r_0 \leqslant 10^{-6}$. The effects shown therefore are a mixture of those shown in preceding figures.

Near $\tau_c = 1$, B_ν, as given by Equation (III-4), is changing rapidly with τ_c in the violet and much less rapidly in the red. As noted in Equation (III-6), this differential gradient in B_ν increases J_c above B_ν in the violet and depresses J_c below B_ν in the red. For strict radiative equilibrium in the continuum, the integral $\int J_c \, d\nu$ must equal $\int B_\nu \, d\nu$, but this is an average condition that does not hold monochromatically.

Deep in an atmosphere, near $\tau_c = 1$, the lines tend to merge with the continuum and thus to approach J_c. We expect, therefore, that, in the presence of a temperature gradient, the behavior of S_L near $\tau_c = 1$ will depend markedly on wavelength. Such an effect is illustrated in Figure III-3a for the case $\varepsilon = 10^{-4}$, $r_0 = 10^{-4}$, and $a = 10^{-2}$. Under these conditions $\delta \gg \varepsilon$. Although the effects shown in Figure III-3a are of small amplitude relative to the overall behavior of S_L, they are nevertheless of profound importance in some applications, particularly in the line blanketing effect to be discussed in Chapter VII.

It is to be expected, also, that the behavior of S_L near $\tau_c = 1$ will be influenced by the values of ε and δ. If ε is large, S_L will lie closer to B, and if ε is small, S_L will lie closer to J_c. The effect of δ is more complicated as is shown by Figure III-3b. For large values of r_0, $\delta \approx 1$, S_L tends to follow J_c closely. As r_0 decreases, S_L first moves close to $B(r_0 = 10^{-2})$, then back towards $J_c(r_0 = 10^{-4})$, then, finally, back to $B(r_0 = 10^{-8})$. This behavior of S_L is evidently related to the saturation of the Doppler core occurring mainly between $r_0 = 10^{-1}$ and 10^{-4} and the development of the line wings for $r_0 < 10^{-4}$.

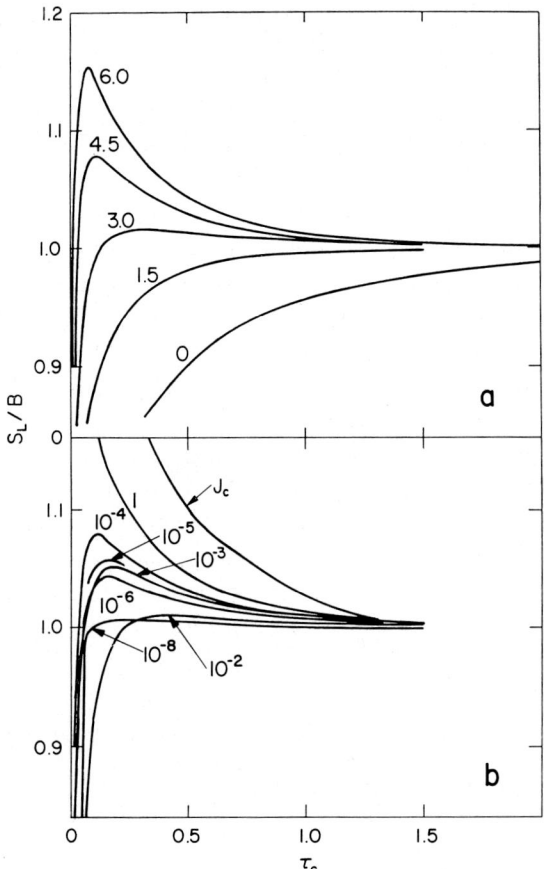

Fig. III-3. The behavior of S_L/B near $\tau_c = 1$ as a function of: (a) $-\beta$, where $B = 1 + \beta\tau_c$, and (b) $-r_0$, with $\beta = 4.5$. In part a, $\varepsilon = 10^{-4}$, $r_0 = 10^{-4}$ and $a = 10^{-2}$ and in part b, $\varepsilon = 10^{-4}$ and $a = 10^{-2}$.

A second type of temperature structure of interest in astrophysics is the case of a temperature rise in the outer part of the atmosphere, i.e., a chromosphere. To represent this temperature structure we adopt as a form for B_ν

$$B_\nu = B_\nu(0)\left(1 + \beta_\nu\tau_c + A_\nu e^{-C\tau_c}\right). \tag{III-8}$$

A_ν provides a measure of the magnitude of the temperature rise and C provides a measure of its location, i.e., of the chromospheric thickness. A rise in T from 4600 to 6500° near $\tau_c = 10^{-4}$ would give $A_\nu \approx 10$ and $C \approx 10^4$ at $\lambda 4000$.

Several studies of the effect of temperature models giving values of B_ν similar to Equation (III-8) have been made in relation to the H and K lines of Ca II. Jefferies and Thomas (1960) first noted that the H_2 and K_2 emission reversals in the H and K lines could be explained, qualitatively, by such a model if A_ν and C were properly chosen. Later workers such as Avrett and Hummer (1965), Dumont (1967), Athay and Skumanich (1968a,b,c), Linsky (1968), Linsky and Avrett (1970) Athay (1970), and Beebe (1971) investigated the qualitative aspects of the problem in some detail. Most of these later studies work with models in which values of T are assigned at each τ or height point rather than the simple functional form given by Equation (III-8). For the sake of parameterizing the nature of the effects, however, it is useful to retain B_ν in the form given by Equation (III-8).

A chromospheric temperature rise may be manifested by a corresponding increase in S_L or by no change in S_L depending upon the relationship between the optical depth where the temperature rise occurs and the thermalization depth. The thickness of the chromosphere in τ_0 is given by $(Cr_0)^{-1}$. If $(Cr_0)^{-1} \ll \tau_{th}$, S_L will show only weak effects from the rise in B, but if $(Cr_0)^{-1} > \tau_{th}$ strong effects will occur. Since C is defined with respect to τ_c, $(Cr_0)^{-1}$ may be much greater than τ_{th} for certain strong lines and much less than τ_{th} for weaker lines.

When there is significant coupling between S_L and B at $\tau_0 \approx (Cr_0)^{-1}$, S_L will tend to exhibit a maximum somewhere near $(Cr_0)^{-1}$. A maximum occurs because S_L first rises in an attempt to follow B_ν then decreases again at smaller τ_c as the coupling between S_L and B diminishes, i.e., as τ_0 becomes smaller with respect to τ_{th}. The amplitude of the maximum in S_L will depend basically upon four parameters: A_ν, C, r_0, and τ_{th}. For the solar H and K lines r_0 and δ are both small compared to ε.

TABLE III-I

Scaling Laws for S_L for B_ν given by Equation (III-9) with $B(0) = 1$

ϕ_ν	Doppler ($a < \varepsilon$)	Voigt ($1 > a > \varepsilon$)	Dispersion (Lorentz) $a > 1$
τ_{th}	ε^{-1}	$a\varepsilon^{-2}$	ε^{-2}
S_L depth	$S_L(0), \ S_L(\max)$	$S_L(0), \ S_L(\max)$	$S_L(0), \ S_L(\max)$
thin $(Cr_0)^{-1} < 1$		$\varepsilon^{1/2}, \ \varepsilon A$	
Effectively Thin $\tau_{th} > (Cr_0)^{-1} > 1$	$\dfrac{\varepsilon A}{(Cr_0)^{1/2}}, \ \dfrac{\varepsilon A}{Cr_0}$	$\begin{cases} \dfrac{Cr_0}{a} > 1 \\ \dfrac{\varepsilon A}{(Cr_0)^{1/2}}, \ \dfrac{\varepsilon A}{Cr_0} \\ \dfrac{Cr_0}{a} < 1 \\ \dfrac{\varepsilon A}{(Cr_0 a)^{1/4}}, \ \dfrac{\varepsilon A}{(Cr_0 a)^{1/2}} \end{cases}$	$\dfrac{\varepsilon A}{(Cr_0)^{1/4}}, \ \dfrac{\varepsilon A}{(Cr_0)^{1/2}}$
Effectively Thick $(Cr_0)^{-1} > \tau_{th}$	$\varepsilon^{1/2} A, \ A$	$\varepsilon^{1/2} A, \ A$	$\varepsilon^{1/2} A, \ A$
$S_L(\tau_0)$ 0 to max	$\sim \tau_0^{1/2}$	$\sim \tau_0^{1/2}, \ \tau_0 < a^{-1}$ $\sim \tau_0^{1/4}, \ \tau_0 > a^{-1}$	$\sim \tau_0^{1/4}$

Hence τ_{th} will be given by a/ε^2, and both a and ε will appear as parameters influencing the maximum in S_L.

Skumanich (1967) has derived approximate scaling laws to exhibit the effect of the various parameters on S_L. His results are summarized in Table III-1 for three types of profiles. The Voigt profile is of most interest for astrophysical applications. Note the similarity of these scaling laws to those given by Equations (II-83) to (II-85).

The solar chromosphere is effectively thin in the Ca II H and K lines. Most of those who have studied the H and K lines conclude that $(Cr_0)^{-1} > a^{-1}$, but some (Engvold, 1967; Zirker, 1968) have argued that $(Cr_0)^{-1} < a^{-1}$. For the case $(Cr_0)^{-1} > a^{-1}$, we note from Table III-1 that any combination of parameters that gives the same value for the quantity $\varepsilon A (Cr_0 a)^{-1/2}$ will give essentially the same maximum value of S_L.

Skumanich's scaling laws are derived for constant Doppler width, i.e., constant ϕ_v. Depth variations of ϕ_v alter these laws somewhat. More importantly, depth

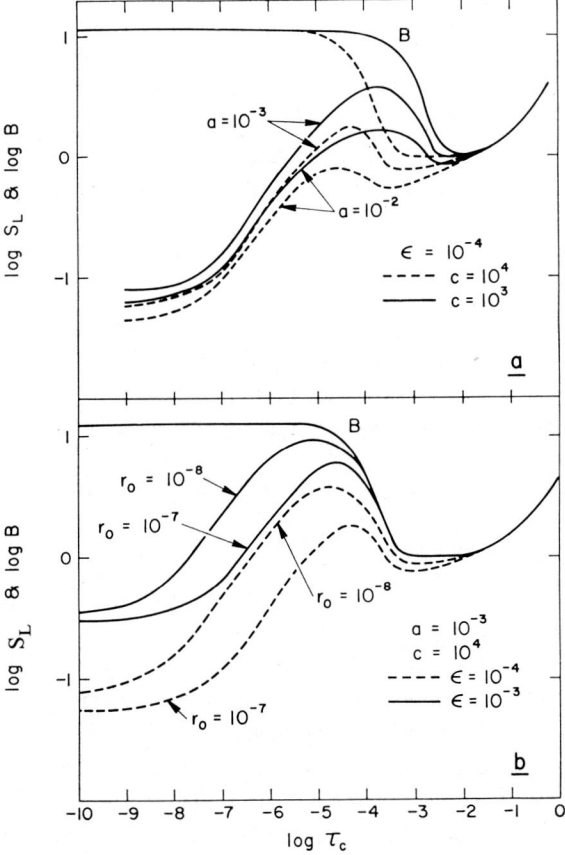

Fig. III-4. The effect of a chromosphere on S_L for: (a) – different values of chromospheric thickness and different values of a, and (b) – different values of ε and different values of r_0.

variations of ϕ_v change the form of the observed line profile for a given $S_L(\tau_0)$. Thus, the scaling laws should not be applied to the observed profile without regard for the fact that I_v and S_L do not necessarily behave in the same way. More will be said about this point in Chapter V.

Figures III-4a and III-4b exhibit the behavior of S_L for two different values of C and a, and for two different values of ε and r_0. Note that for the cases shown in these two figures $(Cr_0/a) \leqslant 1$.

We note at this point that the rule that the surface value of S_L is given by $(\varepsilon+\delta)^{1/2} B$, must be used with some caution. The relevant value of B is the one toward which S_L is saturating, not B at $\tau=0$. Thus, it is the value of B at about a thermalization depth in the atmosphere that fixes the surface value of S_L.

3. The Influence of a Gradient in ε

Within the boundary layer $\tau_0 < \tau_{th}$ the proportionality between S_L and τ_0 is fixed mainly by the form of ϕ_v. Hummer (1964) has shown that the most likely process of photon escape from layers below $\tau_0 = 1$ is by a single scattering event with a large shift in Δv. In other words, a photon undergoes many scatterings near line center without moving very far then, by chance, is scattered from near line center into the wings where it makes a single large step. It is the shape of ϕ_v that determines the probability for a photon to scatter from line center into the wings, so it is not too surprising to find that the functional dependence of S_L on τ_0 is fixed primarily by the form of ϕ_v.

Surface values of S_L, on the other hand, are fixed by ε and δ and depend upon ϕ_v only through the dependence of δ on ϕ_v. However, the thermalization depth, τ_{th}, depends upon ϕ_v explicitly as well as upon ε and δ, as was shown in Chapter II.

It follows from the above considerations that the effect of a gradient in ε will be manifested in the surface values of S_L and in τ_{th}. Both the surface value of S_L and τ_{th} depend upon the mean number of steps in the photon random walk, which is given by ε^{-1} when $\varepsilon > \delta$.

Avrett (1965) investigated the effects of changing ε according to the relation

$$\varepsilon = 10^{-3}(1 - 0.99\, e^{-k\tau_0}). \qquad (III-9)$$

His results are shown in Figure III-5 together with the results for $\varepsilon = 10^{-3}$ and $\varepsilon = 10^{-5}$. The curves for variable ε are very similar to curves with constant ε and lie intermediate to the curves $\varepsilon = 10^{-3}$ and 10^{-5}. Since ε is proportional to electron density and is essentially independent of temperature, steep gradients in ε are not likely to occur in most astrophysical problems.

The effects shown in Figure III-5 will be somewhat different if there is a gradient in B in the regions where ε is changing. However, the effect of a gradient in ε is always to pick some average value of ε intermediate to the extreme ranges. A gradient of B coupled with a gradient in ε will change the effective average of ε, but will not change the basic character of the S_L/B curve.

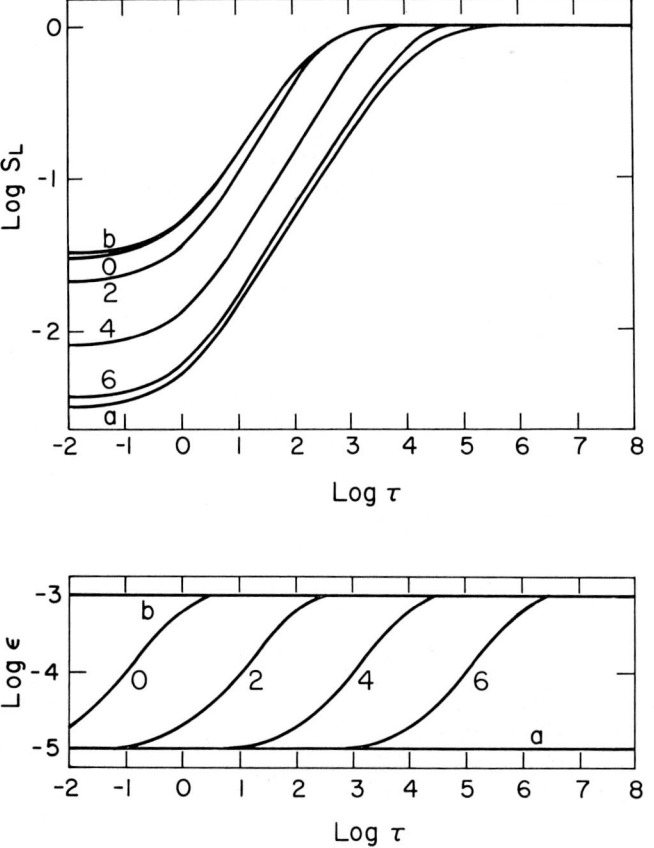

Fig. III-5. The effect on S_L of a gradient in ε (Avrett, 1965; courtesy Smithsonian Institution Astrophysical Observatory).

4. The Influence of a Gradient in r_0

Gradients in r_0 produce changes in S_L that are in some respects analogous to the effects of gradients in ε and in other respects very different. At small τ_0 the primary effect should come from the effective change in δ and produce effects analogous to those produced by a gradient in ε. However, we note from Figures III-3a and III-3b that the behavior of S_L near $\tau_c = 1$ depends sensitively upon both r_0 and upon the gradient in B. Thus, the effect upon S_L of a gradient in r_0 will depend upon the gradient in B, particularly near $\tau_c = 1$.

Temperature gradients in stellar atmospheres lead to changing conditions of excitation and ionization, which will produce gradients in r_0. These gradients can be very large in some cases and small in others.

Expected variations in r_0 for different lines in a particular model atmosphere are too diverse to attempt any systematic illustration of all of the expected effects. If

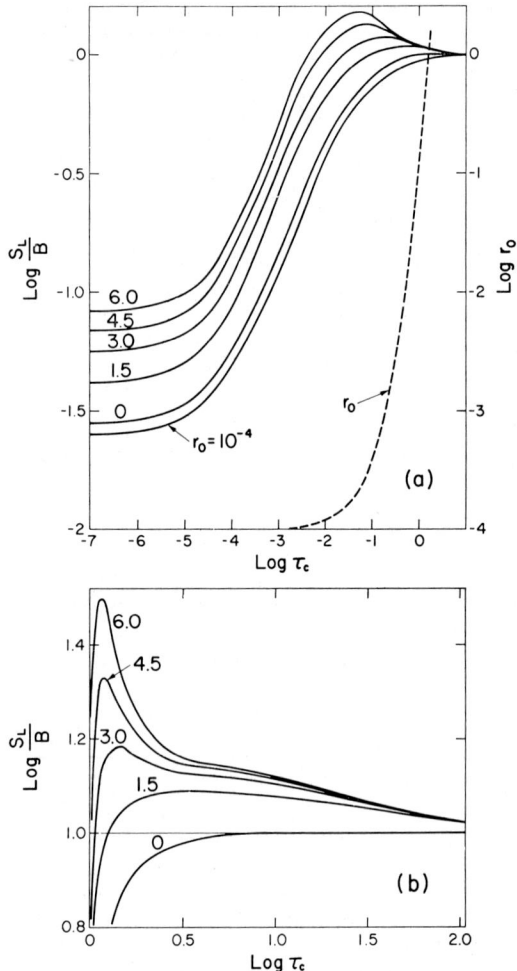

Fig. III-6. The effect on S_L of a gradient in r_0 coupled with a gradient in B. The region near $\tau_c = 1$ is shown in more detail in part b. Values of β ($B = 1 + \beta \tau_c$) are indicated on the different curves; $\varepsilon = 10^{-4}$ and $a = 10^{-2}$.

a star has a temperature minimum, as the Sun does, high excitation lines will show a relative maximum in r_0 in the region of temperature minimum whereas low excitation lines will show a relative minimum in r_0 in this same region.

To illustrate the nature of the effects of gradients in r_0 and their coupling with the gradient in B, we consider r_0 to be of the form

$$r_0 = 10^{-4}(1 + 3\tau_c)^5. \tag{III-10}$$

This produces a sharp rise in r_0 beginning at $\tau_c \approx 0.1$ and mimics in an approximate way the postulated variation of r_0 in a Schuster-Schwarzschild type model atmosphere. Figures III-6a and III-6b show the behavior of S_L for the above form for r_0 and for $\varepsilon = 10^{-4}$, $a = 10^{-2}$, and $B = 1 + \beta\tau_c$. Values of β are indicated on the individual curves. For comparison, Figure III-6a shows a curve for $r_0 = 10^{-4}$ and $\beta = 0$.

The primary effect shown in Figure III-6a for $\tau_c < 10^{-2}$ is a vertical translation of the curves as would be produced by simply increasing δ. (Note that for the parameters chosen $\delta \gg \varepsilon$). Between $\beta = 0$ and $\beta = 6.0$, the apparent increase in δ is by a factor of about 8.8. Evidently the steeper gradient in B samples the effective value of r_0 at larger values of τ_c.

Near $\tau_c = 10^{-2}$ to 1 an important change occurs in the character of S_L. This character is illustrated more clearly in Figure III-6b, which is exactly analogous to Figure III-3a except for the gradient in r_0. The maxima in S_L/B are much more pronounced in Figure III-6b than in Figure III-3a. The increased value of r_0 for $\tau_c > .1$ has driven S_L nearer J_c near $\tau_c = 0.1$ and has driven $S_L > J_c$ near $\tau_c = 1$. This latter result comes from the fact that the spectral line has turned into an emission line as measured in J_c at the local depth in the atmosphere. Because the line is in emission $\int J_\nu \phi_\nu \, d\nu > J_c$. Hence $S_L > J_c$ and $S_L > B$.

The specific behavior of S_L near $\tau_c = 1$ is not of profound importance so far as line profiles are concerned. An effect such as shown in Figures III-6b and III-3a would produce weak emission wings on the Fraunhofer lines. Such wings would be enhanced toward the violet end of the spectrum and for lines on the linear portion of the curve of growth, $r_0 > 1$, and near the top of the shoulder of the curve of growth, $r_0 \approx 10^{-4}$. A much more important result arising from the behavior of S_L near $\tau_c = 1$ is found in the blanketing effect of the lines, to be discussed in Chapter VII. There we will find that effects such as those shown in Figures III-3a and III-6b are of paramount importance.

5. The Influence of a Gradient in ϕ_ν

In a stellar atmosphere the Doppler width of a spectral line varies with depth. Also, if Stark or collision broadening are important in the damping wings, the Voigt

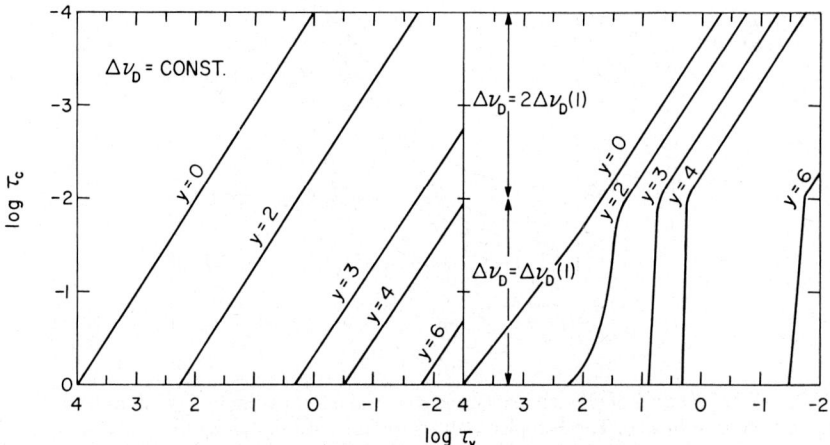

Fig. III-7. Illustration of the effect of a sharp gradient in $\Delta\nu_D$ on the relationship between τ_y and τ_c.

parameter a will vary with depth. Since ϕ_ν is a function of the Doppler width and the a parameter, it follows that ϕ_ν is a depth dependent parameter.

Depth dependence in ϕ_ν can seriously distort the τ grid at frequencies near the edge of the Doppler core. This effect is shown schematically in Figure III-7 for the case where the Doppler width increases sharply by a factor of 2 near $\tau_c = 10^{-2}$. At line center and very far in the line wings the effects on τ_y are relatively small. Near $y = 4$, however, τ_y is nearly constant through much of the lower atmosphere.

Consider the effect of the increase in $\Delta\nu_D$ illustrated in Figure III-7 upon photons at $\tau_c = 10^{-1}$. In the case of constant $\Delta\nu_D$ a photon at $\tau_c = 10^{-1}$ that diffuses to $y = 3$ finds that $\tau_y \approx 0.2$ and the photon readily escapes the atmosphere. On the other hand, in the case where $\Delta\nu_D$ increases by a factor two near $\tau_c = 10^{-2}$ a photon at $\tau_c = 10^{-1}$ that has diffused the same distance in $\Delta\nu$ as in the previous example now finds that

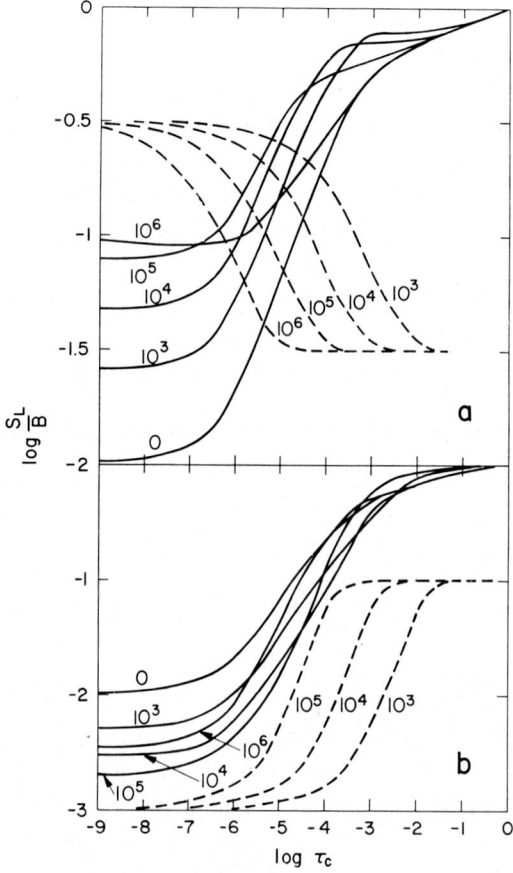

Fig. III-8. The effect on S of (a) – an increase in $\Delta\nu_D$ of the form given by Equation (III-11) and (b) – of a similar decrease in $\Delta\nu_D$. The Doppler width variations are shown by the dashed curves labeled with the corresponding value of χ from Equation (III-11). In all cases the Doppler width changes by a factor of 3 over the total height range.

$\tau_y \approx 6$ and it does not escape. To find $\tau_y < 1$ in the latter case the photon must diffuse to about $y = 5$, which is much less probable than diffusing to $y = 3$. Thus, the increase in Δv_D in the higher layers greatly inhibits the escape of photons in the lower layers.

Figure III-8a shows plots of S_L/B for cases where Δv_D is of the form

$$\Delta v_D = 1 + D e^{-(\chi \tau_c)^{1/2}} \tag{III-11}$$

with $D = 2$. Curves are shown for $\chi = 10^3$, 10^4, 10^5, and 10^6 and for $\varepsilon = 10^{-4}$, $r_0 = 10^{-6}$, and $a = 10^{-3}$. The curve with $\Delta v_D = $ constant is labeled 0 in the Figures. In this particular curve wing transfer dominates for $\tau_c > 10^{-3}$.

Strong effects on S_L/B are noted in all cases. The inhibition of photon escape in the lower layers pushes S_L nearer to B. In the lower layers the effect on S_L is strongest for lowest value of χ, i.e., for the cases where the rise in Δv_D occurs at the largest τ_c. Just the opposite is true in the outer layers. Rises in Δv_D at the smallest τ_c produce the largest increase in the surface values of S_L. Note that the effects are large near $\tau_0 = 1$ ($\tau_c = 10^{-6}$), amounting to as much as a factor of 10 increase in S_L. Everywhere in the atmosphere the outward gradients in Δv_D push S_L closer to B.

Figure III-8b displays the effect of decreasing Δv_D by a factor of 3. The other parameters are the same as in Figure III-8a. Again, we note that at large τ_c, S_L has increased, although the effect is less pronounced than in Figure III-8a for the same gradient and amplitude of the changes in Δv_D. However the increase in S_L at large τ_c is now for a different reason. Without the gradient in $\Delta \lambda_D$ the transfer of radiation for $\tau_c > 10^{-3}$ occurs predominantly in the wings. By decreasing the value of Δv_D in the outer layers we have now reduced the opacity of the higher layers in the near wing and edge of the Doppler core. As a result, the region of dominant transfer in the Doppler core extends to deeper layers.

In the external layers where $\tau_c < 10^{-3}$, S_L is depressed even further below B than for the case of constant Δv_D. This is because we have made photon escape much easier thereby reducing the mean value of J over the line.

Since gradients in Doppler width are very effective in changing the values of S_L in the outer layers of the atmosphere, they are important in influencing the central intensities of strong lines. We will return to a discussion of this point in Chapter V.

6. The Influence of a Moving Atmosphere

In the preceding section, we found that height gradients in the Doppler width may strongly influence S_L. A second type of gradient in ϕ_v results if the atmosphere is in differential motion, i.e., some parts of the atmosphere are rising or subsiding with respect to other parts. This type of motion produces a displacement of the center frequency of ϕ_v from one part of the atmosphere to another. However, the effect of bodily shifting ϕ_v in frequency is very different from the broadening or narrowing of ϕ_v discussed in the preceding section. A broadening of ϕ_v in the upper atmosphere inhibits the escape of photons from lower layers on both sides of the line. On the other hand, a shift in ϕ_v to one side of the line inhibits photon escape on that side of the line but encourages photon escape on the other. It is logical to conclude, therefore, that differential systematic motions are generally much less effective in changing S_L than are random motions which broaden ϕ_ν.

When differential motions are present in an atmosphere a new difficulty arises in computing S_L. The central position of ϕ_ν becomes a function of μ as well as of height. If the motions are vertical, ϕ_ν is shifted by an amount $\Delta\nu = \nu(v/c)\mu$, where v is the vertical velocity.

Because of the angular dependence of ϕ_ν, we can no longer write $d\tau_\nu = -\kappa_\nu\mu\,ds$ as we have in the equation of transfer. Thus, the quantities r_0, ϕ_ν, and τ_ν are each angularly dependent, and we must introduce the angular space as a discretized space in the numerical solutions of the equations. This rapidly increases the dimensions of the problem to be solved, but not in a hopeless way.

Line formation in a differentially moving atmosphere has been discussed by

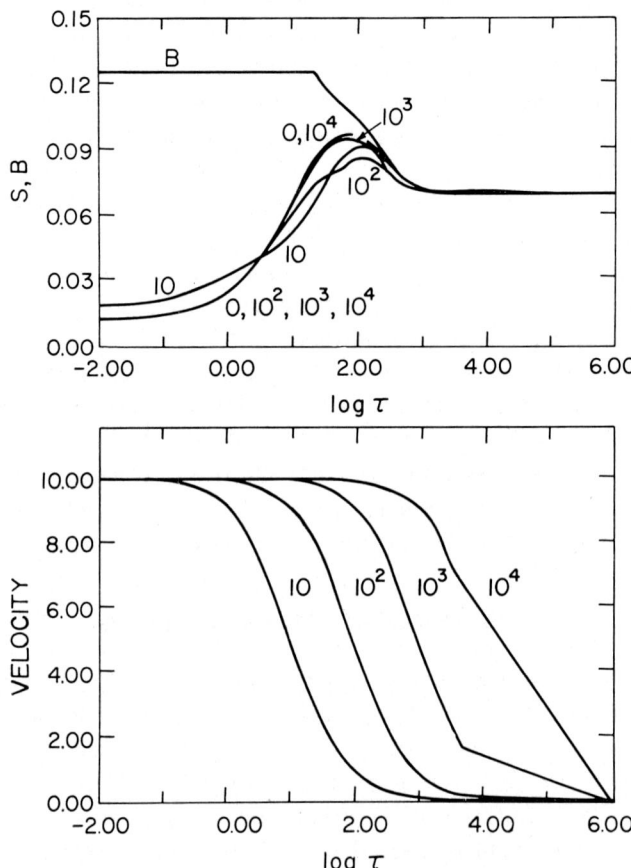

Fig. III-9. The effect on S_L of a differentially moving atmosphere with the velocity profiles illustrated in the lower panel (Kalkofen, 1970; *Extended Atmosphere Stars*, Courtesy U.S. National Bureau of Standards, Department of Commerce).

Magnan (1968), Kulander (1968), Hummer and Rybicki (1968), and Kalkofen (1970), each using different techniques for solving for S_L. The method used by Kalkofen (1970) parallels most nearly the integral equation method discussed in Chapter II for stationary atmospheres.

Figure III-9, taken from Kalkofen's results, shows the influence on S_L of the four velocity profiles shown in the figure. The profile of S_L for the stationary atmosphere is shown for comparison. Five angular points were used in the angular quadrature. Because of symmetry the five points represent the angular division of a single quadrant.

The results in Figure III-9 show that S_L is not very sensitive to radial motion of this type. None of the cases considered shows large deviations from the stationary atmosphere solution. The work of others bears this out.

Velocity gradients have very large effects on the line profiles computed from S_L even if S_L itself is unchanged. Because these effects are large, the relatively small changes in S_L are of secondary importance. It appears therefore that for approximate analyses of line profiles from moving atmospheres one could solve for S_L for the stationary atmosphere and include the motion only in the computation of the line profile. For precise analyses, however, S_L must be computed for the moving atmosphere.

7. The Influence of Frequency Redistribution

In the preceding we have assumed S_L to be independent of frequency by setting $\psi_v \equiv \Phi_v$. The actual shapes of ϕ_v and ψ_v depend upon the distributions of electrons in the absorbing and emitting states in the rest frame of the atoms, upon the velocity distribution of the atoms and upon the angular and frequency redistribution in the scattering process. An absorption event from atomic state l to state u of a photon at frequency v traveling in direction **n** may be followed by a number of alternative possibilities in the re-emission process. The excited electron in state u may return to a state l' or it may move to state u' before returning to state l or l'. Any such internal redistribution of electrons shifts the emitted photon to a new frequency v'. An additional redistribution in frequency between absorption and emission arises from the Doppler effect. Generally the scattered photon will be re-emitted in a direction **n'** that differs from **n** and the velocity component of the atom in direction **n'** will generally differ from that in direction **n**. As a result, the emitted photon will appear at a new frequency v' in the frame of reference of an external observer. This latter type of frequency redistribution, of course, depends, on the velocity distribution and on the angular redistribution of the scattering process.

In thermodynamic equilibrium each absorption at v and **n** must be balanced by a re-emission at v and **n**, which ensures that ψ_v and ϕ_v are identically equal. An equivalent way of bringing equality between ψ_v and ϕ_v is to assume that each of the redistribution processes is complete, i.e., that electrons in state u are completely redistributed into all states u' before re-emission, and that the re-emitted photon may appear in any new direction **n'** with equal probability. In other words, we assume that the distribution of electrons in states l and u has their thermodynamic equilibrium values and that there are no coherency effects between the emitted and absorbed photons. If either of these assumptions fail, as they will in general, the two functions ψ_ν and ϕ_ν will differ and their ratio will be a function of frequency. This, in turn, will influence the coupling between S_L and B and hence will influence the value of S_L.

The effect of having ψ_ν different from ϕ_ν can be partially understood through an inspection of Equation (II-47). The quantity ψ_ν enters Equation (II-47) explicitly in δ' and in H_ν. If we write $dH_\nu/d\tau_\nu$ as $J_\nu - S_\nu$, in accordance with Equation (II-43), and if we further write J_ν as $\Lambda_{\tau_\nu} S_\nu$, where Λ_{τ_ν} is the mean intensity operator defined by

$$J_\nu = \Lambda_{\tau_\nu} S_\nu = \tfrac{1}{2} \int_0^\infty S_\nu E_1(|t - \tau_\nu|) \, dt \qquad \text{(III-12)}$$

(cf., Kourganoff, 1963), Equation (II-47) is of the form

$$S_{L0} = \frac{\varepsilon + \delta}{\varepsilon + \delta'} B + \frac{1}{\varepsilon + \delta'} \int_0^\infty \Phi_\nu (\Lambda_{\tau_\nu} - I) S_\nu \, d\nu. \qquad \text{(III-13)}$$

We recall from Equation (I-10) that

$$S_\nu = \frac{r_0}{\phi_\nu + r_0} B + \frac{\phi_\nu}{\phi_\nu + r_0} S_L,$$

which we may rewrite as

$$S_\nu = \frac{r_0}{\phi_\nu + r_0} B + \frac{\phi_\nu}{\phi_\nu + r_0} \frac{\psi_\nu}{\Phi_\nu} S_{L0},$$

or, as

$$S_\nu = \frac{r_0}{\phi_\nu + r_0} B + \left(1 - \frac{r_0}{\phi_\nu + r_0}\right) \frac{\psi_\nu}{\Phi_\nu} S_{L0}. \qquad \text{(III-14)}$$

For ease of illustration consider that Φ_ν is constant with depth so that Φ_ν may commute with the operator $(\Lambda_{\tau_\nu} - I)$ in Equation (III-13). Equations (III-13) and (III-14) then combine to yield

$$S_{L0} = \frac{\varepsilon + \delta}{\varepsilon + \delta'} B + \frac{1}{\varepsilon + \delta'} \int_0^\infty (\Lambda \tau_\nu - I)$$
$$\times \left[\frac{r_0 \Phi_\nu}{\phi_\nu + r_0} B + \left(1 - \frac{r_0}{\phi_\nu + r_0}\right) \psi_\nu S_{L0}\right] d\nu. \qquad \text{(III-15)}$$

Thus, it is $\psi_\nu S_{L0}$ that enters $dH_\nu/d\tau_\nu$ rather than $\Phi_\nu S_{L0}$. Since it is the term containing $dH_\nu/d\tau_\nu$ in Equation (II-47) that makes S_{L0} differ from B, differences between ψ_ν

and Φ_ν may substantially alter the value of S_{L0}. Note, in particular, that the depth dependence of S_{L0} will differ somewhat from the case where $\psi_\nu \equiv \Phi_\nu$. This arises simply from the fact that the difference between ψ_ν and Φ_ν changes the relative importance of S_L and B. Since ψ_ν/Φ_ν must approach unity at all frequencies at depths in the atmosphere sufficiently great that S_L approaches B, it follows that Ψ_ν/Φ_ν is necessarily depth dependent. This depth dependence adds further change in the depth dependence of S_{L0} from the case where $\psi_\nu \equiv \Phi_\nu$.

Differences between δ and δ' may also contribute to the changes in S_{L0} provided $\delta \geqslant \varepsilon$. However, for $\varepsilon \geqslant 10^{-4}$ we expect $\delta \geqslant \varepsilon$ only when $r_0 \gtrsim 10^{-5}$, and in this case most of the contribution to δ and δ' comes from the Doppler core. As we shall see somewhat later differences between ψ_ν and Φ_ν tend to be small in the Doppler core but may be large in the line wings. Thus, for values of $\varepsilon \gtrsim 10^{-4}$ we do not expect large effects arising from differences between δ and δ'. For much smaller ε, however, such effects could become important.

We further note that each of the terms in Equation (II-47) or Equation (III-13) is frequency independent regardless of whether $\psi_\nu \equiv \Phi_\nu$ or not. Since

$$S_L = \frac{\psi_\nu}{\Phi_\nu} S_{L0},$$

it follows that the frequency dependence of S_L is given directly by ψ_ν/Φ_ν. We again remind the reader, however, that the depth dependence of S_L will also depend upon the ratio ψ_ν/Φ_ν and that ψ_ν/Φ_ν approaches unity as S_{L0} approaches B. The numerical magnitude of these effects can be assessed only through a more quantitative treatment.

Frequency redistribution (or frequency non-coherency) arising from radiation broadening of energy levels has been discussed by Weisskopf and Wigner (1930a, b), Weisskopf (1933), Spitzer (1936), and Woolley (1938). The effects of collision broadening have been discussed by Zanstra (1941), Spitzer (1944), and Edmonds (1955). Efforts to incorporate these effects in the solution of the transfer equation including Doppler broadening have been made by Henyey (1940), Unno (1952a, b, 1955), Warwick (1955), Thomas (1957), Field (1959), Jefferies and White (1960), Hummer (1962, 1965, 1969), Hearn (1964a, b), Sobolev (1965a, b), Finn (1967), Mathis (1968), Auer (1968), and Avery and House (1968). In the interest of brevity we give here only a brief review of this latter phase of the work of frequency redistribution.

Following Warwick (1955) and Hummer (1962), we define the probability that a photon at a frequency between ν and $\nu+d\nu$ traveling in direction $\mathbf{n}$ within solid angle $d\omega$ is absorbed and re-emitted at a frequency between ν' and $\nu'+d\nu'$ and in direction $\mathbf{n}'$ within solid angle $d\omega'$ as

$$R(\nu, \mathbf{n}; \nu'\mathbf{n}') \, d\nu \, d\omega \, d\nu' \, d\omega'.$$

Integration over ν' and ω' gives the absorption probability Φ_ν

$$\Phi_\nu \, d\nu \, d\omega = 4\pi \, d\nu \, d\omega \iint R(\nu, \mathbf{n}; \nu'\mathbf{n}') \, d\nu' \, d\omega'. \tag{III-16}$$

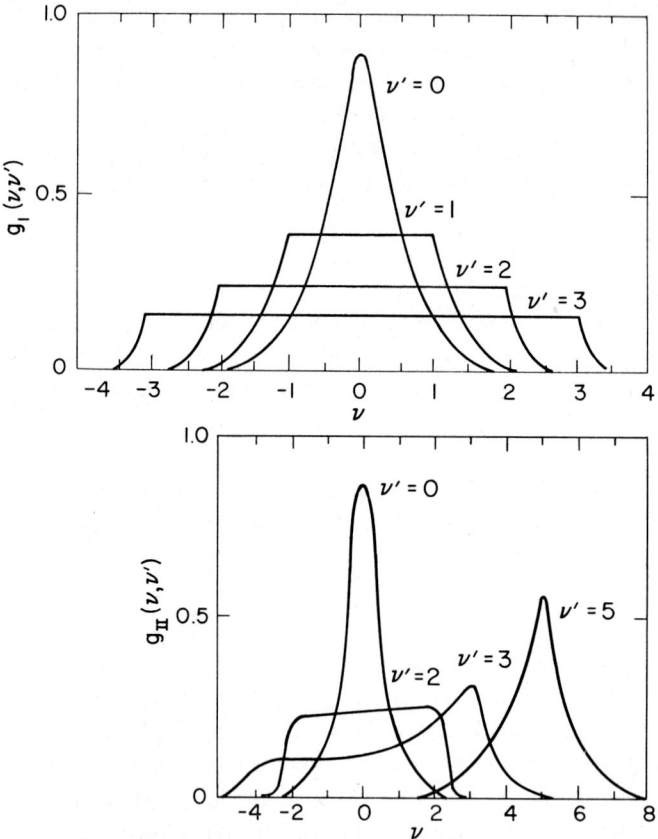

Fig. III-10. The g_I and g_{II} functions (Jefferies and White, 1960; *Astrophysical Journal*, copyright 1960 by the University of Chicago, All rights reserved). (The R_I and R_{II} functions in the text correspond to g_I and g_{II}.)

The normalization conditions are such that

$$\iiint\int R(v, \mathbf{n}; v', \mathbf{n}')\, dv\, d\omega\, dv'\, d\omega' = 1$$

and

$$\iint \Phi_v\, dv\, d\omega = 4\pi.$$

A lengthy discussion of the scattering function is beyond our intent here. Instead, we consider only some of the end results.

Frequency coherence and non-coherence are closely interrelated to angular coherence and non-coherence since it is the angular non-coherence that gives rise to the Doppler redistribution in frequency. Nevertheless, we shall discuss the effects of non-coherency in frequence as those arising within the rest frame of the atom, which are separate from the angular effects.

If we assume complete non-coherence in angular scattering, the scattering probability function reduces to

$$R(v, v') \, dv \, dv'$$

with

$$\Phi_v \, dv = dv \int R(v, v') \, dv' \qquad \text{(III-17)}$$

and

$$\int \Phi_v \, dv = 1.$$

For zero natural line width Unno (1952a) (see also Hummer (1962)) finds

$$R_{\mathrm{I}}(y, y') = \frac{1}{\sqrt{\pi}} \int_{\bar{y}}^{\infty} e^{-t^2} \, dt \qquad \text{(III-18)}$$

where y is $\Delta v / \Delta v_D$ and $\bar{y}$ denotes the larger of $|y|$ and $|y'|$. For the case of radiation damping Unno (1952b) and Sobolev (1965a) (see also Hummer (1962)) find

$$R_{\mathrm{II}}(y, y') = \pi^{-3/2} \int_{1/2|\bar{y}-\underline{y}|}^{\infty} e^{-u^2} \left[\tan^{-1} \frac{\underline{y}+u}{a} - \tan^{-1} \frac{\underline{y}-u}{a} \right] du \qquad \text{(III-19)}$$

where $\underline{y}$ denotes the lesser of $|y|$ and $|y'|$. The functions $R_{\mathrm{I}}(y, y')/\phi(y')$ and $R_{\mathrm{II}}(y, y')/\phi(y')$ are shown in Figures III-10a and III-10b. Equations (III-18) and (III-19) are valid only in the case of coherence in the rest frame of the atom. For complete non-coherence in the rest frame of the atom, Hummer (1962) gives

$$R_{\mathrm{III}}(y, y') = \pi^{-5/2} \int_0^{\infty} e^{-u^2} \left[\tan^{-1} \frac{y+u}{a} - \tan^{-1} \frac{y-u}{a} \right]$$

$$\times \left[\tan^{-1} \frac{y'+u}{a} - \tan^{-1} \frac{y'-u}{a} \right] du. \qquad \text{(III-20)}$$

The difference between the functions R_{I}, R_{II}, and R_{III} can be understood in relatively simple terms. In the case of R_{I} it is assumed that the atom can absorb at $y \neq 0$ only by virtue of its motion. Once having absorbed at this frequency, the atom re-emits with equal probability into all angles and therefore re-emits with equal probability at all frequencies between y and $-y$. Absorption at $y = 0$ means either that the atom is at rest or that it is moving in a direction at right angles to the incident photon. If it is the latter, the re-emitted photon may still Doppler shift to $y \neq 0$.

Atoms with some level broadening, due to natural width or perturbations, may absorb at $y \neq 0$ even if they are at rest. At large values of y, ($y \gtrsim 3$), absorption occurs mainly by atoms near the mean velocity rather than by atoms with unusually high velocity. Hence, in the case of the R_{II} function the Doppler redistribution at large y' consists of a relatively narrow redistribution around the absorbed frequency y. Near y=0 (y ≲ 2) absorption occurs mainly by moving atoms whose central absorption lies near y. The situation is similar, therefore, to case I of no level broadening. These effects are clearly evident in Figures III-10a and III-10b. From the form of R_I and R_{II} for y<3 it is evident that the Doppler effect leads to a strong degree of non-coherence throughout the Doppler core of the line, as noted by Thomas (1957). In the wings, however, the scattering is strongly coherent, provided the atom is scattering coherently in its own rest frame.

For complete redistribution in the rest frame of the atom, R_{III}, some measure of non-coherence still remains in the frame of reference in which v and v' are measured. This effect arises from the level width in a manner analogous to case II, i.e., atoms absorbing at y>3 re-emit preferentially near the same value of y. Hummer (1962) has shown that R_{III} does in fact lead to full non-coherence in the v and v' frame in the wings of the line. However, the form of R_{III} in the Doppler core has not been

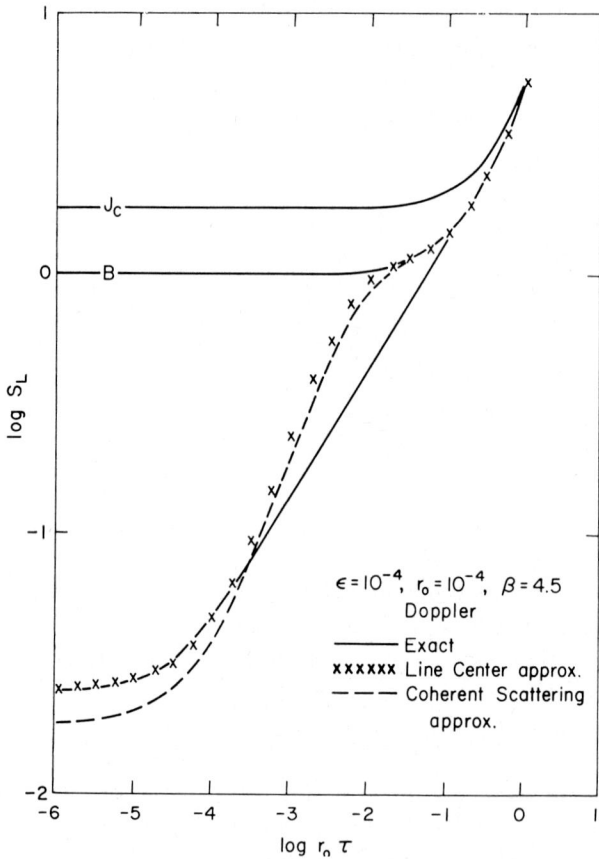

Fig. III-11. Comparison of the solutions for S_L for the coherent scattering approximation and the line center approximation to the solution for S_L for the exact redistribution function (Courtesy A. Skumanich, 1969).

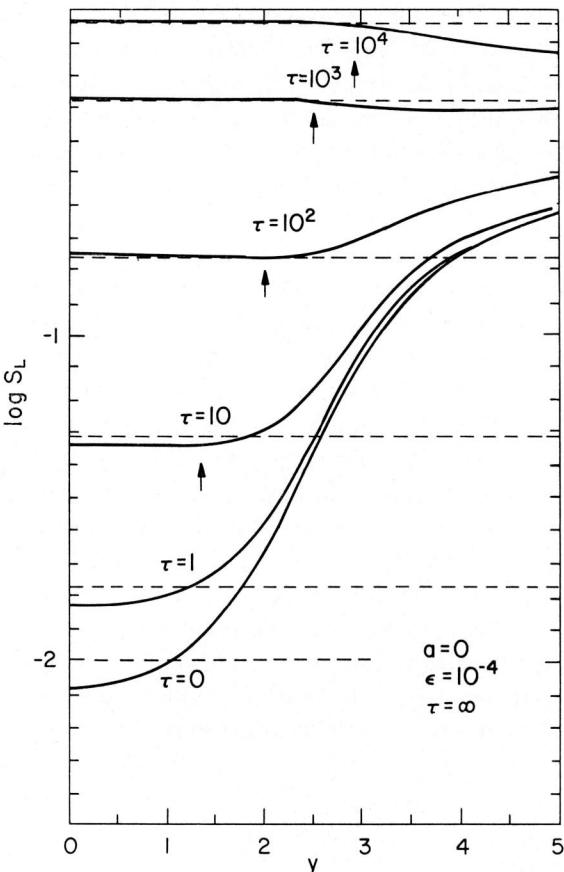

Fig. III-12. Solutions for S_L as functions of y for the redistribution function g. The dashed lines indicate the values of S_L for complete redistribution (Hummer, 1969, courtesy *Monthly Notices Roy. Astronomical Society*, Blackwell Scientific Publications LTD).

studied. It is customarily assumed to give complete non-coherency there also. It is in any case hard to see how this situation could lead to less non-coherency than case II, which is already strongly non-coherent in the Doppler core.

Since there is some non-coherency even in the rest frame of the atom, the true scattering function lies between R_{II} and R_{III}, which is to say only that it lies between coherency and non-coherency in the wings and is strongly coherent in the Doppler core.

Even in the extreme of pure coherent scattering in the v and v' frame, there is not a gross difference in S_L at line center from the non-coherent case. The two cases are compared in Figure III-11 for a Doppler profile from computations by Skumanich (1969).

Hummer (1969) has computed $S_L(v)$ for the redistribution function R_I. His results for a Doppler profile are shown in Figures III-12. The arrows in this figure give the location of $\tau_y = 1$. Within the area defined by $\tau_y \geq 1$ and $\tau_0 \geq 1$, S_L is closely equal to its value for pure non-coherency. This adds considerable confidence to the supposition that for practical purposes the scattering may be treated as non-coherent.

A scattering atom recoils as the photon momentum is changed and the recoil velocity may, in the case of a light atom at low temperature, be a significant fraction of the mean atomic velocity. The effects of recoil have been studied by Field (1959) and Osterbrock (1962). They are of little consequence in stellar atmospheres and will not be discussed further in this text.

Oxenius (1965) has given general arguments to show that the profiles of the spontaneous and stimulated emission coefficients should be identical, i.e., that the profile of stimulated emissions is given by ψ_v rather than ϕ_v, contrary to what we have assumed. His arguments, however, appear to be valid in the rest frame of the atom but not necessarily in the rest frame of the observer. Stimulated emissions are coherent in angle with the direction of the inducing photons. Thus, if the photons come mainly from one direction, as they tend to in the outer layers of stars, the stimulated emissions will be anisotropic. This means that the Doppler redistribution function may be different for stimulated emissions than for spontaneous emissions, hence, that the two emission profiles may differ somewhat. Since stimulated emissions are rarely of importance in astrophysical applications, this point is not worth belaboring further.

We have assumed that the stimulated emission profile is given by ϕ_v. The reason for doing so is directly related to the choice to treat stimulated emissions as negative absorptions. It is not necessary to do this, however. Thus, we could write for the emission coefficient

$$\left(h\nu A_{21} n_2 + h\nu B_{21} n_2 \int J_v \psi_v \, dv \right) \psi_v$$

and for the absorption coefficient

$$h\nu n_1 B_{12} \Phi_v.$$

The source function is then given by

$$S_L = \frac{A_{21} n_2 + B_{21} n_2 \int J_v \psi_v \, dv}{B_{12} n_1} \frac{\psi_v}{\Phi_v},$$

which is equivalent to

$$S_L = \left(\frac{2h\nu^3}{c^2} \frac{b_2}{b_1} e^{-h\nu/kT} + \int J_v \psi_v \, dv \, e^{-h\nu/kT} \right) \frac{\psi_v}{\Phi_v}.$$

If we define a corresponding escape coefficient ϱ as

$$\varrho = 1 - \frac{\int J_v \psi_v \, dv}{S_L (\Phi_v / \psi_v)},$$

we find

$$S_L = \left(\frac{2h\nu^3}{c^2} \frac{1}{(b_1/b_2)e^{+h\nu/kT} - 1 + \varrho}\right)\frac{\psi_\nu}{\Phi_\nu}. \tag{III-21}$$

This differs from the form we have used only in the appearance of the extra term ϱ in the denominator. However, since ϱ is generally small compared to unity the difference is trivial.

8. The Influence of Anisotropic Scattering

As noted in the preceding section, angular scattering is not completely non-coherent. Hummer (1970) has investigated the effect of including the Rayleigh dipole phase function in the scattering term for the case of frequency non-coherence. The normalized phase function is given by

$$g(\mathbf{n}, \mathbf{n}') = \frac{3}{16\pi}[1 + (\mathbf{n}\cdot\mathbf{n}')^2], \tag{III-22}$$

where $\mathbf{n}$ and $\mathbf{n}'$ are the initial and final direction vectors. Since the scattering is anisotropic, S_L is now a function of μ. For a semi-infinite atmosphere with $r_0=0$ and $\varepsilon=10^{-6}$ and a Doppler form for ϕ_ν, Hummer (1970) finds that S_L at $\mu=0.2$ is identical with S_L at $\mu=1$ to three significant figures for all $\tau_0 \geq 1$. S_L at $\mu=1$ is, in turn, equal to the isotropic S_L to the same accuracy. At small τ_0 the angular dependence becomes of some importance but remains relatively small. For example, at $\tau_0 = 10^{-1}$ $S_L(\mu=1)=1.13\times 10^{-3}$, $S_L(\mu=0.2)=1.08\times 10^{-3}$, and $S_L(\text{isotropic})=1.10\times 10^{-3}$. Very similar results were obtained for different ε and r_0 and for a Voigt profile.

Differences in $S_L(\mu)$ at small τ_0 become important in analyses of center-limb data near line center. For the semi-infinite atmosphere, Hummer finds that the Rayleigh dipole phase function leads to a relative depression of the line central intensity at $\mu=0.1$ by about 2% of its isotropic value. In the line wings there is no effect.

Since non-coherence in frequency and non-coherence in angle are interrelated in that it is the non-coherence in angular scattering that produces much of the non-coherence in frequency, Hummer (1970) notes that a combined treatment of dipole scattering with the true frequency redistribution function may show the effects of dipole scattering to be more pronounced than these initial results would suggest.

9. The Influence of a Finite Atmosphere

Although all of the preceding discussion has dealt with semi-infinite atmospheres, this is not a fundamental restriction. Finite atmospheres differ from semi-infinite atmospheres in the degree of coupling between the line source function and the continuum source function, or Planck function. As the total opacity of the atmosphere is reduced the coupling becomes smaller. In many senses the problem of a finite atmosphere is not basically different from the problem of a chromosphere except in the symmetry in τ_0 for the finite atmosphere. Solutions for S_L for finite atmospheres using the flux divergence formulation in Chapter II require only appropriate numerical methods that make note of the fact that S_L reaches a maximum and is symmetric about the maximum on τ scales measured from the nearest border. Thus, for a slab of thickness 10^4, S_L at $\tau_0 = 10^3$ is equal to S_L at $\tau_0 = 9 \times 10^3$ and S_L at $\tau_0 = 10$ is equal to S_L at $\tau_0 = 9.99 \times 10^3$.

A number of authors have considered the behavior of S_L in atmospheres of finite thickness. The case $\varepsilon \gg \delta$ has been discussed by Avrett and Hummer (1965) and the case $\delta \gg \varepsilon$ has been discussed by Hummer (1968). Figure III-13a shows plots of S_L for the case $\varepsilon \gg \delta$ and for different total atmospheric thickness T_0 (Avrett and Hummer, 1965). Figure III-13b shows a similar plot for the case $\delta \gg \varepsilon$.

For the case of $B = $ constant (Figures III-13a and 13b) and for an effectively thin atmosphere, the maximum value of S_L is given approximately by T_0/τ_{th} for the Doppler profile and $(T_0/\tau_{th})^{1/2}$ for the Voigt profile. Surface values of S_L for the corresponding cases are given approximately by $(\varepsilon+\delta)^{1/2} (T_0/\tau_{th})^{1/2}$ and $(\varepsilon+\delta)^{1/2} (T_0/\tau_{th})^{1/4}$. The thermalization lengths are given by $\pi^{-1/2}(\varepsilon+\delta)^{-1}$ and $\pi^{-1/2} a(\varepsilon+\delta)^{-2}$. Thus, the surface value of S_L is given approximately by

$$S_L(0) = \pi^{1/4} (\varepsilon + \delta) T_0^{1/2}, \quad \text{Doppler}, \tag{III-23}$$

and

$$S_L(0) = \pi^{1/8} (\varepsilon + \delta) (T_0/a)^{1/4}, \quad \text{Voigt}, \tag{III-24}$$

The corresponding maximum values of S_L are given by

$$S_L(\text{max}) = \pi^{1/2} (\varepsilon + \delta) T_0, \quad \text{Doppler}, \tag{III-25}$$

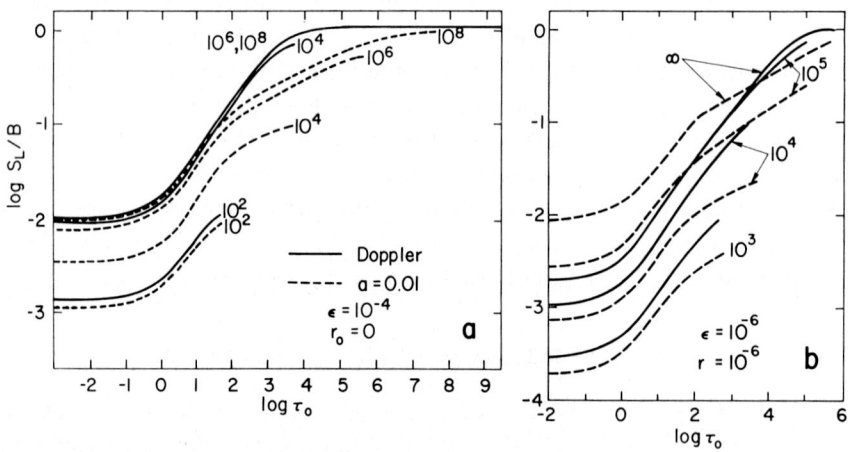

Fig. III-13. Solution for S_L for finite, plane-parallel atmospheres for: (a) – $\varepsilon \gg r_0$ and (b) – $r_0 = \varepsilon$. Results are given for Doppler profiles and for Voigt profiles with $a = 10^{-2}$. The total atmospheric thickness T is indicated on the curves (Avrett and Hummer, 1965 (Figure a); Hummer, 1968 (Figure b); courtesy *Monthly Notices Royal Astronomical Society*, Blackwell Scientific Publications LTD).

and
$$S_L(\max) = \pi^{1/4} (\varepsilon + \delta)(T_0/a)^{1/2}, \quad \text{Voigt}. \tag{III-26}$$

It may be seen by comparison to the chromospheric scaling for the effectively thin case given in Table III-1 that the two sets of laws are the same with the single exception that T_0 is replaced by the chromospheric thickness $(Cr_0)^{-1}$. These scaling laws were first noted by Skumanich (1967), and are in agreement with the values predicted by Equations (II-83) and (II-84).

10. The Influence of Horizontal Structure

It may be seen from inspection of Equation (II-47) that the fundamental quantity determining the ratio of S_L to B is the flux divergence $dH_\nu/d\tau_\nu$ averaged over the line. In a plane parallel atmosphere τ_ν is simply related to the path direction and is a minimum in the vertical direction. As a result of this latter conditions photons escape primarily in the vertical direction. Also, $dH_\nu/d\tau_\nu$ is determined by a single position coordinate. S_L and τ_ν are everywhere the same on a plane surface.

Horizontal structure in an atmosphere in the form of temperature and density inhomogenieties complicate the problem considerably. The temperature and density inhomogenieties will, in general, be accompanied by inhomogenieties in τ_ν so that a surface of constant τ_ν will contain ridges and troughs. Near ridges in τ_ν photons will tend to diffuse laterally into the trough in τ_ν as well as to diffuse upwards. Hence near a ridge $dH_\nu/d\tau_\nu$ will be larger than if the τ_ν surfaces were horizontal and S_L will be correspondingly lower relative to B. In the troughs, on the other hand, photon diffusion in from the sides will partially compensate for photon loss in the vertical direction tending to decrease $dH_\nu/d\tau_\nu$ and to bring S_L closer to B.

In a general way it can be seen that if a given structure has an optical thickness in the horizontal direction, τ_h, that greatly exceeds the thermalization length the structure will have little effect upon S_L, except near the boundaries of the structure. On the other hand, if τ_h is much less than the thermalization length photon diffusion through the sides of the structure will be important in determining $dH_\nu/d\tau_\nu$ and we must expect an effect upon S_L. For strong lines, the thermalization lengths are long. For example, with $\varepsilon = 10^{-4}$ and $a = 10^{-3}$, $\tau_{th} = 10^5$, and if the atmospheric scale height is of the order of 100 km, as in the solar photosphere, the corresponding geometrical length is given by $100 \ln 10^5$ km $= 1150$ km. Thus, we would expect structures with horizontal dimensions of the order of 1150 km (1.6″) or smaller to be very important in determining the character of S_L. In the solar chromosphere τ_{th} increases because of decreasing density and the scale height for τ_0 often increases because of the temperature gradients. For some lines the scale heights are of the order of 500 km and we must expect the possibility at least that structures of even several thousand kilometers in diameter will become important in fixing S_L. Chromospheric spicules and fine mottles and photospheric granulation have characteristic horizontal dimensions less than 1000 km. Thus it seems clear that in these types of structures S_L is partially determined, perhaps dominantly determined, by the geometrical nature of the structure.

In the plane parallel atmosphere τ_v is a function of three variables: the depth, h, direction, μ, and frequency. One of these, μ, is treated analytically, which is made possible by the simple geometry. An atmosphere with horizontal structure becomes a multi-dimensional problem in position and direction coordinates. The optical path length is now a function of both horizontal position and of the direction vector from that position. In addition to h and v, therefore, two coordinates are needed to specify the horizontal position and two coordinates are needed to specify the direction. Thus, τ_v is now a function of six variables. If symmetry is present, some of the variables may be eliminated. Even so, the numerical problem is a great deal larger than the simple plane parallel atmosphere problem.

The methodology for treating multi-dimensional atmospheres is still largely in the developmental stage, although much has been done. To a large extent, the problems that can be treated are limited by computer speed and storage.

Solutions for S_L in multi-dimensional media have been accomplished by four different methods. Wilson (1968, 1969) has proposed an iterative technique, which appears to work well either when S_L does not differ from B by large factors or when the structure is limited to regions where τ_0 is relatively small. Past experience with similar iterative methods attempted in the one dimension case suggests that the method will not work for small values of $\varepsilon (<0.1)$ or for structures at large values of $\tau_0 (>10)$.

Rybicki (1965) has developed a semi-analytic technique for treating multi-dimensional problems employing Green's functions. Both Rybicki (1968) and Jones (1970) have applied the method to restricted types of problems. The method appears to be very useful for restricted types of geometries and has the advantage of using Green's functions which prove instructive.

The flux divergence technique outlined in Chapter II has been extended by Jones and Skumanich (1968) to multi-dimensional problems by a method employing integration along characteristic paths. Although only limited applications of this method have been tested, the flux divergence method shows promise of being adaptable to a wide range of problems.

Monte Carlo techniques are perhaps in some respects the most versatile for treating multi-dimensional media. They suffer primarily from the need to accumulate sufficient photon statistics. The statistics required increase as τ_0 increases and as ε and δ decrease. Thus, Monte Carlo techniques are limited mainly by computer speed. Calculations for two-level atoms have been successfully carried out for τ_0 as large as 10^5 with $\varepsilon = 10^{-5}$, and techniques under investigation promise to extend the range considerably.

The application of Monte Carlo methods in line transfer problems have been developed independently by a number of authors. Reviews of the Monte Carlo method can be found in several texts including Hammersley and Handscomb (1964) and Schreider (1966). A review of the application of the Monte Carlo method to line transfer problems is given by House and Avery (1968).

Three general types of geometries have been studied in the multi-dimensional case: (1) periodic variations in B in one dimension of the form

$$B(\chi) = (1 + \gamma \cos(\pi\chi/L), \tag{III-27}$$

(2) a vertical slab or column of finite thickness imbedded in a semi-infinite atmosphere of different temperature, and

(3) a cylindrical column standing either free or above a radiating surface. In the former two all parameters have been held constant with depth. The latter problem has been studied with variable parameters (Avery, 1969), but we shall discuss only the case where the parameters have been held constant. Results from these three cases discussed in the following are taken from Jones and Skumanich (1968) for the first two and from Avery *et al.* (1969) for the third.

A. PERIODIC STRUCTURE

For the case of a periodic variation in B of the form of Equation (III-27), S_L varies in phase with B but with an amplitude factor relative to B that varies with depth. Thus, if we write

$$S_L(\chi) = \left(\frac{S_L}{B}\right)^* \left(1 + A\gamma \cos \frac{\pi\chi}{L}\right), \tag{III-28}$$

we expect the amplitude factor A to depend upon τ_0 and upon ε, δ and a. The quantity $(S_L/B)^*$ is given by the plane parallel solution, since the equations for S_L are linear in B and solutions may be linearly superposed. If ε, δ and a do not vary with χ, $(S_L/B)^*$ will be a function of τ_0 only. However, if ε, δ, and a vary periodically also, we expect that $(S_L/B)^*$ will be periodic. This case has not been treated, however.

Equations (III-28) and (III-27) state that S_L/B will contain a component given by the plane parallel atmosphere solution but will also contain an added component varying in phase with B. The amplitude of the variation in S_L is $A\gamma$. Solutions for the amplitude factor A (Jones and Skumanich, 1968) have been limited to the case $\varepsilon \gg \delta$ and $a = 0$.

If L is very large in optical length, i.e., $\tau_L \gg \tau_{th}$, we expect that the lateral structure will have little effect upon the solutions. However, if $\tau_L < \tau_{th}$ lateral diffusion of photons might logically be expected to dampen the amplitude of fluctuations in S_L. Figure III-14a shows the variation deep in the atmosphere ($\tau_0 = 50\ \tau_{th}$) of A as a function of ε and τ_L/τ_{th}. Note that saturation to $S_L = B$, i.e., $A = 1$, occurs near $\tau_L = \tau_{th}$, as might be expected from the meaning of τ_{th}. At small values of $\tau_L (\tau_L < 1)$, we find $A = \varepsilon/(1+\varepsilon)$. In other words S_L is increased or decreased in direct proportion to the increase or decrease in the photon generation term εB. This is to be expected since lateral diffusion between adjacent fluctuations in B is complete for $\tau_L \ll 1$ and there is no possibility to build up excess photons in one region.

Between $\tau_L = 1$ and $\tau_L = \tau_{th}$, A is approximately equal to τ_L/τ_{th}. This result is analogous to the result given by Equation (III-25) for the amplitude of S_L for the finite plane parallel atmosphere. We found in this latter case that for a Doppler profile

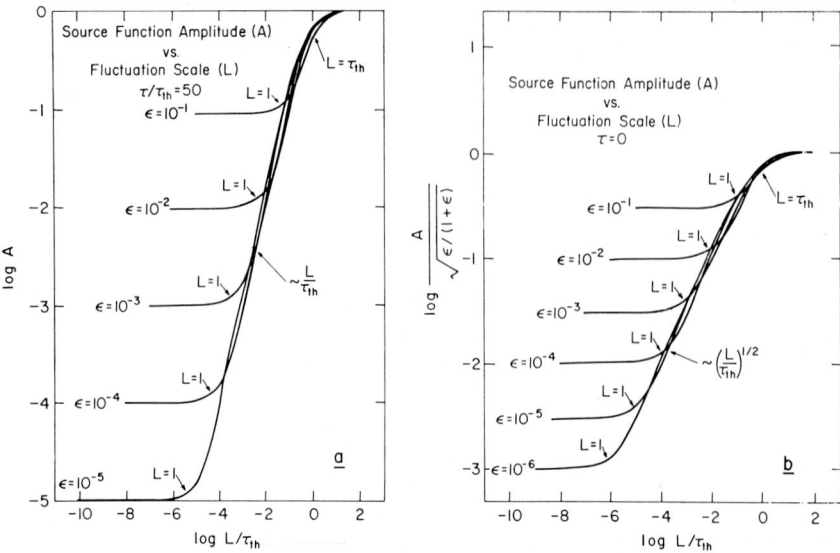

Fig. III-14. The source function amplitude (A) for a one-dimensional sinusoidal variation in B of the form given by Equation (III-27) and for a Doppler profile. Part (a) shows the variations deep in the atmosphere and part (b) shows the surface variations (Jones, 1970 (Thesis); University of Colorado, courtesy National Center for Atmospheric Research Special publication).

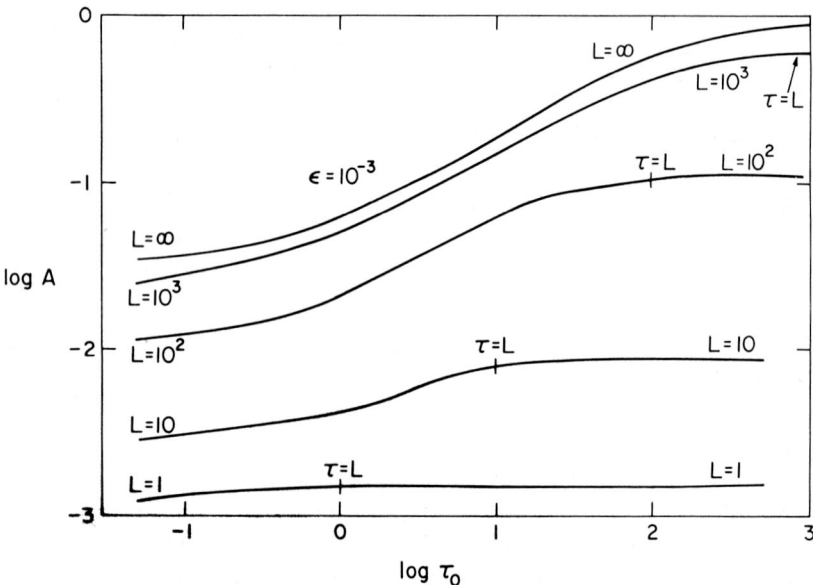

Fig. III-15. Influence of the wavelength of the fluctuation scale on the depth dependence of the source function amplitude for periodic media and for a Doppler profile (Jones, 1970 (Thesis); University of Colorado, courtesy National Center for Atmospheric Research Special publications).

for ϕ_v the amplitude of S_L at the center of the slab was given by $\pi^{1/2}\varepsilon T_0$. For the case shown in Figure III-14a, $\tau_{th} = \pi^{1/2}\varepsilon^{-1}$ so the result $A \approx \tau_L/\tau_{th}$ is just the result $A \approx \pi^{1/2}\varepsilon\tau_L$. Again, by analogy with the finite atmosphere case (Equation (III-22)) we would expect that for a Voigt form for ϕ_v, A would be approximately equal to $(\tau_L/\tau_{th})^{1/2}$, which is equivalent to $\pi^{1/4}\varepsilon(\tau_L/a)^{1/2}$.

Figure III-14b shows a plot similar to Figure III-14a but for the case $\tau_0 = 0$. Note that the ordinate is normalized to the surface value of S_L/B for the one dimensional case. This is done in order to facilitate the analogy with Figure III-14a. Note also, that as in the case of large τ_0, A saturates for $\tau_L > \tau_{th}$ and decreases to $\varepsilon/(1+\varepsilon)$ at $\tau_L < 1$. In the effectively thin region between $\tau_L = 1$ and $\tau_L = \tau_{th}$, we find

$$A \approx \left(\frac{\varepsilon}{1+\varepsilon}\right)^{1/2} \left(\frac{\tau_L}{\tau_{th}}\right)^{1/2} \quad \text{or} \quad A \approx \pi^{1/4} \frac{\varepsilon}{1+\varepsilon} \tau_L^{1/2}.$$

This result again is synonymous with the finite atmosphere result (Equation (III-19)) where $S_L \approx \pi^{1/4}\varepsilon T_0^{1/2}$ for the Doppler profile. The analogy carried to the Voigt profile (Equation (III-20)) would give $A \approx \pi^{1/8}(\varepsilon/(1+\varepsilon))(\tau_L/a)^{1/4}$.

We show in Figure III-15 the variation of A with τ_0 as a function of τ_L for the Doppler profile and $\varepsilon = 10^{-3}$. Note that for $\tau_0 < \tau_L$ the curves are approximately of the same shape as the one dimensional case, $\tau_L = \infty$, differing only by the approximate factor $(\varepsilon\tau_L/(1+\varepsilon\tau_L))^{1/2}$. This suggests that in this optical depth regime the character of the solution is controlled by the nearer boundary $\tau_0 = 0$. For $\tau_0 > \tau_L$, A is approximately constant at $\varepsilon\tau_L/1+\varepsilon\tau_L$ suggesting that photon diffusion to the nearer boundary $\tau_0 = \tau_L$ is now dominating the solution. This result is to be expected since the photon escape probability discussed in Chapter II will always favor the nearer boundary.

B. THE ISOLATED CYLINDER

The transfer of radiation in a cylinder, in similarity with certain imbedding problems, involves two characteristic competing lengths, in this case the radial optical length τ_R and the axial optical length τ_z. When both τ_R and τ_z are large compared to τ_{th}, S_L behaves as it would in the corresponding one dimensional problem, i.e., $S_L = B$ at $\tau \approx \tau_{th}$ and $S_L \approx \varepsilon^{1/2}B$ at the surface. If either τ_R or τ_z is less than τ_{th}, diffusion in the direction of smaller τ dominates over diffusion in the other direction and the amplitude of S_L throughout the cylinder is governed by the smaller of τ_R and τ_z together with the value of ε.

Radial and axial variations of S_L/B are shown in Figure III-16 for a range of values of τ_R and ε. The axial optical length is 100 and the absorption profile is Doppler. Note the similarity between the axial variation of S_L and the variation of S_L for the periodic case shown in Figure III-15. Saturation to constant S_L occurs in the cylinder at $\tau_R \approx \tau_z$ and at $\tau_z < \tau_R S_L$ is decreased in proportion to $\tau_R^{1/2}$. Thus, τ_R replaces τ_L as the competing scale length. An entirely similar effect is present in the radial source function when $\tau_R > \tau_z$.

The scaling laws given by Avery et al. (1969) for cylinders that are effectively thin

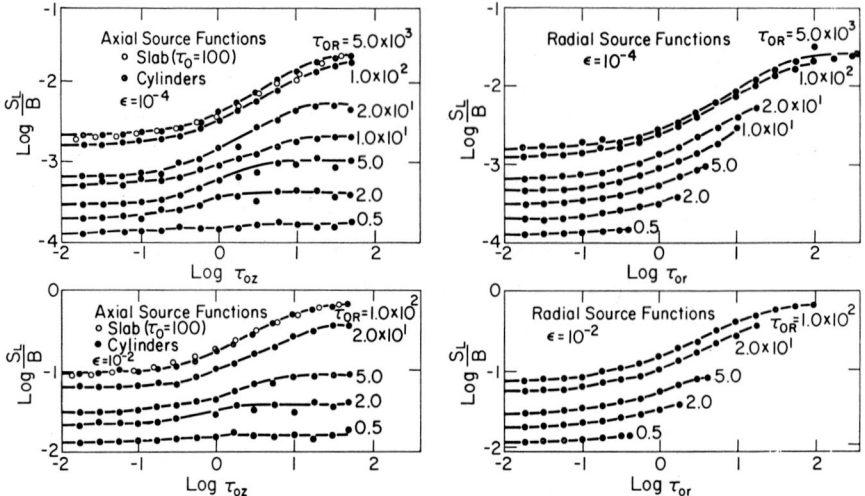

Fig. III-16. Comparison of radial and axial values of the source function in cylinders for different values of ε and different radial optical thickness and for Doppler profiles (Avery et al. 1969; courtesy *Journal of Quantitative Spectroscopy and Radiative Transfer*, Pergamon Press).

in at least one dimension are

$$S_L(0) = \pi^{1/4} \varepsilon \tau_{0\,\text{min}}^{1/2}$$

and

$$S_L(\text{max}) = \pi^{1/2} \varepsilon \tau_{0\,\text{min}}$$

where $\tau_{0\,\text{min}}$ designates the smaller of the quantities τ_R and τ_z. These scaling laws are exactly analogous to those for the finite one-dimensional atmosphere (Equations (III-23) and (III-25)) and the atmosphere that is periodic in B. The only change in the scaling laws occurs in the meaning of $\tau_{0\,\text{min}}$. For the finite atmosphere it is the half thickness T_0 of the slab, for the periodic case it is the wavelength τ_L and for the cylinder it is the smaller of τ_R and τ_z. Again, we note that the surface scaling by $\tau_0^{1/2}$ and the center scaling by τ_0 is a quality of the Doppler profile. For a Voigt profile the scaling should be given by Equations (III-24) and (III-26)

$$S_L(0) = \pi^{1/8} \varepsilon \tau_{0\,\text{min}}^{1/4}$$

and

$$S_L(\text{max}) = \pi^{1/4} \varepsilon (\tau_0/a)^{1/2}$$

provided the analogy with the finite slab remains valid. We see no reason to question the analogy.

C. THE IMBEDDED CYLINDER

A vertical cylindrical structure imbedded in an ambient atmosphere may be expected to exhibit quite different radiation properties from an isolated cylinder. The opacity in the horizontal direction, measured radially outward from the axis of the cylinder does not reach an asymptotic limit at the edge of the cylinder as it does in the isolated cylinder. Instead, the opacity continues to increase indefinitely outside the cylinder. This exterior opacity will have two principle effects: (1) it will inhibit the escape of radiation from the cylinder; and (2) it will cause the effects of the cylinder to diffuse across the borders of the cylinder into the ambient atmosphere. Thus, we expect S_L to be higher in the imbedded cylinder than in the corresponding isolated cylinder and we expect S_L to be above its ambient value for some distance around the cylinder. The diffusion effects of the cylinder might be expected, in fact, to penetrate the ambient medium for a distance approximately equal to the thermalization length.

Figure III-17 illustrates the effect deep in the atmosphere of imbedded cylinders of radii τ_R assuming $\varepsilon = 10^{-4}$, $r_0 = 10^{-6}$ and a Doppler profile for ϕ_ν. These latter quantities have the same values in the ambient medium as in the cylinders. Thus,

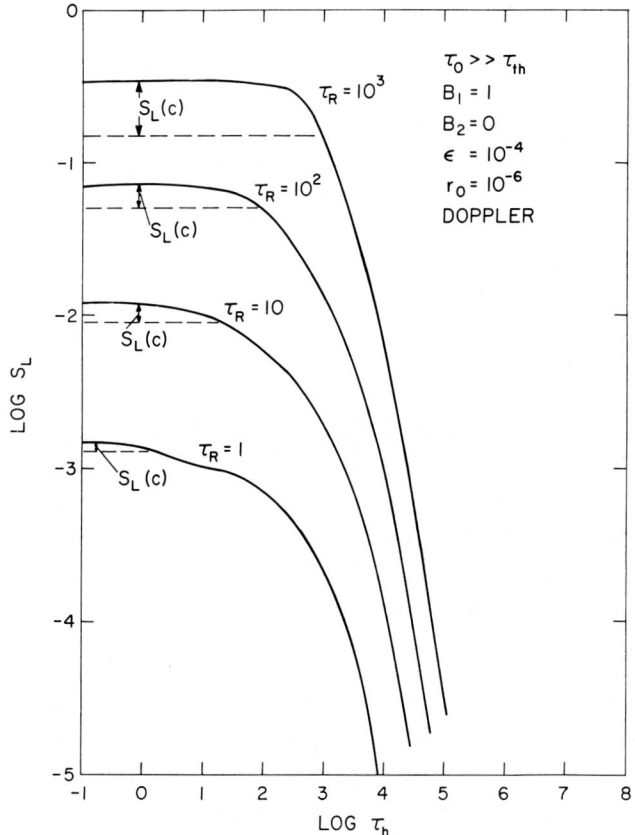

Fig. III-17. Variation of the source functions along the radial direction of imbedded cylinders of optical radii τ_R for depths in the ambient atmosphere larger than τ_{th}. The dashed curves indicate the values of S_L in the equivalent isolated cylinder ($S_L(c)$) (Jones, 1970 (Thesis), University of Colorado, courtesy National Center for Atmospheric Research Special Publication).

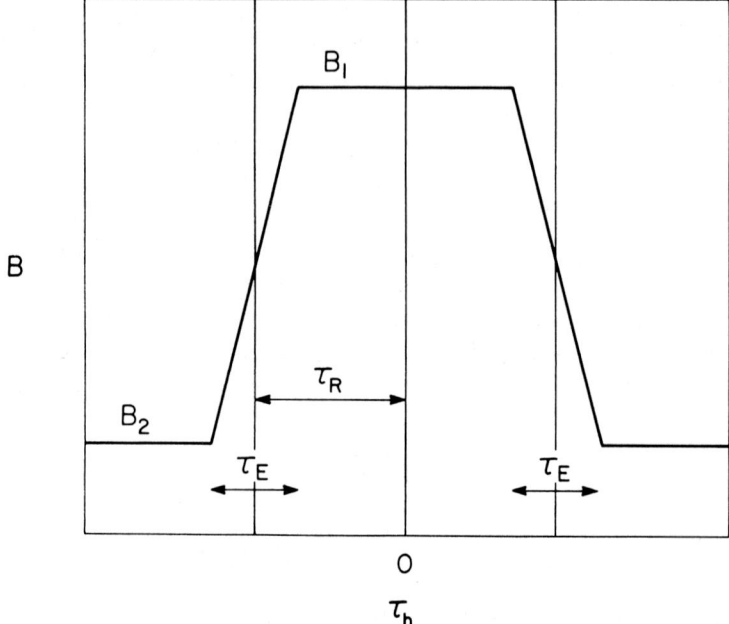

Fig. III-18. The assumed variations of B for the imbedded cylinders shown in Figure (III-17). The ratio of edge thickness τ_E to radial thickness τ_R is 0.1 (Jones, 1970 (Thesis), University of Colorado, courtesy National Center for Atmospheric Research Special Publication).

$\tau_{th} \approx 10^4 \pi^{1/2}$. The assumed variation in B in a horizontal cut through the cylinder is shown in Figure III-18. For the plot in Figure III-17, the ratio of edge thickness, τ_E, to radial thickness, τ_R, is 0.1. B_1 is taken as unity and B_2 as zero. The actual calculations were made with $B_2 = 10^{-4}$, but the solution for S_L in the ambient atmosphere far from the cylinder has been subtracted from S_L in the figure. All of the cases illustrated are effectively thin.

Several effects are notable in Figure III-17. Note firstly that the 'width' of the S_L curves are roughly constant. They each fall to a tenth of their maximum value between τ_h values of approximately 2×10^3 and 4×10^3. At the thermalization distance $\tau_h \approx 10^4$ the value of S_L has fallen to about one percent of its maximum value for each of the curves. To a somewhat surprising degree, the curves for different τ_R are simply displaced vertically from each other. They differ appreciably from a constant shape only in the region of $\tau_h = \tau_R$. This tendency for a constant width of S_L independently of τ_R must result from the fact that the radiation tends to diffuse outwards to approximately τ_{th}. The radius of the cylinder is felt mainly in fixing the amplitude of S_L and the point at which S_L saturates.

We note secondly from Figure III-17 that S_L tends to saturate just inside the edge of the cylinder. This result is perhaps not too surprising in view of the fact that $\tau_R \ll \tau_{th}$, i.e., all of the cylinders are effectively thin.

The values of S_L in Figure III-17 lie everywhere above the values of $S_L(c)$ for the corresponding isolated cylinder. Suppose that S_L is made up of two components, one given by $S_L(c)$ and a second, $S_L(s)$, due to back scattering from the ambient atmosphere. For $S_L(c)$ at the center of the cylinder, we write $S_L(c, \max) = \pi^{1/2}\varepsilon\tau_R$. Then

$$S_L(s, \max) = S_L(\max) - S_L(c, \max)$$
$$= S_L(\max) - \pi^{1/2}\varepsilon\tau_R. \qquad \text{(III-29)}$$

Values of $S_L(s, \max)$ are indicated by the dashed lines in Figure III-17. Note that the dashed lines intersect the S_L curves near $\tau_h = \tau_R$, i.e., just at the edge of the cylinder. The actual values of S_L at the edge of the cylinder are much larger than $\varepsilon\tau_R^{1/2}$, which is the expected value of $S_L(c)$ for the isolated cylinder. Hence the equality of S_L and $S_L(s, \max)$ at the edge of the cylinder means simply that, at this point, the ambient medium is back scattering just the amount $S_L(s, \max)$. Since $S_L(s, \max)$ is defined as the back scatter at the center of the cylinder it follows that $S_L(s)$ is constant inside the cylinder. This is to be expected from the fact that $\tau_R \ll \tau_{th}$. Inside the cylinders, therefore, $S_L = \text{constant} + S_L(c, \tau_R)$, where the constant is $S_L(s)$.

The above interpretation would suggest that S_L should remain nearly constant around $\tau_h = \tau_R$ then rise gradually to $S_L(\max)$ at $\tau_h = 0$. This is not what Figure III-17 shows, however. Instead S_L seems to saturate just inside the cylinder. The explanation for this is to be found in our oversimplification of the problem. The cylinder itself contributes to the buildup of S_L more effectively when surrounded by an opaque medium than when it is isolated. In other words, it adds to the back scatter term.

The ratio of $S_L(s)$ to $S_L(c, \max)$ varies from about 0.9 for $\tau_R = 10^3$ to about 7 for $\tau_R = 1$. Corresponding values of $S_L(\max)/S_L(c, \max)$ are about 2 and 8. To understand this behavior it is helpful to reverse the optical depth coordinate τ_h and view the cylinder from the ambient medium. What we are interested in is the effect of the ambient atmosphere in building up radiation within the cylinder. If the cylinder has a thickness τ_R that greatly exceeds the thermalization length, we will find $S_L = B_1 = S_L(c, \max)$ throughout most of the cylinder. At the border of the cylinder itself $S_L(s)$ will be less than or equal to B_1, but $S_L(s)$ will not penetrate the cylinder much beyond its border. Hence, at the center of the cylinder we should have $S_L(\max)/S_L(c, \max) = 1$ and $S_L(s)/S_L(c, \max) \approx 0$.

If we now begin to decrease τ_R below τ_{th}, back radiation from the ambient atmosphere will penetrate throughout the cylinder. Thus, $S_L(s)$ will no longer be zero at the center of the cylinder. As τ_R decreases still further $S_L(s)$ will grow relative to $S_L(c, \max)$, as is observed.

The area defined by $\int S_L \, d\tau_h$ is of interest in connection with the conservative properties of the system. Numerical integration of the curves in Figure III-17 gives the following results in units of $S_L(\max)$:

τ_R	1	10	10^2	10^3
Area	600	700	1000	1700

These areas are approximately constant at 10^3, which is $\tau_{th}^{3/4}$. They suggest the result that $\int S_L \, d\tau_h \approx S_L(\max) \, \tau^{3/4} \approx$ constant. By inspection of Figure III-17, we find that the constant is approximately $\tau_R B_1$. This is not a surprising result, since it states only that $\varepsilon \int S_L \, d\tau_h \approx \varepsilon B_1 \tau_R$, which represents simply the conservation of energy. At large optical depths, $\tau_0 \gg \tau_{th}$, which is the case under consideration, energy is not diffusing vertically and the conservation condition should hold in any direction.

The value of $S_L(c, \max)$ is given by $(\tau_R/\tau_{th}) B_1$. Thus the preceding result for $S_L(\max)$ gives

$$\frac{S_L(\max)}{S_L(c, \max)} \approx \tau_{th}^{1/4}.$$

To test the consistency of this result, we plot in Figure III-19 the values of $S_L(\max)/S_L(c, \max)$ vs. τ_R obtained directly from the curves in Figure III-17. Note that the curve drawn through the point appears to be approaching an asymptote of approximately $\tau_{th}^{1/4} \approx 11$. The results therefore seem consistent.

To a rather good approximation the points in Figure III-19 can be represented by

$$\frac{S_L(\max)}{S_L(c, \max)} = \tau_{th}^{1/4} \exp - (\tfrac{1}{3}\tau_R^{0.3}).$$

This scaling law is purely heuristic in that the dependence on τ_R is rather slow and could take a variety of forms. That $S_L(\max)/S_L(c, \max)$ should depend upon τ_R can be understood in a qualitative way if we consider the penetration of back scattered radiation into cylinders of different τ_R. For very large τ_R, no back radiation will penetrate and $S_L(\max)/S_L(c, \max)$ will equal unity (both will equal B_1). As τ_R is decreased to τ_{th} back radiation will penetrate from the ambient atmosphere. At

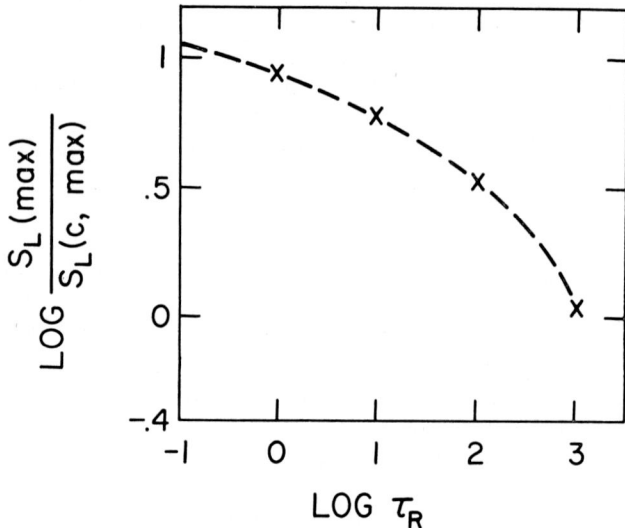

Fig. III-19. The ratio of $S_L(\max)$ to $S_L(c, \max)$ vs. the radial optical thickness of the cylinder.

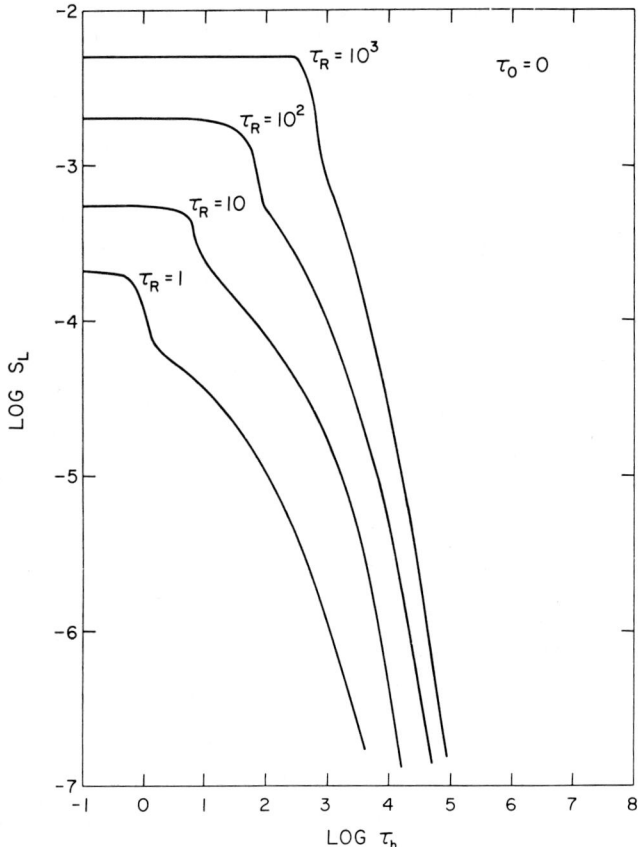

Fig. III-20. Illustration of the radial variation of S_L at the surface of the atmosphere ($\tau_0 = 0$) for the same imbedded cylinders as are illustrated in Figure III-17 (Jones, 1970 (Thesis), University of Colorado, courtesy National Center for Atmospheric Research Special Publication).

$\tau_R < 1$ the cylinder has the appearance of a geometrical line source to the ambient atmosphere. It loses this line character when $\tau_R \gg 1$ and geometric effects from the size of the source become important. We note from Figure III-19 that S_L (max) ≈ 2 S_L (c, max), i.e., S_L (s) $\approx S_L$ (c, max), for $\tau_R = 10^{2.5}$. This is roughly consistent with the preceding result that the characteristic diffusion length for the energy is $\tau_{th}^{3/4}$ rather than τ_{th}.

It is of interest to note in connection with the dependence of S_L (max)/S_L (c, max) on τ_{th} that the limit $\tau_{th} \to 1$ corresponds to perfect reflection by the ambient atmosphere, hence the result S_L (max)/S_L (c, max) $= B_1/B_1 = 1$. At the opposite extreme of $\tau_{th} \to \infty$ the ambient atmosphere acts as a perfect transmitter. However, in this case both S_L (s) and S_L (c, max) go to zero. If τ_{th} remains finite within the cylinder, i.e., $\varepsilon > 0$, but is infinite outside the problem is entirely equivalent to the isolated cylinder problem, and we will again find S_L (max)/S_L (c, max) $= 1$. It is not clear how the scaling

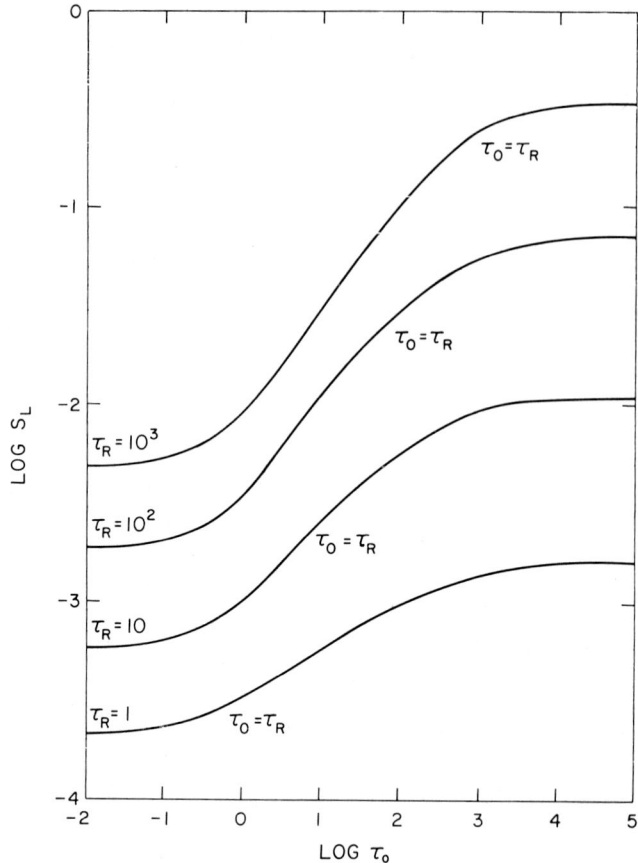

Fig. III-21. The axial variation of S_L for the imbedded cylinder illustrated in Figures III-17 and III-20 (Jones, 1970 (Thesis), courtesy University of Colorado, National Center for Atmospheric Research Special Publication).

law applies in this case. Evidently, the scaling with $\tau_{th}^{1/4}$ cannot continue for indefinitely large τ_{th} in the ambient atmosphere.

Figure III-20 exhibits the variation of surface values of S_L with τ_h for the same imbedded cylinders as in Figure III-17. Note that boundary effects at the vertical surface $\tau_0 = 0$ are now dominating the problem. The maximum values of S_L are given to good approximation by the corresponding values for the isolated cylinder, viz., $\pi^{1/4} \varepsilon \tau_R^{1/2}$. Lateral diffusion is still present but is considerably less enhanced. The widths of the S_L curves at a tenth maximum are now given approximately by 10 τ_R.

The strikingly different behavior of S_L at large τ_0 as opposed to small τ_0 is of considerable interest. It implies, as noted by Jones (1970), that a bright vertical column in the solar atmosphere with $\tau_R < \tau_{th}$, if viewed end on, would have one characteristic width at the center of a line and a considerably broader width in the line wings. The width near $\tau_0 = 1$ would be only somewhat larger than τ_R whereas at large τ_0 (line wings) the width would be set by τ_{th}.

Figure III-21 shows the axial variation of S_L. For $\tau_0 < \tau_R$ the curves are essentially the same as for the isolated cylinders. Rather than saturate to constant S_L for $\tau_0 > \tau_R$, as they do for the isolated cylinder, the buildup in S_L continues until $\tau_0 \approx \tau_{th}$. The reasons for this continued buildup of S_L are implicit in the discussion of Figure III-17.

The results shown in Figures III-17, III-20, and III-21 were obtained by Jones (1970), who expresses some concern about their reliability. They were obtained using the flux divergence approach. Initial attempts by Jones (1970) to verify the results using Rybicki's method gave discordant results inside the cylinder. Later computations by Jones (1970), however, corrected some erroneous numerical procedures and brought the two sets of results into agreement. The Green's functions employed in Rybicki's method exhibit the behavior of S_L in semi-analytic form and should be of help in deriving scaling laws. Such applications are in progress but results are not available at this writing.

References

Athay, R. G.: 1970, *Solar Phys.* **11**, 347.
Athay, R. G. and Skumanich, A.: 1968a, *Solar Phys.* **3**, 181.
Athay, R. G. and Skumanich, A.: 1968b, *Astrophys. J.* **152**, 141.
Athay, R. G. and Skumanich, A.: 1968c, *Solar Phys.* **4**, 176.
Auer, L. H.: 1968, *Astrophys. J.* **153**, 783.
Avrett, E. H.: 1965, *Proceedings 2nd Harvard-Smithsonian Conf. on Stellar Atm.*, Smithsonian Astrophys. Obs. Spec. Report No. 174.
Avrett, E. H. and Hummer, D. G.: 1965, *Monthly Notices Roy. Astron. Soc.* **130**, 295.
Avery, L. W.: 1969, *A Monte Carlo Calculation of Radiative Transfer in Cylinders with Application to Solar Spicules*, Thesis, Univ. of Colorado, Boulder.
Avery, L. W. and House, L. L.: 1968, *Astrophys. J.* **152**, 493.
Avery, L. W., House, L. L., and Skumanich, A.: 1969, *J. Quant. Spectr. Radiative Transfer* **9**, 519.
Beebe, H.: 1971, *Solar Phys.* **17**, 304.
Dumont, S.: 1967, *Ann. Astrophys.* **30**, 421.
Edmonds, F. N.: 1955, *Astrophys. J.* **121**, 418.
Engvold, O.: 1967, *Solar Phys.* **2**, 234.
Field, G. B.: 1959, *Astrophys. J.* **129**, 551.
Finn, G. D.: 1967, *Astrophys. J.* **147**, 1085.
Hammersley, J. M. and Handscomb, D. C.: 1964, *Monte Carlo Methods*, John Wiley, New York.
Hearn, A. H.: 1964a, *Proc. Phys. Soc.* **84**, 11.
Hearn, A. H.: 1964b, *Culham Lab. Report CLM-R* 36.
Henyey, L. G.: 1940, *Proc. Nat. Acad. Sci.* **26**, 50.
House, L. L. and Avery, L. W.: 1968, *Resonance Lines in Astrophysics*, National Center for Atmospheric Research, Boulder.
Hummer, D. G.: 1962, *Monthly Notices Roy. Astron. Soc.* **125**, 21.
Hummer, D. G.: 1964, *Astrophys. J.* **140**, 276.
Hummer, D. G.: 1965, *Proceedings 2nd Harvard-Smithsonian Conf. on Stellar Atm.*, Smithsonian Astrophys. Obs. Spec. Report No. 174.
Hummer, D. G.: 1968, *Monthly Notices Roy. Astron. Soc.* **138**, 73.
Hummer, D. G.: 1969, *Monthly Notices Roy. Astron. Soc.* **145**, 95.
Hummer, D. G.: 1970, *Astrophys. Letters* **5**, 1.
Hummer, D. G. and Rybicki, G.: 1968, *Astrophys. J. Letters* **153**, L107.
Jefferies, J. T. and Thomas, R. N.: 1958, *Astrophys. J.* **127**, 667.
Jefferies, J. T. and Thomas, R. N.: 1960, *Astrophys. J.* **131**, 695.
Jefferies, J. T. and White, O. R.: 1960, *Astrophys. J.* **132**, 767.
Jones, H. P.: 1970, *Line Formation in Multi-Dimensional Media*, Thesis, Univ. of Colorado, Boulder.

Jones, H. P. and Skumanich, A.: 1968, *Resonance Lines in Astrophysics*, National Center for Atmospheric Research, Boulder.
Kalkofen, W.: 1970, Extended Atmosphere Star, NBS Special Publ. **332**, 120.
Kourganoff, V.: 1963, *Basic Methods in Transfer Problems*, Dover, New York.
Kulander, J. L.: 1968, *J. Quant. Spectr. Radiative Transfer* **8**, 273.
Linsky, J. L.: 1968, *Smithsonian Astrophys. Obs. Spec. Report No. 274.*
Linsky, J. L. and Avrett, E. H.: 1970, *Publ. Astron. Soc. Pacific* **82**, 169.
Magnan, C.: 1968, *Astrophys. Letters* **2**, 213.
Mathis, J. S.: 1968, *Resonance Lines in Astrophysics*, National Center for Atmospheric Research, Boulder.
Osterbrock, D. E.: 1962, *Astrophys. J.* **135**, 195.
Oxenius, J.: 1965, *J. Quant. Spectr. Radiative Transfer* **5**, 771.
Rybicki, G. B.: 1965, *Transfer of Radiation in Stochastic Media*, Thesis, Harvard Univ., Cambridge.
Rybicki, G. B.: 1968, Results given in Wilson, 1968.
Schreider, Y. A.: 1966, *The Monte Carlo Methods*, Pergamon, New York.
Sobolev, V. V.: 1965a, *Vestnik LCV* **5**, 85.
Sobolev, V. V.: 1965b, *Vestnik LCV* **11**, 99.
Spitzer, L.: 1936, *Monthly Notices Roy. Astron. Soc.* **96**, 794.
Spitzer, L.: 1944, *Astrophys. J.* **99**, 1.
Skumanich, A.: 1967, *Astron. J.* **72**, 828.
Skumanich, A.: 1969, private communication.
Thomas, R. N.: 1957, *Astrophys. J.* **129**, 551.
Unno, W.: 1952a, *Publ. Astron. Soc. Japan* **3**, 158.
Unno, W.: 1952b, *Publ. Astron. Soc. Japan* **4**, 100.
Unno, W.: 1955, *Publ. Astron. Soc. Japan* **7**, 81.
Warwick, J. W.: 1955, *Astrophys. J.* **121**, 190.
Weisskopf, V.: 1933, *Z. Physik* **85**, 451.
Weisskopf, V. and Wigner, E.: 1930a, *Z. Phys.* **63**, 54.
Weisskopf, V. and Wigner, E.: 1930b, *Z. Phys.* **65**, 18.
Wilson, P. R.: 1968, *Astrophys. J.* **151**, 1029.
Wilson, P. R.: 1969, *Astrophys. J.* **155**, 715.
Woolley, R. v.d. R.: 1938, *Monthly Notices Roy. Astron. Soc.* **98**, 624.
Zanstra, H.: 1941, *Monthly Notices Roy. Astron. Soc.* **101**, 273.
Zirker, J. B.: 1968, *Solar Phys.* **3**, 164.

CHAPTER IV

THE MULTILEVEL CASE: TWO OR MORE LINES

1. General Comments

In the discussion of the two-level problem in Chapter III, several perturbing influences were considered. Most of these can be incorporated in the solution for S_L in a straightforward way without numerical or conceptual difficulty. Only the frequency and angular dependence of S_L, differentially moving atmospheres and non-planar geometries present substantial added numerical difficulties.

A further perturbation on S_L arises when additional energy levels are added to the model atom. The added energy levels provide new sources and sinks for photons, which in some cases may exceed in importance the sinks and sources already represented by ε and δ. Multilevel effects arising from interlocking between different spectral lines of the same atom may be taken into account without conceptual difficulty. They increase considerably the numerical scope of the problem though not in a prohibitive way.

One point should be made clear at the outset: the addition of more energy levels with their added photon sinks and sources does not necessarily lead to a closer approximation to LTE. Often the reverse is true. The added energy levels provide new escape routes for photons that may substantially increase the thermalization length in a particular line. This may be the major effect of the added levels, in which case the departure of S_L from B extends deeper in the atmosphere. Conversely, the added levels may lead to more rapid thermalization with closer agreement between S_L and B, or to closer coupling between S_L and J_c which, indirectly, gives closer agreement between S_L and B. It is not usually obvious, in any particular case, in which direction the major effects will go.

In multilevel problems it becomes important to distinguish between the degradation length and the thermalization length. It is still the thermalization length that determines the convergence of S_L and B at large τ_0. However, in the multilevel problem there are photon sink and source terms that are non-thermal. Photon energy often changes from one line to another without coupling to the thermal reservoir. Such photon degradation effects may inhibit the build-up of photons in a given line by providing an easy escape route for the photons. This results in a cooperative behavior of the lines and links the individual source functions together. In the words of Jefferies (1968): "The generalization from coherent to non-coherent scattering made necessary the consideration of the simultaneous transfer of photons at all frequencies in the line along with the parallel recognition that the photons belonged to the whole line and no longer to a particular frequency. Correspondingly, when we generalize our discussion to a multilevel atom, we should be prepared ... to take account of the fact that the photons no longer pertain to a particular line but in a real sense belong to the radiation fields in the totality of transitions emitted by the atom."

The cooperative behavior of photons from interlocked lines tends to couple the thermalization length for the different lines. However, the degradation lengths remain properties of the individual transitions. A good example of this distinction between degradation and thermalization lengths is provided by the Lyman-β line of hydrogen. Lyman-β photons branch into Balmer-α and Lyman-α photons with a relative probability of $A_{32}/A_{31}=0.8$. Thus, for each 2.2 scattering events in Lyman-β one of the photons degrades to Balmer-α and Lyman-α.

Because the opacity in Balmer-α is orders of magnitude lower than in Lyman-β the Balmer-α photon will very likely escape the scattering environment in those regions of the atmosphere where the Lyman lines are formed. Thus, the degradation of the Lyman-β photons is irreversible. The effect of this short degradation length is to inhibit any buildup of photons in the Lyman-β line; they always have an easy escape route. This suppresses the $\int J_\nu \Phi \, d\nu$ term in the Lyman-β source function and gives the source function an abnormally low value, as will be illustrated numerically in Section 8 of this chapter.

The easy escape of Lyman-β photons does not influence in any appreciable way the thermalization length of either Lyman-α or Lyman-β. For a model hydrogen atom with three energy levels the three lines thermalize together in a cooperative way (cf., Cuny, 1967; Jefferies, 1968). If the atomic parameters and relative energies of the three energy levels are changed, Lyman-β and Balmer-α may pair together with a common thermalization depth whereas Lyman-α may have quite a different thermalization depth (cf., Section 8 of this chapter).

Multilevel line transfer problems are not basically more difficult to treat than the two-level problem except in numerical procedures. Each individual source function in the multilevel problem is of the same general form as it is for the two-level problem. It is true in the multilevel problem that ε^* and $\varepsilon^\dagger$ are more complicated functions of atomic parameters than is ε and that ε^* and $\varepsilon^\dagger$ depend upon the full set of source functions which are initially unknown. Nevertheless, ε^* and $\varepsilon^\dagger$ may be reduced to numerical form through the use of the escape coefficients in the steady state equations. The only essential complication, therefore, is that the solutions must be extended to the full set of lines either by an iterative technique or by a simultaneous solution.

The simplest extension of the two-level problem to the multilevel case is through an iterative technique. To iterate the equations one need only follow the simple scheme of:

(i) Assume an initial set of ϱ_{ji}, usually $\varrho_{ji}=0$.
(ii) Compute ε^*_{ji} and $\varepsilon^\dagger_{ji}$ from the resulting rate coefficients.
(iii) Solve the steady state equations for n_j or b_j and compute τ_{ij} and δ_{ij}.
(iv) Solve Equation (II-46) for S_L for each transition.
(v) Use this set of S_L and solve Equation (II-88) for a new set of ϱ_{ji}.

(vi) Repeat steps (ii) through (v) until convergence is achieved.

In most cases convergence is achieved in relatively few iterations (four to six). However, I iterations for L lines represents a total of LI individual solutions for S_L and even a simple multilevel problem may easily require more than an order of magnitude increase in computer time over the simple two level case if done by this method.

The results presented in the remainder of this chapter have each been obtained using the above iterative techniques. In addition to being costly in computer time the iterative technique often encounters an added difficulty of non-consistency in the final converged solutions. This problem is discussed in Section 2 of this chapter.

Excellent methods for solving the coupled transfer and steady state equations for multilevel problems that avoid the two major difficulties of the iterative technique (time and consistency) have been developed by Kalkofen (1968), Auer and Mihalas (1969), Skumanich and Domenico (1970), and Rybicki (1970). The latter three of these methods are outlined in Chapter VIII.

In reality is it not entirely fair to compare multilevel problems to the simple two-level problem discussed in the preceding chapter. All of the problems considered there were for specified values of r_0, which implies that the population is known for the ground state of the transition. If one is discussing a resonance line of an ionized metal in a star of solar type, it is perhaps safe to assume that nearly all of the atoms of that specie are in the ground state of the ion. In this case r_0 is relatively well-known and the problems of Chapter III have direct application. It is not usually the case, however, that r_0 is known initially. Hence r_0 must be solved for simultaneously with S_L. To solve for r_0 it is necessary to solve for the state of ionization of the atom and this cannot be done without including transitions to and from the continuum. The simplest general problem therefore, for all except the resonance lines of the dominant ion, is that of two bound levels plus continuum together with their interlocking transitions.

If the degree of ionization of the atom being considered is independent of the population of the upper level of the line transition, the degree of ionization is uncoupled from S_L and can be treated separately. Thus, r_0 can be determined independently of S_L. In most cases this is not a sufficiently good approximation, however, and r_0 and S_L are coupled. A good example is provided by the Lyman continuum problem. To solve it one must first guess at the ionization of hydrogen in order to obtain τ_0 and r_0. Using this τ_0 and r_0, he then solves the transfer equation to obtain b_1 and a new ionization of hydrogen. The next iteration provides a new b_1 and a third estimate of the ionization. This process is repeated until convergence is achieved. In the case of a line, the continuum transitions may be optically thin so that the continuum radiation field is known and only the line transfer problem need be solved. Even in this case, however, S_L must still be iterated against r_0 using the steady state equations in conjunction with the S_L equation.

In Thomas' (1957) discussion of the line source function, he defined the two-level case as two bound levels plus continuum and he treated the continuum transitions as being optically thin. As noted above, this does not eliminate the need to iterate S_L

against r_0. Furthermore, it introduces continuum transition rates into ε^* and $\varepsilon^\dagger$. More recent usage defines the two-level case as that in which S_L and r_0 are uncoupled, as in Chapter III.

The three level transfer problem has been rather extensively discussed by Avrett and Kalkofen (1968). Much of the following is guided by their work as well as by the multilevel work of Jefferies (1960, 1968) and Finn and Jefferies (1968, 1969). However, each of the following multilevel examples has been computed independently by the author, except where noted otherwise.

Multilevel problems provide a large array of possible configurations and interactions and it is not practical to treat them with the same sense of thoroughness that has been lavished on the two-level atom. The presence of extra energy levels and their coupling effects do not change the basic behavior of the line source function. Coupling to other spectral lines produces cooperative effects that change the thermalization lengths and the detailed behavior of S_L. In the following, we seek to illustrate some of the coupling effects for selected, simple configurations.

Although coupling effects are perhaps more readily apparent when all parameters of the problems are held constant with depth, we choose at this point to use a model atmosphere that is more realistic. This has the advantage of giving the reader a better concept of how S_L behaves in a real atmosphere without markedly obscuring the coupling effects we wish to illustrate. Since the primary reason for adding more energy levels is to introduce more realism into the computations of S_L, the use of a realistic model atmosphere runs parallel with the use of more energy levels.

One effect of coupling between different spectral lines is to partially destroy the simple scaling laws of the two-level atom. The use of a realistic model atmosphere with gradients in B, ε, r_0 and Φ_y further destroy the scaling laws. Thus, the surface values of S_L are no longer given by $(\varepsilon+\delta)^{1/2}$ at the surface and the slope of S_L/B is not generally given by $\tau_0^{1/2}$ for the case of Doppler profiles. The reader should be aware that gradients are present and should note that not all of the differences between S_L in this chapter and S_L in Chapter III for corresponding parameters are due to coupling. Where possible, the effects of coupling and of the model atmosphere will be noted separately. The model atmosphere for which the computations are made is summarized in Table IV-1. Each of the illustrations in this chapter, except those drawn from previous publications, use the model in Table IV-1 and they use a Doppler from for Φ_y. In most of the illustrations, we give values of $\log \tau_{5000}$ to allow the reader to relate the results to the model atmosphere in Table 1V-1. However these are not necessarily the values of τ_c used for computing r_0 and δ. It is advantageous in some cases, to be able to change r_0 without changing the model. This allows us to compare the effects of $\varepsilon \gg \delta$ vs. $\delta \gg \varepsilon$ without changing the thermalization length, for example.

A further change introduced at this point is to work with specific rate coefficients rather than ratios of rates, such as are represented by ε. With each of the illustrations we will give the relevant spontaneous transition probabilities and collision cross-sections. The ε's, δ's and coupling parameters vary with depth in the atmosphere. How-

TABLE IV-1
Model Atmosphere Parameters (Microturbulence = 0)

Height (km)	τ_{5000} (H^-)	T	N_H	N_e
1980	1.9×10^{-11}	9800	6.8×10^{10}	3.9×10^{10}
1965	9.3×10^{-11}	8800	8.1×10^{10}	4.1×10^{10}
1870	1.0×10^{-9}	8000	1.2×10^{11}	5.7×10^{10}
1650	9.0×10^{-9}	7300	2.7×10^{11}	9.2×10^{11}
1350	7.5×10^{-8}	6750	1.1×10^{12}	1.4×10^{11}
1000	6.9×10^{-7}	6250	6.9×10^{12}	1.7×10^{11}
700	5.5×10^{-6}	5900	4.3×10^{13}	2.3×10^{11}
500	3.1×10^{-5}	5400	2.7×10^{14}	2.4×10^{11}
450	5.4×10^{-5}	5000	6.5×10^{14}	2.0×10^{11}
400	1.2×10^{-4}	4700	1.5×10^{15}	2.4×10^{11}
300	6.5×10^{-4}	4350	4.4×10^{15}	5.0×10^{11}
180	3.4×10^{-3}	4450	6.4×10^{15}	6.8×10^{11}
0	2.0×10^{-2}	4600	1.6×10^{16}	1.6×10^{12}
-100	6.7×10^{-2}	4920	4.0×10^{16}	4.0×10^{12}
-170	1.9×10^{-1}	5430	7.5×10^{16}	7.5×10^{12}
-240	5.0×10^{-1}	6100	1.3×10^{17}	1.3×10^{13}
-340	1.8	7000	2.7×10^{17}	2.7×10^{13}

ever, values of ε and δ will be indicated at representative optical depths. Since we are here dealing with multilevel problems, it is necessary to solve simultaneously for the S_L's and δ's.

2. Consistency Checks

In most problems involving numerical iterative techniques there is a question as to whether the solutions obtained are valid. The present problem is no exception.

For a model atom with M energy levels the level populations are uniquely determined by M-1 independent source functions and an absolute population for one energy level. In general, there are more than M-1 allowed transitions between the energy levels and each transition has an associated source function. Thus, some of the source functions are redundant. Their radiation fields must be included in the solutions of the equilibrium equations, however.

The problem of redundancy is readily seen for the case of three energy levels with three permitted transitions. In this case one of the source functions will be redundant. The 2-1 and 3-2 source functions are given by (neglecting stimulated emissions);

$$S_{21} = \frac{2h\nu_1^3}{c^2} e^{-h\nu_{21}/kT} \frac{b_2}{b_1}, \qquad \text{(IV-1)}$$

and

$$S_{31} = \frac{2h\nu_{31}^3}{c^2} e^{-h\nu_{31}/kT} \frac{b_3}{b_1}. \qquad \text{(IV-2)}$$

Since,

$$S_{32} = \frac{2h\nu_{32}^3}{c^2} e^{-h\nu_{32}/kT} \frac{b_3}{b_2}, \qquad \text{(IV-3)}$$

we have

$$S_{32} = \frac{2h\nu_{32}^3}{c^2} \left(\frac{\nu_{21}}{\nu_{31}}\right)^3 \frac{S_{31}}{S_{21}}. \qquad \text{(IV-4)}$$

Solutions of the coupled equilibrium equations and transfer equations give S_{21}, S_{31}, S_{32}, ϱ_{21}, ϱ_{31}, ϱ_{32}, b_1, b_2 and b_3. It is possible therefore to test the self-consistency of the solutions by a variety of methods. A convenient test for each transition is given by

$$\frac{S_{ji}}{B_{ji}} \frac{b_i}{b_j} = 1, \qquad \text{(IV-5)}$$

or, if stimulated emissions are included,

$$\frac{S_{ji}}{B_{ji}} \frac{b_i}{b_j} \frac{1 - \beta b_j/b_i}{1 - \beta} = 1. \qquad \text{(IV-6)}$$

A converged set of solutions obtained by iteration has the property that ϱ_{ji} and b_j satisfy the equilibrium equations closely and S_{ji}, ϱ_{ji} and b_{ji} satisfy the radiative transfer equations closely. It does not necessarily follow that S_{ji} satisfies the equilibrium equations. The tests indicated by Equations (IV-5) and (IV-6) will detect such a failure.

Coupling between the equilibrium equations and the transfer equations is established through ϱ_{ji}. The simplest computation of ϱ_{ji} is via Equation (II-88)

$$\varrho_{ji} = \varepsilon_{ji}^* \frac{B_{ji}}{S_{ji}} - \varepsilon_{ji}^\dagger, \qquad \text{(IV-7)}$$

where B_{ji}/S_{ji} comes from the transfer equations and ε_{ji}^* and $\varepsilon_{ji}^\dagger$ come from the equilibrium equations. This often provides sufficiently accurate coupling and leads to consistent solutions.

An alternative way of computing ϱ_{ji} that sometimes markedly improves the consistency is to use the defining equation

$$\varrho_{ji} = 1 - \frac{\int J_\nu \phi_\nu \, d\nu}{S_{ji}}, \qquad \text{(IV-8)}$$

where $\int J_\nu \Phi_\nu \, d\nu$ is computed from the local value of S_{ji} from the equation

$$\int J_\nu \phi_\nu \, d\nu = (1 + \varepsilon_{ji}^\dagger) S_{ji} - \varepsilon_{ji}^* B_{ji} \qquad \text{(IV-9)}$$

but where S_{ji} in Equation (IV-8) is computed from $S_{ji} = (b_j/b_i) B_{ji}$ using the value of b_j/b_i obtained from the equilibrium equations. This method was proposed by Avrett (1968) to be used at small values of τ_0 when consistency could not be obtained using

Equation (IV-7). It leads to substantial improvement in the rate of convergence to consistent solutions in many cases.

A third method of computing ϱ_{ji} utilizes the consistency conditions in one or more lines directly. As an illustration of this method consider the 3-level case with three transitions. The equilibrium equation for the ground level is

$$b_1 W_1 (C_{12} + C_{13}) = b_2 W_2 (C_{21} + A_{21}\varrho_{21}) + b_3 W_3 (C_{31} + A_{31}\varrho_{31}).$$
(IV-10)

We solve this for ϱ_{31} to obtain

$$\varrho_{31} = \frac{b_1}{b_3} \frac{W_1}{W_3} \frac{C_{12} + C_{13}}{A_{31}} - \frac{b_2}{b_3} \frac{W_2}{W_3} \frac{C_{21} + A_{21}\varrho_{21}}{A_{31}} - \varepsilon_{31}.$$
(IV-11)

Consistency requires that $b_1/b_3 = B_{31}/S_{31}$ and $b_2/b_3 = B_{32}/S_{32}$. The introduction of these equalities into Equation (IV-11) will force the 3-1 and 3-2 solutions towards consistency, particularly if a corresponding equation is used for ϱ_{32}. Equation (IV-7) is used to evaluate ϱ_{21}, which provides the initial coupling between the equilibrium and transfer equation.

If the consistency is quite poor, forcing equality of b_1/b_3 with B_{31}/S_{31} and b_2/b_3 with B_{32}/S_{32} may be too harsh a procedure. A partial forcing of consistency is achieved by setting

$$\frac{b_1}{b_3} = \frac{b_1}{b_3} \left(\frac{b_3}{b_1} \frac{B_{31}}{S_{31}} \right)^x$$

and

$$\frac{b_2}{b_3} = \frac{b_2}{b_3} \left(\frac{b_3}{b_2} \frac{B_{32}}{S_{32}} \right)^x.$$

This gives

$$\varrho_{31} = \frac{b_1}{b_3} \left(\frac{b_3}{b_1} \frac{B_{31}}{S_{31}} \right)^x \frac{W_1}{W_3} \frac{C_{12} + C_{13}}{A_{31}}$$
$$- \frac{b_2}{b_3} \left(\frac{b_3}{b_1} \frac{B_{32}}{S_{32}} \right)^x \frac{W_2}{W_3} \frac{(C_{21} + A_{21}\varrho_{21})}{A_{31}} - \varepsilon_{31}.$$
(IV-12)

The corresponding equation for ϱ_{32} is

$$\varrho_{32} = \frac{b_2}{b_3} \left(\frac{b_3}{b_2} \frac{B_{32}}{S_{32}} \right)^x \frac{W_2}{W_3} \frac{(C_{23} + C_{21} + A_{21}\varrho_{21})}{A_{32}}$$
$$- \frac{b_1}{b_3} \left(\frac{b_3}{b_1} \frac{B_{31}}{S_{31}} \right)^x \frac{W_1}{W_3} \frac{C_{12}}{A_{32}} - \varepsilon_{32}.$$
(IV-13)

By setting x equal to, say, $\frac{1}{4}$ the requirement for consistency is gradually forced upon successive iterations. It is necessary to compute at least one of the ϱ_{ji} from S/B as in Equation (IV-7). However, if two of the transitions are forced to consistency the third is necessarily consistent.

After the ϱ_{ji} are computed from Equations (IV-7), (IV-12), and (IV-13) they are substituted into the equilibrium equations to obtain a new set of b_j and a new set of ε^*, $\varepsilon^\dagger$ and r_0.

Still a fourth method for improving consistency involves the use of S_{ji} to compute b_j/b_i via Equation (IV-6). These values of b_j/b_i (together with one additional b_j) can be used in the steady state equations to obtain the new values of ϱ_{ji}, ε^*_{ji} and $\varepsilon^\dagger_{ji}$ for the next iteration. The one additional b_j required, say, b_i, can be obtained in the usual way by solving Equation (IV-7) for ϱ_{ji} and using these ϱ_{ji} in the steady state equations to solve for b_j. In other words one uses the steady state equations twice: once with prescribed ϱ_{ji} (Equation (IV-7)) to obtain one of the b_j and once with prescribed b_j/b_i (Equation (IV-6)) to obtain ϱ_{ji}. The new set of b_j and ϱ_{ji} obtained in this way are sufficient to determine new τ_{ji} and ε^*_{ji} and $\varepsilon^\dagger_{ji}$.

Multilevel problems have not yet been examined in enough detail to offer general rules for the best methods for computing ϱ, ε^* and $\varepsilon^\dagger$. One normally starts with the simplest methods, say, the use of the lower level equations to compute ε^* and $\varepsilon^\dagger$ and Equation (IV-7) to compute ϱ. If this fails to give consistency, consistent solutions may often be obtained simply by switching to the full set of equilibrium equations for the computation of ε^* and $\varepsilon^\dagger$. Further improvement is found by computing ϱ following one of the alternate procedures outlined in the above discussion. Although each of these methods improves consistency, none guarantees it.

In spectral configurations involving, say, three to five spectral lines, satisfactory solutions can be quite readily obtained using iterative techniques with the simple two-level formulation and appropriate ε^* and $\varepsilon^\dagger$. When many lines are required or when many computations are to be done, however, use should be made of the improved techniques outlined in Chapter VIII.

3. Two Levels Plus Continuum

Addition of a continuum level to the two-level problem provides one additional source term and one additional sink term in S_L. The added source term arises from ionization from level 1 followed by recombinations to level 2. Similarly, the added sink term takes the reverse route; ionization from level 2 followed by recombination to level 1.

It is commonly the case in stellar atmospheres that collisional ionizations and recombinations occur much less frequently than radiative ionizations and radiative recombinations. Furthermore, the free-bound continua do not usually have opacities that are large relative to the more general continuum absorption, which means that the radiative intensities are formed in the deeper atmpsphere. In the line forming layers for many resonance lines, therefore, the atmosphere is transparent to the photoionizing radiation and the radiation intensity is uncoupled from the local temperature. This means that the photoionization rates and, hence, both the added sink and source terms are essentially independent of the local temperature. In the two level case this is not true. The source term is proportional to the Planck function.

It might be expected, therefore, that the added continuum level will, under certain conditions, change the character of S_L markedly.

To illustrate this distinction more clearly, we consider the form of $\varepsilon^\dagger$ and ε^* for two bound levels and a continuum. The steady-state equations are:

$$(P_{12} + P_{1c}) b_1 - P_{21} b_2 - P_{c1} = 0$$

and
$$-P_{12} b_1 + (P_{21} + P_{2c}) b_2 - P_{c2} = 0,$$
(IV-14)

from which we find

$$\frac{b_2}{b_1} = \frac{P_{12}(P_{c1} + P_{c2}) + P_{1c} P_{c2}}{P_{21}(P_{c1} + P_{c2}) + P_{2c} P_{c1}}.$$
(IV-15)

Define

$$p_{c1} = \frac{P_{c1}}{P_{c1} + P_{c2}}$$
(IV-16)

and

$$p_{c2} = \frac{P_{c2}}{P_{c1} + P_{c2}}.$$
(IV-17)

We then have

$$\frac{b_2}{b_1} = \frac{P_{12} + P_{1c} p_{c2}}{P_{21} + P_{2c} p_{c1}}$$

$$= \frac{W_1}{W_2} \frac{B_{12} \int J_\nu \Phi_\nu \, d\nu + C_{12} + R_{1c} p_{c2}}{A_{21} + C_{21} + R_{2c} p_{c1}}$$

$$= \frac{\dfrac{1}{B} \int J_\nu \Phi_\nu \, d\nu + \varepsilon + \dfrac{W_1}{W_2} \dfrac{R_{1c}}{A_{21}} p_{c2}}{1 + \varepsilon + \dfrac{R_{2c}}{A_{21}} p_{c1}},$$
(IV-18)

where we have used $(W_1/W_2)(B_{12}/A_{21}) = 1/B$ and $\varepsilon = C_{21}/A_{21}$. Stimulated emissions are neglected for convenience. The source function for the line is given by

$$S_L = \frac{b_2}{b_1} B = \frac{\int J_\nu \Phi_\nu \, d\nu + \varepsilon B + \dfrac{2h\nu^3}{c^2} \dfrac{\omega_1}{\omega_2} \dfrac{R_{1c}}{A_{21}} p_{c2}}{1 + \varepsilon + \dfrac{R_{2c}}{A_{21}} p_{c1}}.$$
(IV-19)

We rewrite this, following the notation of Thomas (1957) and Jefferies (1968), as

$$S_L = \frac{\int J_\nu \phi_\nu \, d\nu + \varepsilon B + \eta B^*}{1 + \varepsilon + \eta}.$$
(IV-20)

Note that the ratio of the photoionization source to the collision source is given by

$$\frac{\eta B^*}{\varepsilon B} = \frac{R_{1c}}{C_{12}} p_{c2}, \qquad \text{(IV-21)}$$

and the ratio of the photoionization sink to the collisional sink is given by

$$\eta/\varepsilon = (R_{2c}/C_{21}) p_{c1}. \qquad \text{(IV-22)}$$

To the approximation that collisional ionization may be neglected, the ionization rates are given by

$$R_{ic} = 4\pi \int (J_\nu \alpha_\nu / h\nu)\, d\nu, \qquad \text{(IV-23)}$$

where α_ν is the atomic absorption cross-section for photoionization. These rates are independent of density and essentially independent of temperature in the higher layers of the atmosphere where J_ν is approximately constant. The relative recombination probabilities, p_{ci}, are slow functions of temperature and independent of density. Thus, the added sink and source terms due to the continuum are essentially constant in the line forming layer. By contrast, ε is directly proportional to electron density (Chapter II), and the source term, εB, is strongly temperature dependent as well.

Thomas (1957) noted the different character of the two source terms and argued on this basis that resonance lines should fall into one of two classes: collisionally controlled lines for which $\varepsilon B \gg \eta B^*$ and $\varepsilon \gg \eta$ and photoelectrically controlled lines in which the reverse inequalities hold.

The photoionization rates are strongly influenced by the distance of the energy levels from the continuum: If the levels are relatively near the continuum J_ν will be large and R_{1c} and R_{2c} will be relatively large. However, if the levels are far from the continuum J_ν will be small and R_{1c} and R_{2c} will be small.

For an adopted collisional excitation cross-section of πa_0^2 the collisional excitation rate is

$$\begin{aligned}
C_{12} &= \pi a_0^2 \int_{v_0}^{\infty} v n_e(v)\, dv \\
&= \pi a_0^2 \int_{v_0}^{\infty} 4\pi n_e \left(\frac{m}{2\pi kT}\right)^{3/2} v^3 e^{-mv^2/2kT}\, dv \\
&= 5.5 \times 10^{-11} n_e T^{1/2} \left(\frac{1.44}{\lambda T} + 1\right) e^{-1.44/\lambda T}. \qquad \text{(IV-24)}
\end{aligned}$$

For $n_e = 10^{11}$, $T = 6000°$, and $\lambda = 4000$ Å, this equation gives $C_{12} = 7.5$ s^{-1}. The corresponding value of $C_{21} = (\tilde{\omega}_1/\tilde{\omega}_2) e^{1.44/\lambda T} C_{12} = 3 \times 10^3$ s^{-1} for $\tilde{\omega}_1 = \tilde{\omega}_2$.

Photoionization rates are given by Equation (IV-23). For ease of illustration we set $J_\nu = B_\nu(T_R)$ and $\alpha_\nu = \alpha_{\nu 0}(\nu_0/\nu)^2$. We then have, neglecting stimulated emission in

$B_\nu(T_R)$,

$$R_{ic} = \frac{4\pi\alpha_{\nu_0}}{h} \frac{2h}{c^2} \int_{\nu_0}^{\infty} e^{-h\nu/kT_R} \, d\nu$$

$$= \frac{8\pi\alpha_{\nu_0}}{c^2} \nu_0^2 \frac{kT_R}{h} e^{-h\nu_0/kT_R}$$

$$= 5.2 \times 10^{11} \frac{\alpha_{\nu_0} T_R}{\lambda_0^2} e^{-1.44/\lambda_0 T_R}. \tag{IV-25}$$

Equation (IV-25) may be used with given values of T_R and α_{ν_0} to find the values of λ_0 at which $R_{1c} = C_{12}$ and $R_{2c} = C_{21}$. This, in turn, will allow us to state which types of atoms will have $R_{1c} > C_{12}$ and $R_{2c} > C_{21}$ and vice versa.

For $T_R = 6000°$ and $\alpha_{\nu_0} = 0.1 \, \pi a_0^2$, Equation (IV-25) gives $R_{1c} \geqslant 7.5 \text{ s}^{-1}$ at $\lambda_{0,1} \geqslant 1450$ Å and $R_{2c} \geqslant 3 \times 10^3 \text{ s}^{-1}$ at $\lambda_{0,2} \geqslant 2500$ Å. These wavelengths correspond to ionization potentials of $\leqslant 8.5$ and $\leqslant 5.0$ eV, respectively. The neutral metal resonance lines and Hα have ionization potentials satisfying these criteria and therefore are the most likely candidates for the class of photoelectrically controlled lines. For the resonance lines of ionized metals, however, the ionization potentials lie considerably above these limits and it is reasonable to place these lines in the collisionally controlled category. This distinction between neutral metal lines and ionized metal lines was pointed out by Thomas (1957).

Thomas further argued that the different character of the source and sink terms in the source functions for the collisionally controlled and photoelectrically controlled lines would produce a fundamentally different character in the line profiles. Since in the case of collisional control the ε terms are much larger than the photoionizing terms, the collision case is entirely analogous to the strict two-level problem. Collisions within the line transition are the main source and sink terms. Lines in this category should exhibit the properties of S_L as discussed in Chapter III. In particular, a very strong line of this category should respond to a chromospheric rise in T by showing a maximum in S_L in the low chromosphere. The line itself should be self-reversed as are the H and K lines of Ca II and the corresponding lines of Mg II.

In the case of photoelectrically controlled lines the main source and sink terms are constant, independent of T. Thomas (1957) concluded from this that these lines would not respond to a chromospheric temperature rise, which is true if $\eta B^* \gg \varepsilon B$ since S_L will follow the constant driving term ηB^* much more closely than it follows the weaker term εB. It should be noted from Equation (IV-21), however, that the condition $\eta B^* \gg \varepsilon B$ requires that $(R_{1c}/C_{12}) p_{c2} \gg 1$. The quantity p_{c2} is often small, of the order of 0.1 so that R_{1c} often must exceed C_{12} by more than a factor of 10 before the line is truly photoelectrically controlled.

A typical ionization potential for a neutral metal is, say, 6.5 eV. In the numerical examples cited above with $Q_{12} = \pi a_0^2$, $\alpha_{\nu_0} = 0.1 \, \pi a_0^2$ and $T_R = 6000°$, we found $R_{1c} = C_{12}$ for an ionization potential of 8.5 eV and a line of upper excitation potential 3.1 eV

(4000 Å). If the ionization potential is reduced to 6.5 eV and other parameters are held constant, R_{1c} increases to 40 C_{12}. This would perhaps satisfy the condition for photoelectric control. However, we have considerably biased the results in favor of this conclusion, at least for the solar case. A more appropriate value of T_R at 1900 Å (6.5 eV) in the solar spectrum is 5000°, which decreases R_{1c} by about a factor of 13. Furthermore, we have used a relatively large value for α_{v_0} and a relatively small value for Q_{12} for many metals. It is not clear, therefore, whether any of the solar lines do, in fact, qualify as photoelectrically controlled. Of the solar lines that have been considered in some detail, the most likely candidates for this classification would be the Na D lines and Hα. These have ionization potentials from the lower levels of 5.14 and 3.4 eV respectively. However, for the Na D lines $Q_{12} \approx 10\ \pi a_0^2$ and $\alpha_{v_0} \approx 0.0014\ \pi a_0^2$. Also, the longer wavelength of the Na D lines increases the collision rate C_{12} still further. The net result of all these factors leaves the Na D lines collisionally controlled (Johnson, 1964); Athay and Canfield, 1969). Similarly, in Hα the excitation cross-section is of the order of 100 πa_0^2 and the longer wavelength of the line further increases the collisional excitation rate by another factor of ten over the reference case. Thus,

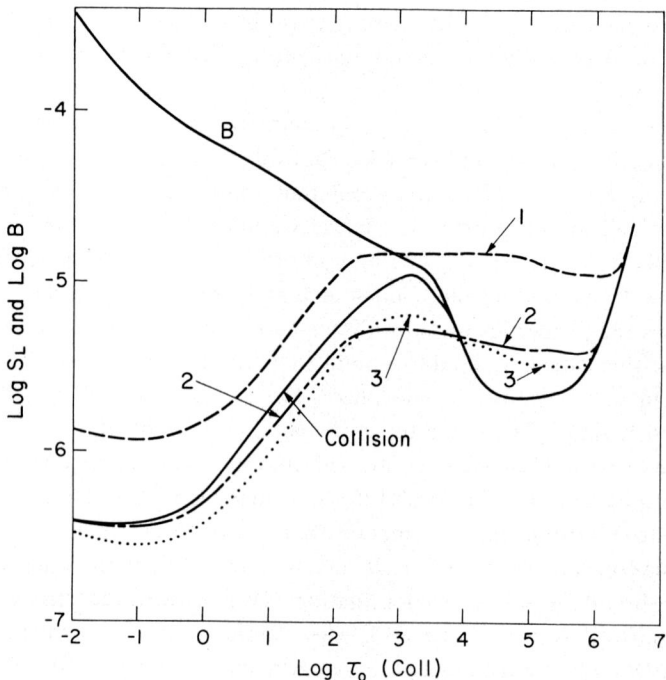

Fig. IV-1. The influence of photo-ionization on S_L for a chromospheric temperature profile. Curve 1 – $T_r = 6000°$, $\eta B^* = 4.2\ \varepsilon B$ and $\eta = 12\varepsilon$; curve 2 – $T_r = 5000°$, $\eta B^* = 6.6\ \varepsilon B$ and $\eta = 10.5\varepsilon$; curve 3 – $T_r = 5000°$, $\eta B^* = 1.4\ \varepsilon B$ and $\eta = 4\varepsilon$. Values of ε and B are given at τ_0 (coll) = 10^3. The line has a wavelength of 4000 Å and $\varepsilon \gg \delta$. The profile is Doppler.

for $n_e = 10^{11}$, $T = 6000°$, $\lambda = 6563$ Å and $Q = 100\ \pi a_0^2$ Equation (IV-24) gives a collisional excitation rate of approximately 5×10^3 s^{-1}. The photoionization rate in the Balmer continuum of hydrogen is of the order of 2×10^4 s^{-1} and the factor equivalent to p_{c2} is about 0.4. Thus $\eta B^*/\varepsilon B$ is of the order of unity.

We illustrate in Figure IV-1 four source functions and their response to a chromospheric temperature rise. One is collisionally controlled with $\eta B^* \ll \varepsilon B$ and $\eta \ll \varepsilon$. The other three have $\eta B > \varepsilon B$ and $\eta > \varepsilon$ at $\log \tau_0 (\text{coll.}) = 3$ by the amounts indicated in the figure. For the example chosen, the upper level of the line lies midway between the ground level and the continuum. In this case the wavelength in B^* is the same as the wavelength of the line and S_L saturates to $B(T_R)$ in the photoelectrically controlled case. Note that for $\eta B^* > 4\ \varepsilon B$ there is relatively little response of S_L either to the chromospheric rise in T or the drop in T in the upper photosphere. For $\eta B^* = 1.4\ \varepsilon B$, S_L rises above B^* at its maximum values by an amount given by

$$S_{\max} = \frac{\varepsilon B}{\varepsilon B + \eta B^*} B + \frac{\eta B^*}{\varepsilon B + \eta B^*} B^*, \tag{IV-26}$$

i.e., in direct proportion to the weighted average of the source terms.

Both εB and ε vary with depth in the example in Figure IV-1. The change in ε is gradual from 5×10^{-4} near the minimum in B to 2×10^{-4} near the maximum in S_L and to 4×10^{-5} near $\tau_0 = 1$.

The type of distinction noted by Thomas (1957) between collisionally controlled and photoelectrically controlled lines is an important consideration in the interpretation of line profiles and in the use of lines as diagnostic tools. Lines with large ηB^* are to be avoided if one is seeking information on the temperature minimum or the chromospheric rise in temperature.

Actually, there are several parameters that influence the ability of a line to respond to temperature structure at a given τ. These include r_0 and a as well as ηB^* and ε. For $\tau_c \leqslant 10^{-4}$ we expect only the very strong lines to respond to changes in B. The only lines in the solar spectrum that are known to exhibit the reversal phenomenon associated with the chromosphere are the Ca II and Mg II lines. But these are precisely the lines with the smallest values of r_0 and, on that basis alone, they should show the strongest effects. Current chromospheric models place the temperature minimum near $\tau_{5000} = 10^{-4}$. There are no lines in the visual solar spectrum other than H and K that have sufficient opacity above $\tau_{5000} = 10^{-4}$ to exhibit a maximum in their source functions. If iron is as abundant as recent data suggest, the resonance lines of Fe II are likely candidates to show self reversal. They have not yet been sufficiently well observed, however.

Many other strong lines have their centers formed in the chromosphere. For these lines the chromospheric rise in temperature may appreciably raise the central intensities without going to the extreme of producing a reversal. None of the strong lines studied show clear evidence of photoelectric control. However, only a few lines have been considered in any detail (cf., Chapter V).

4. Upper Level Multiplets

The essential properties of multiplet behavior can be demonstrated for doublet levels. Higher multiplicity increases the number of possible interactions but does not introduce any basic changes in the effects of multiplet structure. Doublets are easier to handle computationally than higher order multiplets and the doublet problem has the added advantage of having only a single coupling parameter, which is convenient for illustration. Although we restrict the discussion of multiplets in this section to doublets, generalization of the results to higher multiplicity is usually rather obvious.

Upper level doublets, such as the Na D lines, the Ca II H and K lines or the Mg II resonance doublet, provide a good starting point to illustrate coupling effects. We number the two upper levels 2 and 3 and the lower level 1. Initially, we take the excitation energies of the doublet levels as having the same value. This means that the two lines have the same wavelength and eliminates the need to consider the effects of wavelength differences. We treat τ_v separately for each line so that photons in one line cannot be absorbed in the other.

The 3-2 transition is taken to be optically forbidden, but not collisionally forbidden. Thus, the direct coupling between levels 2 and 3 will be by collisions. Cross-sections for such collisions are often large (Chapman and Krauss, 1966) and may involve atom-atom collisions as well as electron-atom collisions. For the sake of illustration, we consider only electron-atom collisions.

Coupling between levels 2 and 3 is not restricted to the direct route of course. Electrons may go from level 2 to level 3 via some fourth level in the sequence 2-4-3 and its reverse. Level 4 may be either a bound level or the continuum. Coupling of this type is often strong and will be illustrated in this section for the case of coupling through the continuum. Coupling through a fourth bound level is illustrated in Section 10.

To understand the effects of the coupling it is helpful to consider the equilibrium equations. Consider the 3-1 equilibrium. Equation (IV-14), with subscript c replaced by 3, gives

$$\frac{b_3}{b_1} = \frac{P_{13}(P_{21} + P_{23}) + P_{12}P_{23}}{P_{31}(P_{21} + P_{23}) + P_{32}P_{21}}$$

$$= \frac{P_{13} + P_{12}p_{23}}{P_{31} + P_{32}p_{21}}. \tag{IV-27}$$

The source function is

$$S_{31} = \frac{\int J_v \Phi_v \, dv + \varepsilon_{31}^* B}{1 + \varepsilon_{31}^\dagger}, \tag{IV-28}$$

where

$$\varepsilon_{31}^* = \varepsilon_{31} + \frac{P_{12}}{W_3 A_{31}} p_{23} \tag{IV-29}$$

and

$$\varepsilon_{31}^\dagger = \varepsilon_{31} + \frac{P_{32}}{W_3 A_{31}} p_{21}. \qquad \text{(IV-30)}$$

The influence of the coupling on S_L is determined by the ratio $(\varepsilon^*/\varepsilon)_{31}$, provided $\varepsilon \gg \delta$. Equation (IV-29) gives

$$\left(\frac{\varepsilon^*}{\varepsilon}\right)_{31} = 1 + \frac{P_{12}}{W_3 C_{31}} p_{23}. \qquad \text{(IV-31)}$$

To a sufficiently good approximation, the value of P_{12} is given by $W_1 B_{12} \int J_\nu \Phi_\nu \, d\nu$, which, in turn, is given approximately by $(b_2/b_1) W_2 A_{21}$. Since we might expect $A_{21} > C_{23}$, the probability p_{23} is given approximately by C_{23}/A_{21}. Hence, we have

$$\left(\frac{\varepsilon^*}{\varepsilon}\right)_{31} = 1 + \frac{W_2}{W_3} \frac{b_2}{b_1} \frac{C_{23}}{C_{31}}, \qquad \text{(IV-32)}$$

$$= 1 + \frac{b_2}{b_1} \frac{C_{32}}{C_{31}}.$$

The corresponding quantity for the 2-1 transition is obtained by transposing subscripts 2 and 3.

Note that when C_{23} and C_{32} are zero ε^* and ε are equal and the two source functions are independent. Each acts exactly as in the equivalent two-level case. Suppose, for illustration, that $\varepsilon_{21} = \varepsilon_{31}$, that $C_{23} = C_{32} = 0$ and that $\tau_{31} = n\tau_{21}$, also that ϕ_y is Doppler for both lines and that $\varepsilon \gg \delta$. Solutions to the two-level problem in an atmosphere with constant B, ε and ϕ_y then give the following results: $S_{21}/B = S_{31}/B = \varepsilon^{1/2}$ at $\tau_0 = 0$, $S_{21}/B = S_{31}/B$ at $\tau_0(21) = \tau_0(31)$, and $S_{21} = B$ at $\tau_0(21) = \varepsilon^{-1}$ and $S_{31} = B$ at $\tau_0(31) = \varepsilon^{-1}$. Since the profile of ϕ_y is Doppler, we have $S/B \propto \tau_0^{1/2}$ for $1 \lesssim \tau_0 \ll \varepsilon^{-1}$. Hence, within this range of τ_0 the two values of S_L have the ratio $S_{31}/S_{21} = [\tau_0(31)/\tau(21)]^{1/2} = n^{1/2}$. If there are gradients in ε and B, the effective values of εB are different for the two lines because of the differences in $\tau_0(21)$ and $\tau_0(31)$ and the ratio of S_{31}/S_{21} will differ somewhat from $n^{1/2}$. However, the basic character of the results will not change, viz., without coupling S_{21} and S_{31} differ at all depths other than $\tau_0 = 0$ and $\tau_0 > \varepsilon^{-1}$. Any amount of coupling between levels 2 and 3 will reduce the difference between S_{21} and S_{31}.

The equilibrium equations give for the ratio b_2/b_3

$$\frac{b_2}{b_3} = \frac{P_{32} + P_{31} p_{12}}{P_{23} + P_{21} p_{13}}. \qquad \text{(IV-33)}$$

Strict equality of the two source functions occurs when $b_2/b_1 = b_3/b_1$, i.e., when $b_2/b_3 = 1$. A sufficient condition for $b_2 = b_3$ is $P_{32} \gg P_{31} p_{12}$ and $P_{23} \gg P_{21} p_{13}$, or, since p_{13} and p_{12} are unity or smaller, $P_{32} \gg P_{31}$ and $P_{23} \gg P_{21}$, as first noted by Waddell (1963). These inequalities may be rewritten as $C_{32} \gg A_{31}$ and $C_{23} \gg A_{21}$. Equation

(IV-33) then gives

$$\frac{b_2}{b_3} = \frac{W_3 C_{32}}{W_2 C_{23}} = 1.$$

In upper level doublets we usually have $R_{31} = R_{21}$. Also, P_{12} is proportional to $B_{12} \int J_\nu \Phi_\nu \, d\nu$ and P_{13} is proportional to $B_{13} \int J_\nu \Phi_\nu \, d\nu$. If the lines are strong the two values of $\int J_\nu \Phi_\nu \, d\nu$ approach values that are approximately equal. Thus, P_{12}/P_{13}

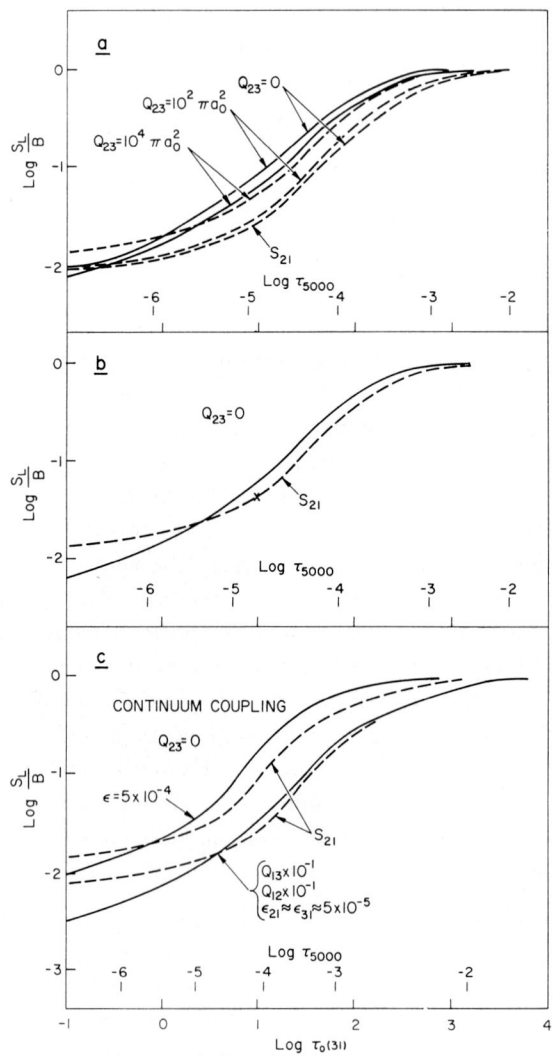

Fig. IV-2. Comparison of upper level doublet source functions at $\lambda 5000$. (a) – collisional coupling between upper levels. $A_{21} = A_{31} = 10^8$, $Q_{12} = \pi a_0^2$, $Q_{13} = 10 \ \pi a_0^2$, $\tilde{\omega}_2 = 1$, $\tilde{\omega}_3 = 10$, $\varepsilon_{21} = \varepsilon_{31} = 5 \times 10^{-4}$, $\varepsilon \gg \delta$. (b) – Same as (a) except $\delta \gg \varepsilon$, $\delta_{21} \approx 6 \times 10^{-3}$ and $\delta_{31} \approx 6 \times 10^{-4}$. (c) – Same as (a) but with a continuum level added, $\alpha_{\nu 0} = 10^{-17}$ cm^{-2} and $T_r = 6000°$.

approaches B_{12}/B_{13} or $\tilde{\omega}_2/\tilde{\omega}_3$. In general $\tilde{\omega}_2/\tilde{\omega}_3$ differs from unity. The condition $C_{32} \gg A_{31}$ and $C_{23} \gg A_{21}$ is therefore a necessary condition for strict equality between S_{21} and S_{31} (Waddell, 1963). On the other hand, the coupling effects between S_{21} and S_{31} bring about approximate equality for much less stringent conditions on C_{23} and C_{32}. The effects of intermediate coupling between levels 2 and 3 are illustrated in Figure IV-2a for a strong resonance doublet. For upper level doublets the absorption oscillator strengths are in the ratio $f_{12}/f_{13} = \tilde{\omega}_3/\tilde{\omega}_2$. The collisional excitation cross-sections are in the same ratio. De-excitation cross-sections and spontaneous transition probabilities are each the same in the two lines, however. These equalities result from the fact that the emission oscillator strengths are the same for the two lines. Because of the equality between the two de-excitation cross-sections on the one hand and the two transition probabilities on the other, the two lines have identical values of ε.

In the limit of strong coupling, Equation (IV-30) gives

$$\frac{\varepsilon_{31}^\dagger}{\varepsilon_{21}^\dagger} = \frac{P_{32}/\tilde{\omega}_3 P_{23}}{P_{23}/\tilde{\omega}_2 P_{32}} = \frac{\tilde{\omega}_2}{\tilde{\omega}_3} \frac{P_{32}^2}{P_{23}^2}$$

$$= \frac{\tilde{\omega}_2}{\tilde{\omega}_3} \left(\frac{\tilde{\omega}_3}{\tilde{\omega}_2}\right)^2 \left(\frac{C_{32}}{C_{23}}\right)^2 = \frac{\tilde{\omega}_2}{\tilde{\omega}_3} \left(\frac{\tilde{\omega}_3}{\tilde{\omega}_2}\right)^2 \left(\frac{\tilde{\omega}_2}{\tilde{\omega}_3}\right)^2$$

$$= \frac{\tilde{\omega}_2}{\tilde{\omega}_3}.$$

Since the thermalization length is given by $\varepsilon^{\dagger-1}$, the thermalization length is increased in one line and decreased in the other as a result of the coupling. This effect is evident in Figure IV-2a.

Note that for $Q_{23} = 10^4 \pi a_0^2$, which gives $\varepsilon^*/\varepsilon \approx 1.1$, the coupling between levels 2 and 3 is fairly complete. For $\tau_0(31) > 1$ the two source functions differ by less than or about 25% for this case where $\tau_0(31) = 10\,\tau_0(21)$. For a smaller τ_0 ratio, the coupling would be closer. At small τ_0, the two source functions at first diverge as the coupling is increased. This results from the fact that level 2 is gaining electrons from level 3 whereas level 3 is losing electrons to level 2. The balance is not complete. When Q_{23} is made sufficiently large the two levels come into balance and the two source functions again approach each other. These effects were first noted by Avrett and Kalkofen (1968).

Note also from Figure IV-2a that the two source functions cross near $\tau_0 = 1$. This tends to make the two lines appear to have the same source functions near line center even when the coupling is relatively weak.

In weaker multiplets it is very probable that δ will exceed ε in each of the lines. This alters the relative values of the source functions because the thermalization length is now set by δ^{-1} and the δ's are necessarily different for the different lines. For Doppler profiles $\delta_{21}/\delta_{31} \approx \tilde{\omega}_3/\tilde{\omega}_2$. Even in the uncoupled case, therefore, both the surface values of S_L and the thermalization lengths will differ for the two lines. Since the weakest line will have the largest value of δ, we expect the weak line source function to increase relative to the strong line source function.

Figure IV-2b illustrates the effect of $\delta \gg \varepsilon$ for the same atom as in Figure IV-2a except that Q_{13} and Q_{12} have each been reduced a factor of ten and r_0 has been increased so that δ_{31} has approximately the same value as ε had in Figure IV-2a.

We note from Figure IV-2b that the two source functions have the same relative appearance as the strongly coupled source functions in Figure IV-2a even though $Q_{23}=0$ for the example shown in Figure IV-2b. In fact the two source functions appear to be about as stongly coupled as when $Q_{23}=10^3 \ \pi a_0^2$ and $\varepsilon \gg \delta$. This is only an apparent similarity in coupling, however, and the two cases are not really analogous. Calculations made with $Q_{23}=10^4 \ \pi a_0^2$ and other parameters the same as in Figure IV-2b change S_{21} and S_{31}, in fact, by about one percent only. The condition $\delta \gg \varepsilon$ is completely dominating S_L.

For the case $\delta \gg \varepsilon$, the effective driving term in S_L is δB rather than $\varepsilon^* B$, and the effect of the coupling between levels 2 and 3 is measured by ε^*/δ. This may be written as $(\varepsilon^*/\varepsilon)/(\delta/\varepsilon)$. For the example in Figure IV-2b, $\varepsilon^*/\varepsilon$ is the same as in Figure IV-2a. However, we now have $(\delta/\varepsilon)_{31} \approx 10$ and $(\delta/\varepsilon)_{21} \approx 10^2$. Thus, the coupling term due to C_{23} and C_{32} are much weaker relative to the main driving term in S_L in this case than in the preceding example where $\varepsilon \gg \delta$. This explains the smaller response of S_{21} and S_{31} to a given value of Q_{23}.

Referring back to Figure III-1b, we note that the result in Figure IV-2b could have been predicted. The agreement between the values of S_L for different values of r_0 in Figure III-1b is even closer than in Figure IV-2b, a result which can be attributed to the gradients in B, r_0 and Φ_y that are present in the calculations for Figure IV-2b but not in Figure III-1b.

In the case of upper level multiplets the coupling rates between different J states competes with the downward spontaneous transition probabilities, which generally are large quantities. For this reason complete coupling via collisional effects is difficult to achieve. Atom-atom collisions may contribute appreciably to the coupling as may photoionizations and recombinations and coupling transitions via other bound levels.

Figure IV-2c shows the effect on S_{31} and S_{21} of adding photoionizations and recombinations to levels 2 and 3. Results for two different values of ε are shown. The photoionization cross-sections are taken as 10^{-17} cm^2 and the continuum radiation temperature equals 6000°. The continuum lies 6.5 eV above level 1 and 4.0 V above levels 2 and 3. Note that the coupling effect is relatively strong, about the equivalent of $Q_{23}=10^3 \ \pi a_0^2$. Coupling through a fourth bound level lying substantially closer to levels 2 and 3 than the continuum may be even stronger than the continuum coupling (see Section 10).

It appears from the preceding examples that it is relatively easy to bring the source functions for upper level multiplets to within about 10 to 20% of each other, but may be difficult to bring them into equality to within 1 or 2%. It seems likely therefore that complete equality of source functions in such multiplets is not to be expected, but that equality to within, say, 10% may be quite common.

It should be noted that the τ ratio of ten to one in the examples given is rather extreme for multiplets. More commonly adjacent pairs of multiplet lines have opacity ratios of two to three. Also, the transition probabilities of 10^8 s^{-1} used in the examples are larger than those found in many multiplets. Smaller opacity ratios reduce the intrinsic differences between the source functions and close coupling is achieved for smaller values of $\varepsilon^*/\varepsilon$. Similarly, smaller transition probabilities enhance the relative importance of the coupling terms and again reduces the demands on $\varepsilon^*/\varepsilon$ for close coupling.

Separating levels 2 and 3 in excitation energy does not seriously alter the situation in this particular case of an upper level doublet since this does not change markedly the opacity ratio between the two lines. Such a separation will reduce somewhat the coupling rate C_{23} for a given value of Q_{23} (assuming level 2 lies below level 3). However, this is a relatively small effect as long as the energy separation does not become large.

Close coupling between two energy levels that are separated in energy takes on a somewhat different meaning than in the close doublets used in the illustrations. When the energy levels separate the multiplet lines have different wavelengths and different Planck functions. Close coupling still means that $S_{21}/B \approx S_{31}/B$, but not that $S_{21} \approx S_{31}$. This point will be elaborated in Section 5 of this chapter dealing with configurations involving a metastable level.

As an illustration of the practical application to a doublet problem, we show in Figure IV-3a source functions for the Na D lines computed by Athay and Canfield

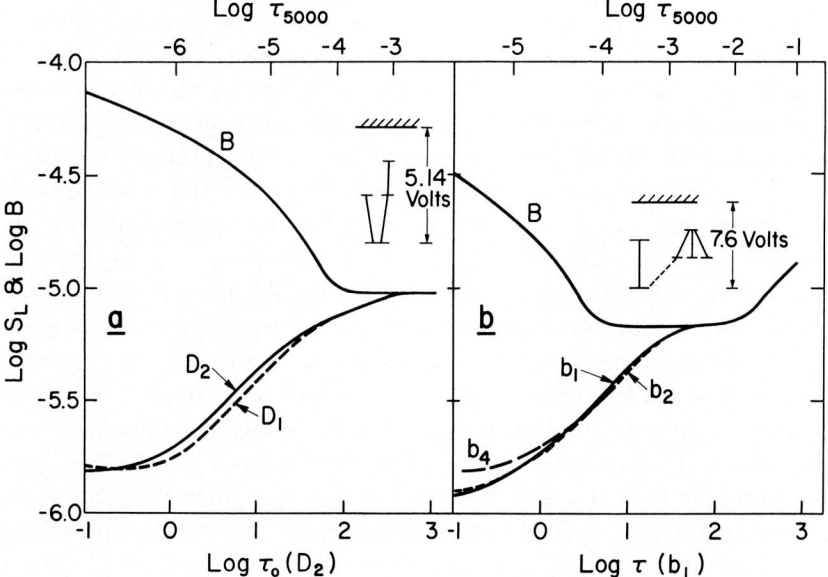

Fig. IV-3. Solutions for the Na D and Mg b lines (Athay and Canfield, 1969; courtesy *Astrophysical Journal*, copyright 1969 by the University of Chicago. All rights reserved).

(1969) for the 4-level sodium atom illustrated in the figure. The intramultiplet collision cross-section for electrons was taken as $10^3 \pi a_0^2$. However, the dominant coupling within the multiplet is via photoionizations and recombinations, and essentially the same coupling is achieved for much smaller values of the intramultiplet collision cross-section. Note that the two source functions agree to within 10% over most of the relevant range in τ_0, i.e., $\tau_0 > 0.1$. Note also that level 4 does not couple the two doublet lines in the model chosen. In the real Na atom the $3d\,^2D$ level does couple to both of the upper levels of the D line multiplet. However, Athay and Canfield encountered numerical instabilities when this added coupling was present. If coupling to this $3d\,^2D$ level were properly included, the two D line source functions would undoubtedly lie even closer together. The $3d\,^2D$ level is important also in the ionization equilibrium and is therefore essential for fixing the τ_0 scale for the D lines.

5. Lower Level Multiplets

Multiplets arising from lower level splitting are basically different in their coupling properties from upper level multiplets. Lower level multiplets have common upper levels. A photon absorbed in any one of the lines places an electron in the common upper level and the excited electron may return spontaneously to any of the lower states, thus emitting a photon in any of the multiplet lines. Photons always freely exchange between the lines therefore and the source functions always exhibit some effects of coupling. The effect is generally quite strong and the source functions will be more nearly equal than for the corresponding upper level multiplets.

For illustration, we again consider the doublet problem. If the statistical weights of the lower levels, here numbered 1 and 2, are $\tilde{\omega}_1$ and $\tilde{\omega}_2$, the transition probabilities will have the ratio $A_{31}/A_{32} = \tilde{\omega}_1/\tilde{\omega}_2$. Also, the collisional de-excitation rates will be in the same ratio $C_{31}/C_{32} = \tilde{\omega}_1/\tilde{\omega}_2$, so the two lines again have common values of ε. The absorption oscillator strengths are generally equal for the two lines. Thus, in close coupling the opacity ratio will be given by $\tau_0(31)/\tau_0(32) = \tilde{\omega}_1/\tilde{\omega}_2$. However, as the coupling becomes weaker the populations of levels 1 and 2 will depart from the Boltzmann ratio, i.e., b_1 will not equal b_2, and the opacity ratio will change from $\tilde{\omega}_1/\tilde{\omega}_2$.

Still a further, and perhaps more important, difference in the coupling properties of the lower level multiplet is found in the size of the term against which the intramultiplet coupling competes. Strong coupling in this case results when the collision rate between the lower levels exceeds the excitation rates to the upper level. Generally, the photoexcitation rates, abbreviated as A_{13} and A_{23}, exceed the collisional excitation rates. Hence, strong coupling requires $C_{12} \gg A_{13}$ and $C_{21} \gg A_{23}$. To a fair degree of approximation the rates A_{23} and A_{13} are given by $A_{23} = (n_3/n_2)A_{32}$ and $A_{13} = (n_3/n_1)A_{31}$. For lines in the visual part of the spectrum and for temperatures near $6000°$ the Boltzmann ratios for n_3/n_1 and n_3/n_2 are of the order of 10^{-2}. Departures from thermodynamic equilibrium further reduce these ratios by factors of 10^{-1} to 10^{-2}. Thus, A_{13} and A_{23} are often several orders of magnitude smaller than A_{31} and A_{32}.

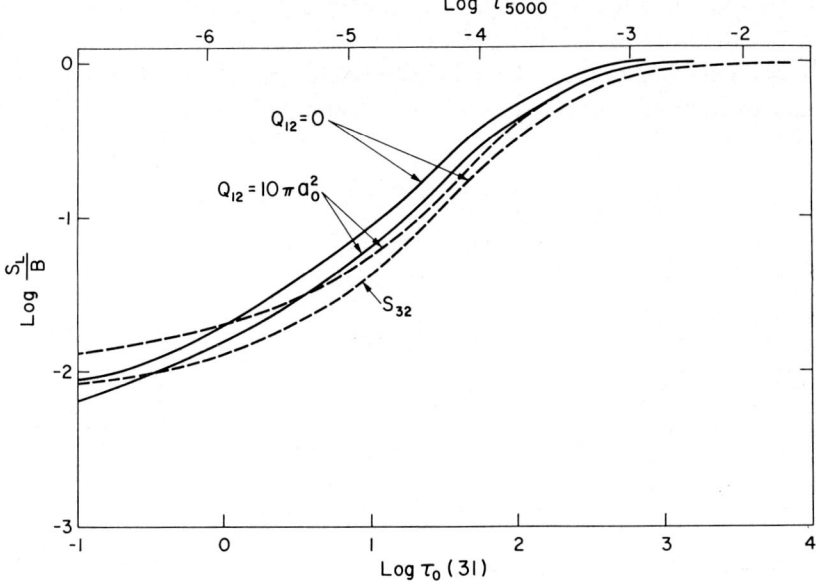

Fig. IV-4. Comparison of lower level doublet source functions at $\lambda 5000$ for $A_{32} = 10^7$, $A_{31} = 10^8$, $Q_{13} = Q_{23} = 10\ \pi a_0^2$, $\tilde{\omega}_1 = 10$, $\tilde{\omega}_2 = 1$, $\varepsilon_{31} \approx \varepsilon_{32} \approx 5 \times 10^{-4}$, $\varepsilon \gg \delta$.

It follows that intramultiplet collision rates will produce close coupling in lower level multiplets for collision cross-sections that are orders of magnitude smaller than those required for close coupling in upper level multiplets.

Figure IV-4 illustrates the coupling effect of $Q_{12} = 10\ \pi a_0^2$ and $Q_{12} = 0$ for a lower level doublet with the same parameters as the upper level doublet illustrated in Figure IV-2a. Note that for $Q_{12} = 10\ \pi a_0^2$ the two source functions are equal to within about 15%, which required Q_{23} in excess of $10^4\ \pi a_0^2$ in the upper level doublet. With $Q_{12} = 0$ the two source functions in Figure IV-4a differ by about a factor of 2 at intermediate values of τ_0.

Again we call attention to the fact that an opacity ratio of ten to one and transition probabilities of $10^8\ \text{s}^{-1}$ are large for multiplets. Also, an intramultiplet collision cross-section of $10\ \pi a_0^2$ is modest. It appears therefore that close coupling between source functions in lower level multiplets will be the rule rather than the exception.

Figure IV-3b shows the source functions for the Mg b triplet computed by Athay and Canfield (1969) for the six level model of the Mg I atom shown in the figure. Intramultiplet collision cross-sections of $10\ \pi a_0^2$ for $J = 0$ to $J = 1$ and $J = 1$ to $J = 2$ and zero for $J = 0$ to $J = 2$ were used. Obviously, the source functions in the b lines are closely coupled in the line forming layers.

Small differences in energy between the lower levels again make relatively little difference to the coupling in lower level multiplets. Such separation will change the opacity ratio somewhat and will have some effect on the collision rates for coupling. The case of large energy separation is treated in the following section.

It is often the case that multiplets arise from splitting in both the upper and lower states. This does not add any fundamental changes over the upper level and lower level cases. Coupling between the individual levels, whether upper or lower, follows the same patterns as indicated in this section and in Section 4.

6. Metastable Levels

Many atoms, such as Ca II, Fe I, and Fe II have low lying metastable levels whose direct transitions to the ground state are optically forbidden but which are coupled

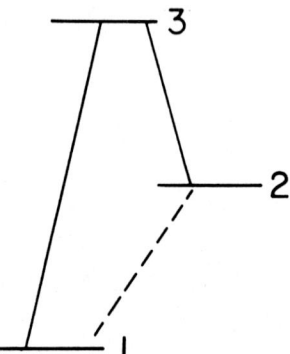

Fig. IV-5. Model atom for metastable configuration. The dashed line is optically forbidden.

to the ground state via collisions and via permitted transitions to a third, higher level. Such a configuration is illustrated in Figure IV-5.

This situation is analogous to the lower level doublet discussed in the preceding section, the principal difference being that the opacity ratio $\tau_0(31)/\tau_0(32)$ will, in general, be much larger than in the close doublet. It does not matter so much that the transitions 3-1 and 3-2 may themselves be multiplets. The rather strong tendency for source function equality within multiplets means that multiplets can be treated as single lines with a reasonable degree of accuracy. This follows from the fact that the different J states are populated very nearly in proportion to their Boltzmann weights and one need not solve the transfer equations for each of these states separately. The single line equivalent of the multiplet can be obtained either by taking just the strongest line with its individual statistical weight, transition probability and collisional excitation cross-section or by taking the entire multiplet with its combined statistical weight and the weighted average of the transition probabilities and collision cross-sections. In using the single line equivalent of the multiplet it must be remembered that small systematic errors will be present since the multiplet source functions are not identical.

With the above comments in mind, we note that the configuration in Figure IV-5 may be regarded as representing such cases as the interaction of the H and K lines (3-1) with the infrared triplet (3-2) of Ca II.

In comparing source functions at different wavelengths, it is necessary to define the basis of comparison. Source function equality can be defined in alternative, and quite different, ways. To illustrate this we again consider just two lines 3-1 and 3-2, and for simplicity of illustration we ignore stimulated emissions. We then have

$$S_{31} = \frac{2h\nu_{31}^3}{c^2} \frac{b_3}{b_1} e^{-h\nu_{31}/kT} = \frac{b_3}{b_1} B_{31}$$

and

$$S_{32} = \frac{2h\nu_{32}^3}{c^2} \frac{b_3}{b_2} e^{-h\nu_{32}/kT} = \frac{b_3}{b_2} B_{32}.$$

Equality of S_{31}/B_{31} and S_{32}/B_{32} means, therefore, that $b_2 = b_1$. This is the sense in which we have used source function equality to this point.

An alternative way of considering source function equality is to define a source function temperature, more commonly called the excitation temperature, by the relation

$$S_L = B(T_{ex}) = \frac{2h\nu^3}{c^2} e^{-h\nu/kT_{ex}}. \tag{IV-34}$$

In other words, we define T_{ex} for the j-i transition by

$$\frac{b_i}{b_j} e^{-h\nu/kT} = e^{-h\nu/kT_{ij}}, \tag{IV-35}$$

where we simplify the notation by setting $T_{ex}(i-j) = T_{ij}$. Equation (IV-35) may be written as

$$\frac{\tilde{\omega}_i}{\tilde{\omega}_j} \frac{b_i}{b_j} e^{-h\nu/kT} = \frac{n_i}{n_j}$$

$$= \frac{\tilde{\omega}_i}{\tilde{\omega}_j} e^{-h\nu/kT_{ij}}, \tag{IV-36}$$

which accounts for the identification of T_{ij} with the excitation temperature defining the relative populations of levels i and j. Equation (IV-34) suggests that source function equality can be defined as the condition $T_{31} = T_{32}$, or, in other words, that the source functions at different wavelengths are defined by a Planck function at a single excitation temperature, which, in general, differs from the kinetic temperature. Since T_{ij} is a function of both b_i/b_j and ν_{ij}, equality of T_{31} and T_{32} necessitates that $b_1 \neq b_2$. If the wavelengths of the lines are the same, this distinction disappears, of course, and the two definitions of source function equality are equivalent.

The effect of strong collisional coupling between levels 1 and 2 in Figure IV-5 is to produce equality of b_2 and b_1, hence equality of S_{21}/B and S_{31}/B but inequality of T_{31} and T_{32}. For $S_L < B$, we will have $T > T_{31} > T_{32}$ at a given depth in the atmosphere.

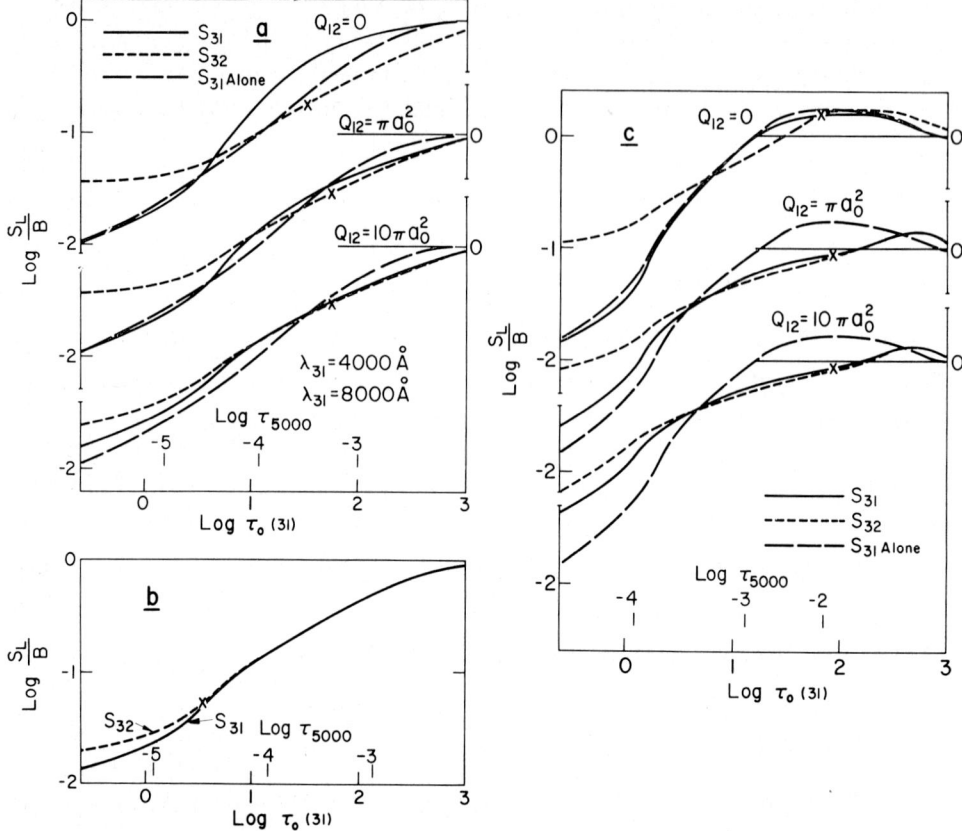

Fig. IV-6. Source functions for the metastable configurations.

(a) – $\lambda_{31} = 4000$, $\lambda_{32} = 8000$, $A_{31} = 10^8$, $A_{32} = 10^7$, $Q_{13} = 10\,\pi a_0^2$, $Q_{23} = \pi a_0^2$, $\varepsilon_{31} \approx 5 \times 10^{-4}$, $\varepsilon_{32} \approx 3 \times 10^{-4}$, $\varepsilon \gg \delta$.

(b) – Same as (a) except $A_{32} = 10^8$ and $Q_{12} = Q_{13} = Q_{23} = 10\,\pi a_0^2$.

(c) – $\lambda_{31} = 2500$, $\lambda_{32} = 5000$, $A_{31} = A_{32} = 10^8$, $Q_{13} = Q_{23} = 10\,\pi a_0^2$, $\varepsilon_{31} \approx 10^{-3}$, $\varepsilon_{32} \approx 5 \times 10^{-4}$, $\varepsilon \gg \delta$.

The crosses indicate the depths at which $\tau_0(32) = 1$.

Figure IV-6a shows the coupling effects for different values of Q_{12} for a case where the opacity ratio between the two lines is moderately large. Figure IV-6b illustrates a case of smaller opacity ratio and Figure IV-6c illustrates the case of both larger opacity ratio and shorter wavelengths. The ε values are kept comparable and $\varepsilon \gg \delta$ in all cases. As in the case of the lower level close doublet the coupling between levels 1 and 2 is close for $Q_{12} \geqslant 10\,\pi a_0^2$. Furthermore b_2 is equal to b_1 within about 20% even for $Q_{12} = \pi a_0^2$.

Values of S_{31} when level 2 is absent are shown in the figures for comparative purposes. One effect of level 2 is to carry the departure between S_L and B deeper into the atmosphere. This is a manifestation of the cooperative behavior of the two lines. They tend to thermalize together at a depth where τ_0 in the weaker line is

TABLE IV-2
Excitation Temperatures from Figures IV-6a and IV-6c
For $Q_{12} = 10 \, \pi a_0^2$

$\tau_0(31)$	T	T_{31}	T_{32}	T_{32} at $\tau_0(32) = \tau_0(31)$
Figure IV-6a				
1	5850	3650	2800	3500
3	5500	3750	2950	3900
10	4800	3750	3100	4350
30	4550	3800	3300	–
100	4400	3950	3600	–
300	4450	4200	4000	–
Figure IV-6c				
1	5100	4250	3850	4300
3	4700	4150	3900	4300
10	4450	4200	4050	5200
30	4350	4250	4050	–
100	4500	4450	4400	–
300	4600	4700	4750	–

at its thermalization depth. In other words, the system thermalizes at the deeper of the thermalization depths, as noted by Avrett and Kalkofen (1968). At small values of τ_0 the coupling raises the value of S_{31} and depresses the value of S_{32}. Level 1 is thus providing a source of electrons for level 2 and level 2 is acting as a sink of electrons for level 1.

In Table IV-2 we give excitation temperatures for the cases $Q_{12} = 10 \, \pi a_0^2$ from Figures IV-6a and IV-6c. At small τ_0 the excitation temperatures are quite different at a given depth. It is of interest to note, however, that at a common optical depth ($\tau_0(32) = \tau_0(31)$) near $\tau_0 = 1$-3 the excitation temperatures are quite similar.

The maximum in S_{31}/B near $\tau_0(31) = 10^2$ in Figure IV-6c is an effect of wavelength, as may be seen by comparison to the results in Figure IV-6b. The maximum occurs, in this case, in the region of decreasing B just below the temperature minimum. At short wavelengths $dB/d\tau$ is large and J_c is considerably above B. Even though we have $\varepsilon \gg \delta$ in Figure IV-6c, the mean radiation intensity $\int J_\nu \phi_\nu \, d\nu$ builds up to values exceeding B and causes S_L to rise above B. The effect is analogous to that evident in Figures III-4 and III-7, although the detailed reasons are somewhat different. S_{32} is forced to rise above B because of its coupling with S_{31}. However as the coupling increases the influence of S_{32} becomes strong in the region of $\tau_0(32) = 1$ and both source functions drop below B as is consonant with the behavior of S_L at $\lambda 5000$.

We show in Figure IV-7 results for $\lambda_{32} = 3000$ Å and $\lambda_{31} = 2500$ Å for three different opacity conditions: one in which $\delta \ll \varepsilon$ for both lines, one in which $\delta \approx \varepsilon$ and one in which $\delta \gg \varepsilon$. The shorter wavelengths of the lines enhances the separation between J_c and B. Strong optical pumping occurs in cases 2 and 3 in both the 3-2 and 3-1 transitions elevating b_3 above both b_2 and b_1. The primary reason for the strong

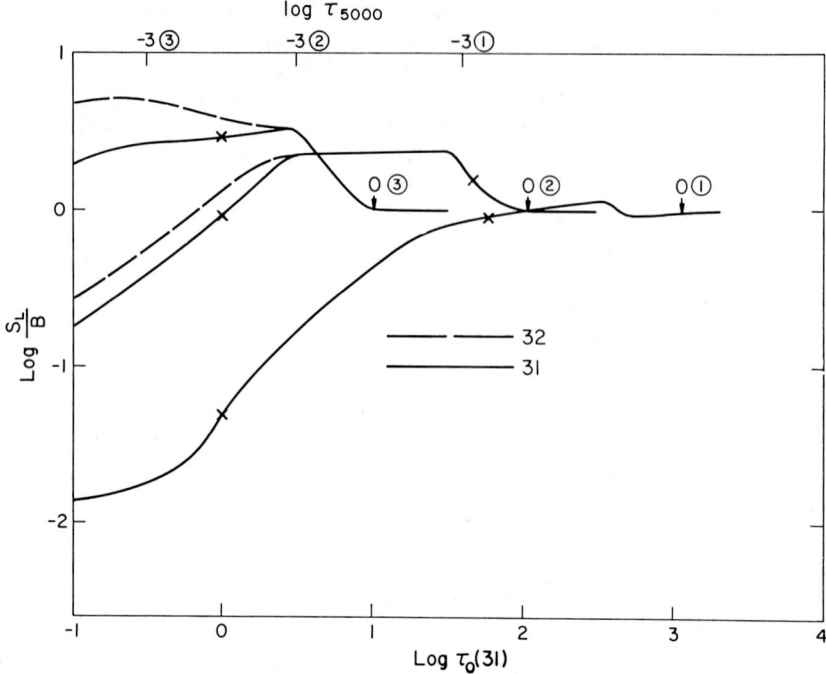

Fig. IV-7. Source functions for the metastable configuration with $\delta \gg \varepsilon$. $\lambda_{31} = 2500$, $\lambda_{32} = 3000$ Å, $A_{31} = 10^8$, $A_{32} = 10^7$, $Q_{13} = Q_{23} = 10\ \pi a_0^2$, $Q_{12} = 3\ \pi a_0^2$.

	δ_{31}	δ_{32}	ε_{31}	ε_{32}
Case 1	0.0002	0.01	0.001	0.03
Case 2	0.001	0.02	0.001	0.1
Case 3	0.03	1	0.003	1

Values of ε and δ are given near $\tau_0 = 1$.

optical pumping, in these cases, is the low opacity in the 3-2 transition. This exposes the 3-2 transition to the mean continuum intensity, which exceeds the Planck function by a relatively large factor. A similar computation made at wavelengths of $\lambda_{32} = 8000$ Å and $\lambda_{31} = 4000$ Å shows that both S_{32}/B and S_{31}/B remain less than unity when $\tau_0(32)$ becomes small. However, S_{32}/B is close to unity, in this case, and this forces S_{31}/B to be much closer to unity than it would be if the 3-1 transition were uncoupled from the 3-2 transition.

TABLE IV-3

Values of S_{31}, T_e and T_{31} at $\tau_0(31) = 1$ in Figure IV-7

Case	S_{31}	T_e	T_{31}	τ_{5000}
1	2×10^{-8}	5200	4100	4×10^{-5}
2	7×10^{-8}	4450	4400	4×10^{-4}
3	2×10^{-7}	4450	4900	3×10^{-3}

This cooperative behavior of coupled transitions and the elevating of one source function by another will be evident in several subsequent illustrations.

Since the abscissa in Figure IV-7 is $\log \tau_0(31)$ rather than τ_{5000}, the values of B are not the same for the three cases at a fixed position on the abscissa. The differences in B account for part of the differences in S/B. At $\tau_0(31)=1$, for example, the three values of S_{31} and the corresponding values of T_e and T_{31} are as given in Table IV-3.

7. Two Lines in Series

Coupling effects between two lines in series, say, 3-2 and 2-1 transitions, are less pronounced than in the preceding examples. This results primarily from the fact that levels 1 and 3 are generally well separated in energy and are generally coupled by collisions less strongly than are levels 1 and 2 and levels 2 and 3. Strong effects can occur, however, if it is assumed that the 1-3 coupling is strong as was shown by Avrett and Kalkofen (1968).

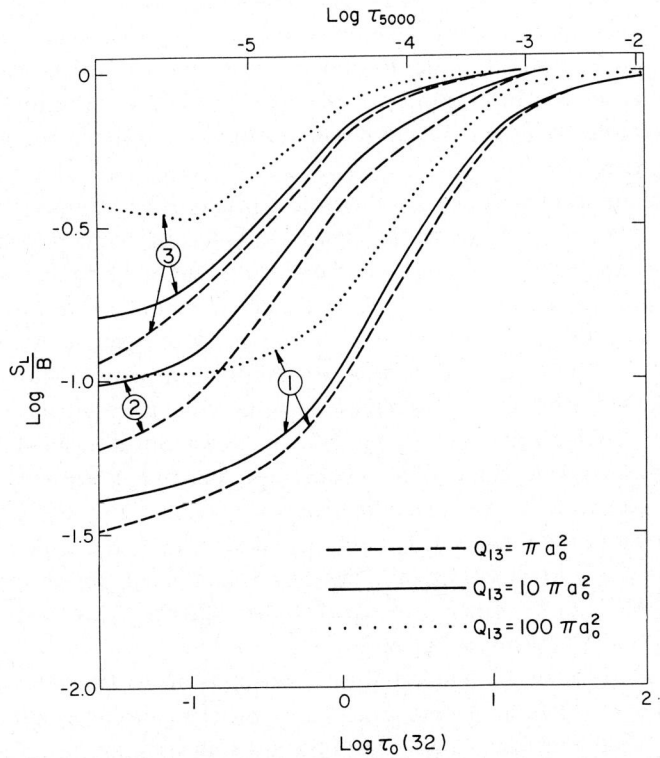

Fig. IV-8. Solutions for the 3-2 source function for two lines in series. $\lambda_{21}=4000$, $\lambda_{32}=6000$, $Q_{12}=10\,\pi a_0^2$, $A_{21}=10^8$.

Case 1 – $A_{32}=10^8$, $Q_{23}=10\,\pi a_0^2$, $\varepsilon_{32}\approx 5\times 10^{-3}$;
Case 2 – $A_{32}=10^7$, $Q_{23}=10\,\pi a_0^2$, $\varepsilon_{32}\approx 5\times 10^{-3}$;
Case 3 – $A_{32}=10^7$, $Q_{23}=100\,\pi a_0^2$, $\varepsilon_{32}\approx 5\times 10^{-2}$.
Values of Q_{13}, A_{32} and Q_{23} have no effect on S_{21}.

Figure IV-8 shows the effect on S_{32} of varying the collision cross-section Q_{13} for three different combinations of A_{32} and Q_{23}. The effects are most pronounced when Q_{23} is smallest (case 2) and when Q_{13} is largest, as might reasonably be expected. None of the cases shown in the figure produced appreciable changes in S_{21}.

Collisional coupling between levels 1 and 3 in the configuration 3-2-1 with the 3-1 transition optically forbidden raises the population of level 3 relative to levels 1 and 2, thus increases S_{32}. The growth in n_3 occurs at the expense of levels 1 and 2. However, because $n_3 \ll n_1$ a large relative change in n_3 may occur even though the relative change in n_1 is small.

8. Three Line Loops

Closed loops of transitions between energy levels such as those represented by Figure IV-5 when the transition 2-1 is permitted have interesting coupling properties. They are again fundamentally different from any of the preceding cases considered. In the metastable configuration for level 2, collisions between levels 1 and 2 compete against the upward excitation rates, R_{13} and R_{23}. When the 2-1 transition is permitted, however, the 2-1 collision rate must compete against A_{21}, which may well be orders of magnitude larger than R_{13} and R_{23}. We expect that ε_{21} will be small and the effects of scattering will be large in S_{21}. By making the 2-1 transition a permitted transition, therefore, we permit b_2 to differ markedly from b_1 in the boundary regions of the atmosphere.

We show in Figures IV-9a and IV-9b results for two different three line loops for the case $\varepsilon \gg \delta$. The loop represented in Figure IV-9a has the same parameters as the metastable configuration in Figure IV-6b, the single exception being that $A_{21} = 10^8 \text{ s}^{-1}$ rather than zero. Similarly, the loop represented in Figure IV-9b has parameters matching those in the lower set of curves in Figure IV-6c again with the exception that $A_{21} = 10^8 \text{ s}^{-1}$ rather than zero. For comparative purposes, Figures IV-9a and IV-9b show the solutions for S_{31} from these corresponding metastable configurations. Note the very striking difference in S_{31} and S_{32} from the metastable case. There is no longer any tendency for S_{31}/B to equal S_{32}/B except in the very deep layers where S_{21} is thermalized. The controlling source functions, in both figures, are S_{32} and S_{21} throughout the range $\tau_0(31) < 100$. At larger $\tau_0(31)$, S_{21} is thermalized and S_{31} couples with S_{32} to cooperate in fixing their values. Changing A_{32} and Q_{23}, for example, changes both S_{32} and S_{31} by equal factors whereas S_{21} remains unchanged. Changing A_{31} and Q_{13} changes S_{31} only.

The dominance of the 3-2 and 2-1 source functions arises from the relative ease with which the 3-1 photons degrade, i.e., the short degradation length for the 3-1 line. Since $A_{32} = A_{31}$, photons absorbed in the 1-3 transition are just as likely to be destroyed by cascade in the 3-2-1 sequence as to reappear in the 3-1 transition. The relatively low opacity in the 3-2 transition, means that the 3-2-1 cascade provides a ready escape route for 3-1 photons. In effect, the 3-1 photons cannot scatter more than once or twice without escaping. This prohibits the buildup of photons in the 3-1 line and severely depresses the 3-1 source function.

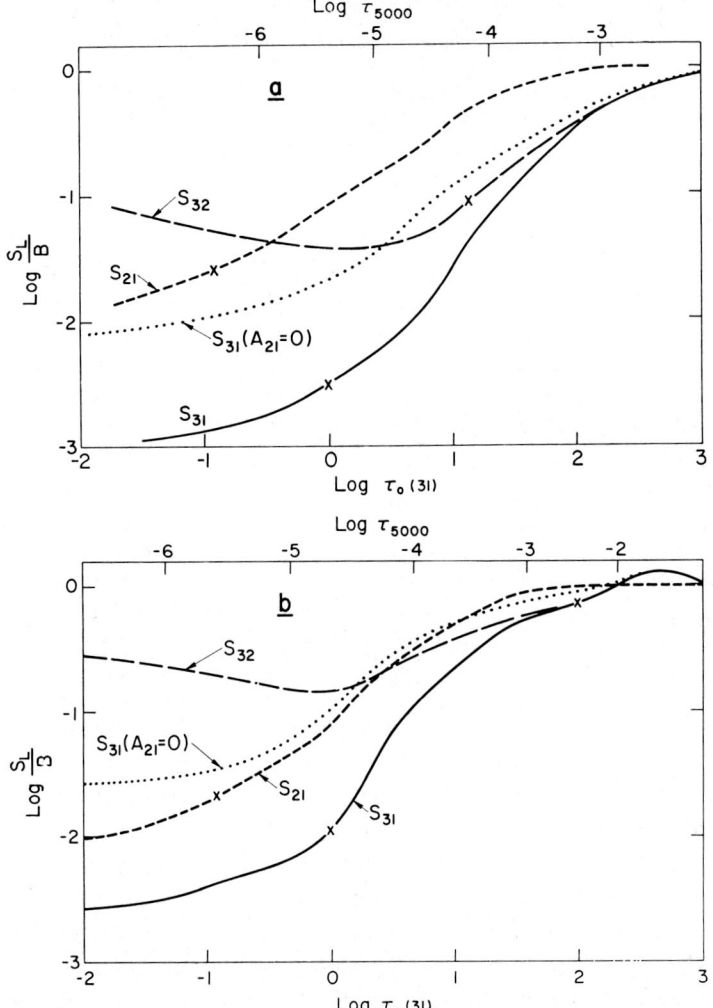

Fig. IV-9. Solutions for the three line closed loop. $A_{32} = A_{31} = A_{21} = 10^8$, $Q_{12} = Q_{13} = Q_{23} = 10 \pi a_0^2$, $\varepsilon_{31} = 10^{-3}$, $\varepsilon_{32} = 3 \times 10^{-4}$, $\varepsilon_{21} = 3 \times 10^{-4}$, $\varepsilon \gg \delta$.

a – $\lambda_{31} = 4000$, $\lambda_{32} = \lambda_{21} = 8000$.
b – $\lambda_{31} = 2500$, $\lambda_{32} = \lambda_{21} = 5000$.
The x's mark the location of $\tau_0 = 1$.

Note that the opposite effect does not happen. Photons absorbed in the 2-3 transition may temporarily be delayed by a 3-1 transition. But, because of the high opacity in the 1-3 transition, the 3-1 photon is immediately reabsorbed and is again likely to give rise to a 3-2 photon to replace the one originally absorbed. The 2-3-1-3-2 cycle merely delays the 2-3-2 scattering, therefore, without destroying the photon. An exception to this occurs when $\tau_0(31)$ becomes of order unity or less, in which

TABLE IV-4
Excitation Temperatures from Figures IV-9a and IV-9b

$\tau_0(31)$	T	T_{21}	T_{31}	T_{32}	T_{21} at $\tau_0(21) = \tau_0(31)$	T_{32} at $\tau_0(32) = \tau_0(31)$
Figure IV-9a						
0.1	6250	2700	2900	3150	–	–
0.3	6150	2950	2950	2950	2500	2800
1	5950	3250	3050	2850	2700	2900
3	5750	3600	3200	2800	2950	3250
10	5100	3950	3350	2900	3250	3650
30	4600	4100	3600	3100	–	–
100	4400	4300	3850	3500	–	–
300	4400	4400	4100	3900	–	–
Figure IV-9b						
0.1	6050	3300	3800	4500	–	–
0.3	5850	3550	3800	4250	3150	3850
1	5650	3800	3900	4100	3300	4050
3	5050	4050	4050	4050	3600	4550
10	4650	4200	4150	3950	3850	–
30	4400	4300	4200	3850	–	–
100	4350	4350	4250	4050	–	–
300	4550	4550	4550	4550	–	–

case 3-1 photons are about as likely to escape as 3-2 photons. By this time, however, $\tau_0(32)$ is so small that little effect is felt in either S_{32} or S_{31}.

In both Figures IV-9a and IV-9b, S_{32}/B increases outward for $\tau_0(31) < 1$. This has its explanation in the fact that $\tau_0(21)$ is larger than $\tau_0(31)$. As a result S_{21}/B continues to decrease at $\tau_0(31) < 1$, whereas S_{31}/B and S_{32}/B are tending to level out. The net effect is to decrease b_2 relative to both b_1 and b_3, which increases S_{32} relative to B.

The main effect of adding the spontaneous transition 2-1, as contrasted with the metastable case, is to provide a more effective escape of electrons from level 2. Level 2 depopulates to level 1 reducing S_{21}/B from near unity to the values shown in the figures. Level 3 also depopulates as a result of the depopulation of level 2. This reduces S_{31}/B by a substantial factor, though not as large as the reduction in S_{21}/B.

Excitation temperatures for the three source functions in Figures IV-9a and IV-9b are given in Table IV-4. Note that T_{32} (hence S_{32}) passes through a minimum near $\tau_0(32) = 1$.

Because of redundancy, equality of any two of the T_{ex} values requires equality with the third. Also, the greater energy difference between levels 3 and 1 requires that T_{31} lie between T_{21} and T_{32}. We note again the general inequality of T_{21}, T_{31}, and T_{32} at a given depth, but again a rather striking tendency for approximately equal values of T_{ex} near $\tau_0 = 1$ in the different lines. The only notable exception to this is for T_{21} in Figure IV-9b.

Each of the preceding illustrations of the 3 line loop is for the case $\varepsilon > \delta$. When $\delta > \varepsilon$ the problem of achieving consistent solutions for S_L becomes considerably more difficult. The values of the S_L's are determined by the δ's in this latter case, whereas the ratios of the b_j's are still determined by the ε's. This results in a weak coupling between the steady state equations and the radiative transfer equations and enhances the problems of inconsistency in iterative solutions. However, consistent solutions can be achieved using the forcing procedures given in Section 2.

Three sets of solutions for the 3 line loop are shown in Figure IV-10. In set 1, the relative abundance of the element, A, is 10^{-9} and $\delta > \varepsilon$ for all three lines. Values of δ and ε given in the caption are at depths $\tau_0 = 1$. For this case, $\tau_0(32)$ is unity near $\tau_{5000} = 1$ and transfer effects in the 3-2 transition are minimal. Note that S_{32}/B is near unity throughout the atmosphere and that S_{31}/B and S_{21}/B are approximately equal. The effects of optical pumping are again evident in case 1 where $S_L/B > 1$.

In set 2, $A = 10^8$ and $\delta \approx \varepsilon$ in all three transitions near $\tau_0 = 1$. Transfer effects in the 3-2 transition are now important. As a result S_{32}/B drops below unity and S_{31}/B drops below S_{21}/B, as in Figures IV-9a and IV-9b. Note here also that the abscissa in

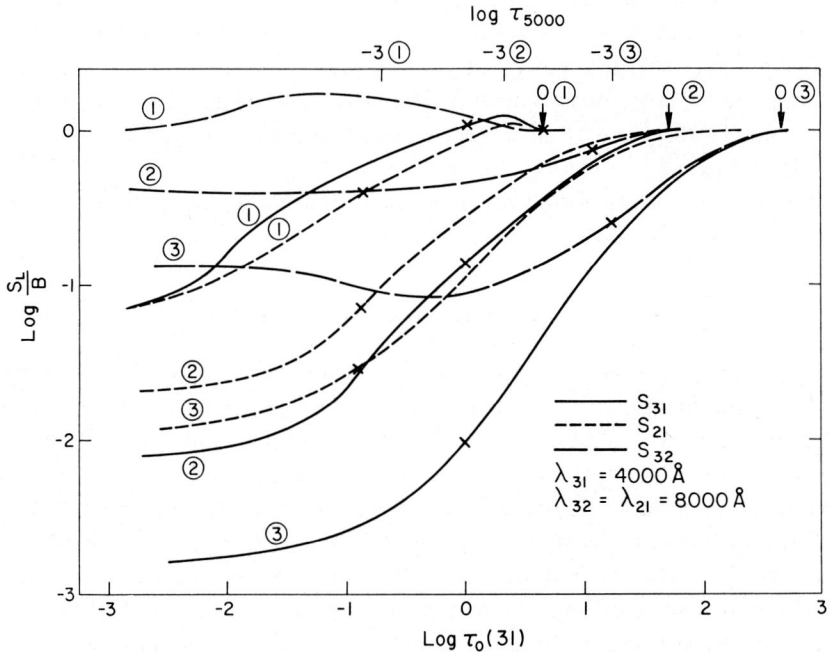

Fig. IV-10. The effect of $\delta \gg \varepsilon$ for the three line closed loop.
$A_{31} = A_{21} = A_{32} = 10^8$, $Q_{13} = Q_{12} = Q_{23} = 10\pi a_0^2$.

	δ_{21}	δ_{31}	δ_{32}	ε_{21}	ε_{31}	ε_{32}
Case 1	0.005	0.05	1	0.0008	0.001	0.08
Case 2	0.0003	0.003	0.06	0.0003	0.003	0.003
Case 3	0.00003	0.0003	0.003	0.0004	0.0006	0.001

Figure IV-10 is $\tau_0(31)$ rather than τ_{5000}, and, hence, that for a given position on the abscissa B is different in each of the three cases. The locations of $\log \tau_{5000} = 0$ and -3 are shown in the figure for the three sets of curves.

In set 3, $A = 10^{-7}$ and $\varepsilon > \delta$. The resultant source functions for the three lines now are very similar to those in Figure IV-9a where $A = 10^{-6}$. As $\tau_0(32)$ becomes larger both S_{32}/B and S_{31}/B become progressively smaller. However, S_{21}/B remains relatively unchanged once $\varepsilon_{21} \gg \delta_{21}$. This may be seen by noting that S_{21}/B for case 3 in Figure IV-10 is very similar to S_{21}/B in Figure IV-9a where the abundance (hence the opacity) is a factor of 10 higher. At $\tau_0(21) = 1$, $\log S_{21}/B = 1.55$ in Figure IV-10, set 3, and -1.6 in Figure IV-9a. By comparison, at $\tau_0(31) = 1$, $\log S_{31}/B = -2.0$ and -2.5 in the same figures.

A computation made with the abundance of case 2 but with $A_{32} = 10^7 \text{ s}^{-1}$, shows results very similar to case 1 for the 3-2 transition, i.e., $S_{32}/B \approx 1$. The 2-1 and 3-1 source functions are closely coupled, as in case 1, but S_{21}/B and S_{31}/B have absolute values close to those for S_{21}/B for case 2. S_{21}/B and S_{31}/B do not differ in this case by more than 15% at any depth. They cross near $\tau_0(21) = 1$ with S_{31}/B slightly larger than S_{21}/B at smaller optical depths.

Three line loops such as those considered in this section generally occur only in hydrogenic configurations or in configurations that approximate the hydrogenic case such as the higher energy states of simple atoms.

Solutions to the Lyman-α, Lyman-β, Hα loop for hydrogen by Cuny (1967) and Athay et al. (1968) give S/B ratios very similar to those shown in Figure IV-9a and Figure IV-10, case 3.

A spectroscopic configuration more common than the three line loops is the four line loop, which we consider next.

9. Four-Line Metastable Loops

A common spectroscopic configuration in complex atoms is one in which one or more low-lying energy levels is metastable with the metastable level (or levels) and the ground state each connected to two or more common upper levels by relatively strong transitions. A simple case of this type is illustrated in Figure IV-11. In some situations of this type all four lines may be permitted in l-s coupling (for example if levels 1 and 2 are s and d states and levels 3 and 4 are both p states) and may have comparable transition probabilities. In other cases one of the transitions may be forbidden in l-s coupling but permitted in more complex coupling. This could result in one of the transitions, e.g., the 4-1 or 4-2 transitions, having a smaller transition probability than the remaining three lines.

The configuration in Figure IV-11 is, in some respects, similar to that in Figure IV-5 and some of the same effects will be present. In particular, the collisional coupling between levels 1 and 2 will tend to make $b_2 = b_1$ and, hence, to make $S_{31}/B_{31} = S_{32}/B_{32}$ and $S_{41}/B_{41} = S_{42}/B_{42}$. Similarly, the collisional coupling between levels 3 and 4 will tend to make $b_3 = b_4$, $S_{31}/B_{31} = S_{41}/B_{41}$ and $S_{32}/B_{32} = S_{42}/B_{42}$. If there were no

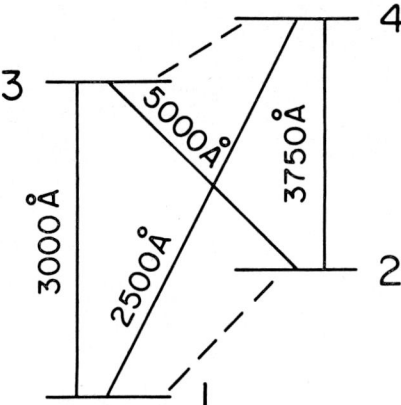

Fig. IV-11. Model atom for the four line metastable case.

competing effects, this would lead to a common value for all four of the quantities S/B.

Two effects are present in configurations of the type shown in Figure IV-11 that tend to destroy the equality between b_1 and b_2 and b_3 and b_4. These two effects are: (1) the wavelengths of the lines are generally different, and (2) the opacities of the lines are generally different. Both the wavelength effects and the opacity effects expose the different lines to radiation fields of different mean intensity and tend to force S_L/B to different values. If the opacity is large in all of the lines, these uncoupling effects are minimized. They become important, however, when the opacity becomes small in one or more lines.

Figure IV-12a shows results for the four lines depicted in Figure IV-11 for two cases in which the opacities differ by a factor of 10^2 relative to the continuum opacity. In the more opaque case (case 1) $\varepsilon \gg \delta$ for each of the lines and in the less opaque case (case 2) the reverse is true. In the examples shown each of the transition probabilities are 10^8 s^{-1} and each of the collisional excitation cross-sections are $10\,\pi a_0^2$ for the permitted transitions and $3\,\pi a_0^2$ for the forbidden transitions. Values of ε and δ near $\tau_0 = 1$ in each of the lines are indicated in the caption.

Note that for case 1 where all the opacities are large the source functions are all strongly coupled. For $\tau_0(31) \ll 1$ the strongest tendency is for $b_3 = b_4$ and for $\tau_0(31) \gg 1$ the strongest tendency is for $b_2 = b_1$. In the low opacity case both $\tau_0(32)$ and $\tau_0(42)$ reach unity near $\tau_{5000} = 1$. This exposes levels 3 and 4 to the strong continuum radiation field in the 2-3 and 2-4 transitions and this forces more electrons into levels 3 and 4 relative to both levels 2 and 1. The 4-2 transition is in the violet and both the 4-2 and 4-1 source functions exhibit the effect of optical pumping quite strongly. Parallel effects are evident in Figure IV-7 for the simpler metastable configuration and in Figure IV-10 for the three line loop. In Figure IV-12a, S_{41}/B and S_{42}/B lie above S_{31}/B and S_{32}/B because the former two transitions lie further in the violet and have somewhat lower opacities.

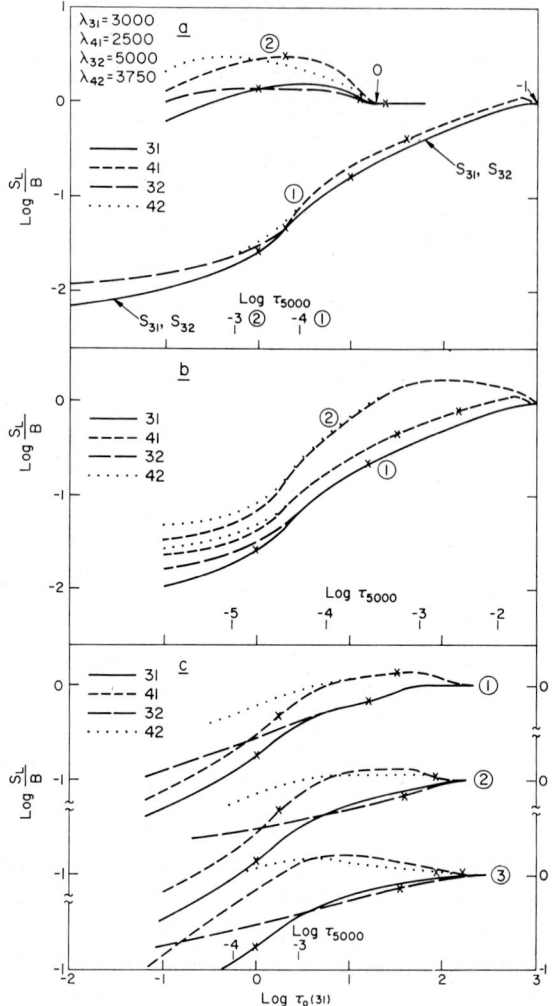

Fig. IV-12. Solutions for the four line metastable case.
(a) — $A_{ji} = 10^8$ (permitted), $Q_{ji} = 10\pi a_0^2$ (permitted) and $Q_{ij} = 3\pi a_0^2$ (forbidden).

	δ_{31}	δ_{41}	δ_{32}	δ_{42}
Case 1	0.0001	0.0002	0.001	0.004
Case 2	0.01	0.03	0.4	1

	ε_{31}	ε_{41}	ε_{32}	ε_{42}
Case 1	0.001	0.001	0.001	0.002
Case 2	0.003	0.003	0.02	0.08

(b) — Case 1 $A_{41} = 10^6$, Case 2 $A_{42} = 10^6$.
(c) — Case 1 $A_{41} = 10^8$, $\lambda_{42} = 3750$
     Case 2 $A_{41} = 10^8$, $\lambda_{42} = 5000$
     Case 3 $A_{41} = 10^6$, $\lambda_{42} = 5000$

The reader should note once again that the abscissa in Figure IV-12a is for a fixed $\tau_0(31)$ rather than a fixed τ_{5000}. Thus, B is different in cases 1 and 2 at a fixed $\tau_0(31)$. At $\tau_0(31)=1$, for example, S_{31} for case 2 exceeds S_{31} for case 1 by a factor of 6.5. The corresponding excitation temperatures for S_{31} are 3900° (case 1) and 4600° (case 2). By comparison, S_{31}/B for case 2 exceeds S_{31}/B for case 1 by a factor of 50.

Note that in going from case 1 to case 2 $\varepsilon_{31}+\delta_{31}$ increases by a factor of ten and $\varepsilon_{41}+\delta_{41}$ increases by a factor of 30. By simple analogy with the two level case we would expect S_{31} and S_{41} to increase by at most a factor of $30^{1/2}$ at small τ_0. Instead they increase by almost a factor of 100 at small τ_0 and S/B rises well above unity near $\tau_0=1$ as a result of the optical pumping effect.

Raising level 2 to a higher excitation energy reduces the opacities in the 3-2 and 4-2 lines, reduces the collisional coupling between levels 1 and 2 and increases the wavelengths of the 3-2 and 4-2 lines. The reduced opacities in the 3-2 and 4-2 lines and the reduced collisional coupling both tend to uncouple the source functions for the four lines. The energy separation between levels 1 and 2 in Figure IV-12a is 1.67 eV. Increasing this separation to 2.5 eV decreases $\tau_0(32)$ and $\tau_0(42)$ by about a factor of four relative to $\tau_0(31)$. For the same values of $\tau_0(31)$ as shown in Figure IV-12a, the same trends are found in the source functions but the differences in $\log S_L/B$ for the four lines are about twice as large as those shown in Figure IV-12a for case 1.

For case 2 at the higher value of χ_2, the differences in $\log S_L/B$ are similar to those in Figure IV-12a, but the detailed behavior of the four curves is different. The curves for S_{31}/B and S_{41}/B are nearly identical to those in Figure IV-12a but both S_{32}/B and S_{42}/B are nearer unity. Thus, the strong optical pumping effect is still present in the 4-1 transition even though the 3-2 and 4-2 transitions occur, respectively, at $\lambda 7500$ and $\lambda 5000$ where $J_c \lesssim B$. This implies that the main pumping effects arise through the cooperative behavior of the lines rather than through individual lines.

To illustrate the optical pumping effect still further, we show in Figure IV-12b comparative cases where A_{41} and A_{42} are alternately changed to 10^6 s^{-1}. This reduces the opacity in these transitions in the respective cases by a factor of 10^2 without changing the opacities in the other lines. Aside from these changes the opacities, wavelength, δ's and ε's are the same as in case 1 in Figure IV-12a.

Case 1 in Figure IV-12b shows the effect of reducing $\tau_0(41)$ and case 2 shows the effect of reducing $\tau_0(42)$. Both S_{32} and S_{31} are independent of $\tau_0(41)$ and $\tau_0(42)$ and have values essentially identical to those in Figure IV-12a, case 1. The reduction in $\tau_0(41)$ by a factor of 10^2 raises both S_{42} and S_{41} by a factor of about 1.25 for $\tau_0(31)\gg 1$ and by a factor of about 2.5 for $\tau_0(31)\ll 1$. At all depths the dominant tendency is for $b_2=b_1$.

In case 1, Figure IV-12b, $\tau_0=1$ for all lines occurs at $\tau_{5000}<0.003$. In case 2, however, $\tau_0(42)=1$ occurs near $\tau_{5000}=2$. The effect of this low value of $\tau_0(42)$ for case 2 on both S_{41} and S_{42} is pronounced. Near $\tau_{5000}=10^{-3}$, S_{41} and S_{42} have increased by about a factor of four and both now lie above B. Even at $\tau_0(41)=1$, S_{41} has increased by a factor of about 2.5 over its value for $A_{42}=10^8$. This strong dependence of the 4-1 source function on the opacity in the 4-2 transition should serve as ample warning that interlocking effects of even weak transitions can sometimes reach major proportions.

Figure IV-12c shows three curves where $\tau_0(31)$ is intermediate to case 1 and 2 in Figure IV-12a. In each of the three cases $\tau_0(31)$ is a factor 10 smaller relative to τ_{5000} than in case 1 in Figure IV-12a. The relative values of τ_0 in the other lines are altered in the three cases by changing the energy of level 2 (cases 2 and 3) and by changing A_{41} (case 3). We note that the tendency for $b_2 = b_1$ at large values of $\tau_0(31)$ remains moderately strong in all three cases but that the tendency for $b_3 = b_4$ at small $\tau_0(31)$ becomes progressively weaker from case 1 to case 3. Also, the effect of optical pumping in level 4 becomes stronger as either $\tau_0(41)$ decreases (case 3) or λ_{42} decreases (case 1). The former effect is felt mainly near $\tau_0(41) = 1$ (as measured in cases 1 and 2) and the latter is felt mainly near $\tau_0(42) = 1$.

Numerous other variations of parameters (such as the various collision rates and the coupling to the free-bound continuum) are of interest but practical limitations exclude their evaluation and discussion at this time.

10. Four-Line Closed Loop

Closed loops of transitions involving four energy levels, as illustrated in Figure IV-13, are of considerable interest and occur frequently in both simple and complex spectra. This particular configuration is analogous to the two-line series discussed in Section 7 of this chapter. There is no particular reason to expect coupling between S_{21} and S_{42}, for example, but we expect strong coupling between S_{21} and S_{31} on the one hand and between S_{42} and S_{43} on the other. In other words, we expect a tendency for equality between b_2 and b_3.

We show in Figure IV-14a two sets of curves for cases in which the line opacities have been changed by about a factor of 10^2 and for the case where $\chi_2 = \chi_3$. Note

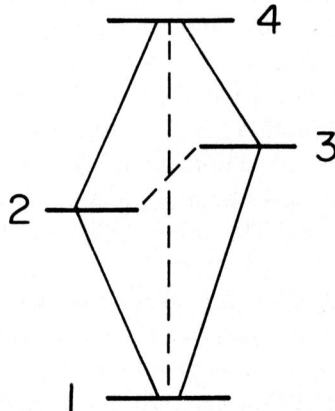

Fig. IV-13. Model atom for four line closed loop.

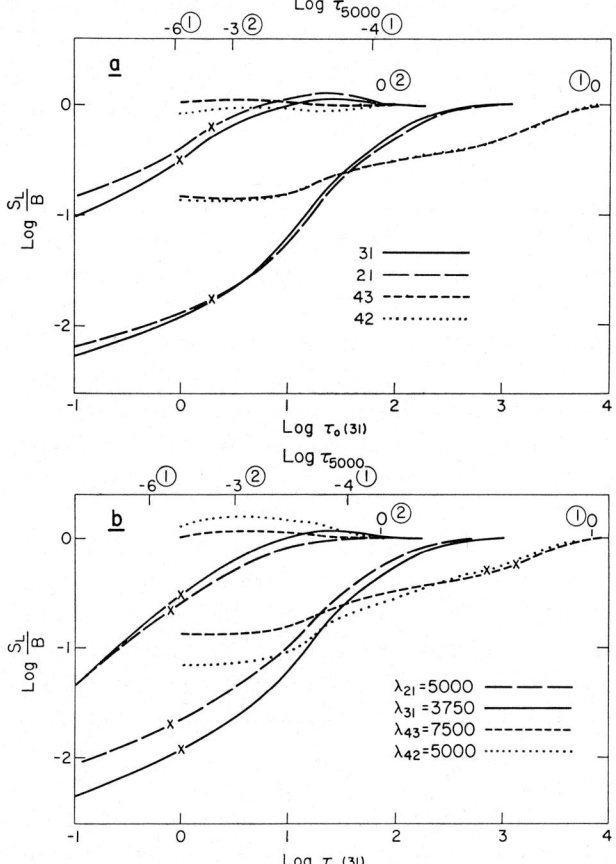

Fig. IV-14. Solutions for four line closed loop.

(a) – $\lambda_{21} = \lambda_{31} = 3750$, $\lambda_{42} = \lambda_{43} = 7500$,
$A_{ji} = 10^8$ (permitted), $Q_{ji} = 10\pi a_0^2$ (permitted)
$Q_{ij} = 3\pi a_0^2$ (forbidden)

	δ_{31}	δ_{21}	δ_{43}	δ_{42}
Case 1	0.00002	0.00004	0.05	0.05
Case 2	0.003	0.005	1	1

	ε_{31}	ε_{21}	ε_{43}	ε_{42}
Case 1	0.0003	0.0007	0.004	0.002
Case 2	0.0005	0.002	0.1	0.05

(b) – Same as (a) except $\lambda_{21} = 5000$, $\lambda_{31} = 3750$, $\lambda_{43} = 7500$, $\lambda_{42} = 5000$.

that in the high opacity case (case 1) the coupling between the two members of each doublet is very close. In the lower opacity case where $\delta \gg \varepsilon$ for all four lines, the coupling is somewhat weaker.

As in the preceding examples in Section 9, Figure IV-12, we note a strong tendency for S/B to increase markedly for all four lines and for S to exceed B in the 2-1 and

3-1 transitions just before thermalization occurs. Again, part of the increase in S/B is expected from the simple two-level analogy. However, note that S_{31}/B is of the order of 0.3 at $\tau_0(31)=1$ in case 2 even though $\varepsilon_{31}+\delta_{31} \approx 0.0035$ and $\varepsilon_{21}+\delta_{21} \approx 0.007$, i.e. $S_{31}/B \gg (\varepsilon_{31}+\delta_{31}+\varepsilon_{21}+\delta_{21})^{1/2}$. By contrast, in case 1, $S_{31}/B < \varepsilon_{31}^{1/2}$ near $\tau_0(31)=1$ and $S_{21}/B < \varepsilon_{21}^{1/2}$ near $\tau_0(21)=1$.

The relative changes in S_{21} and S_{31} in the two cases in Figure IV-14a are similar to those in S_{31} and S_{41} in Figure IV-12a. However, in this former case (Figure IV-12a) the reasons for the strong increase in S_{31} and S_{41} are readily explained as arising from optical pumping in the 3-2 and 4-2 transitions. The results in Figure IV-14a are not so readily explained. If the increase in S_{21} and S_{31} were due to optical pumping in the 4-3 and 4-2 transitions we would expect relatively stronger increases in S_{21} and S_{31} as λ_{42} and λ_{43} were decreased. Instead we find just the opposite effect. By lowering χ_2 and χ_3 to where λ_{42} and λ_{43} are in the violet (3750) and λ_{21} and λ_{31} are in the red (7500) we find, in the equivalent of case 2 (Figure IV-14a) that S_{42} and S_{43} each increases as expected but that S_{21} and S_{31} *decrease* by a corresponding factor. In other words, b_2 and b_3 decrease whereas b_1 and b_4 are essentially unchanged.

The rather strong similarity between Figures IV-12a and IV-14a clearly suggests that optical pumping effects are present in both cases. It is evidently the pumping effects in the 2-1 and 3-1 transitions themselves, however, that increase S_{21} and S_{31} in Figure IV-14a. On the other hand, the effect is not present nearly so strongly when level 4 is missing so the pumping effect does depend upon the coupling of levels 2 and 3 to level 4. Thus, the full explanation of the increase in S_{21} and S_{31} in case 2 remains obscure.

In discussing the upper level doublet in Section 4 of this chapter, we indicated that coupling between the upper levels was strongly increased by the presence of a fourth level connecting to both of the upper levels of the doublet by permitted transitions. The results in Figure IV-14a fully confirm this suggestion, as may be seen by comparing the results for case 1 to the results in Figure IV-2. The coupling in Figure IV-14a, case 1, is closer than that produced by $Q_{23}=10^4 \ \pi a_0^2$ in Figure IV-2. Part of this improvement, however, arises from the lower opacity ratio of the two doublet lines. The ratio is 2:1 in Figure IV-13a as compared to 10:1 in Figure IV-2.

Separating levels 2 and 3 in excitation energy reduces the coupling somewhat between the two levels. This is illustrated in Figure IV-14b where level 2 has been lowered by 0.84 eV. The resulting values of S_{31} and S_{43} are nearly identical to those in the preceding figure, except for $\tau_0(31) \ll 1$ in case 2. However, the change in S_{21}/B from case 1 to case 2 is considerably less than for the same cases in Figure IV-14a. In fact, S_{21} increases at $\tau_0(21)=1$ in Figure IV-14b by only 30% from case 1 to case 2. Thus, the two cases lead to essentially the same value of T_{21}, which is near 3300° at $\tau_0(21)=1$. This contrasting result appears, again, to be due to the effect of wavelength and the balance between J_c and B. In Figure IV-14b $\lambda_{21}=5000$ Å and $J \approx B$, whereas in Figure IV-14a $\lambda_{21}=3750$ Å and $J_c > B$. Thus, at $\tau_0(21)=1$ S_{21}/B is much lower for case 2 in Figure IV-14b than in Figure IV-14a. For case 1, at $\tau_0(21)=1$, S_{21}/B is nearly the same in the two figures.

11. Interlocking Effects on a Strong Line of Fixed ε, r_0 and λ

In the preceding illustrations of interlocking effects we have allowed the interlocking to change the values of r_0 from the comparison cases and, in addition, we have considered the effects of deliberately changing r_0. Many of the effects illustrated, therefore, include the effects of the changes in r_0. This is a realistic approach to the effects of interlocking since such effects will be present in any real situation.

It is instructive, however, to consider the effects of interlocking when r_0 is held relatively constant. To do this we adjust the abundance of the element following each new case considered until the value of r_0 near $\tau_0 = 1$ is the same as in the reference case. In addition, we keep both λ and ε identical to those in the reference case. This allows us to determine the influence of the added energy levels independently of their effects upon r_0. We consider only the strong line case ($\varepsilon \gg \delta$) in order to avoid the influences of optical pumping that are so important when some of the interlocking lines are weak to moderately strong.

Figure IV-15 illustrates 12 configurations of energy levels in each of which one spectral line is marked by a circle. This line has $\lambda = 3750$, $\delta = 10^{-4}$ and $\varepsilon \approx 8 \times 10^{-3}$ near $\tau_0 = 1$ in each of the cases. Each of the permitted transitions has $A_{ji} = 10^8$ s^{-1} and $Q_{ij} = 10 \ \pi a_0^2$. For the forbidden transitions $Q_{ij} = 3 \ \pi a_0^2$. Each of the bound levels is coupled to the continuum at 7.5 eV with photoionization cross-sections of 10^{-17}

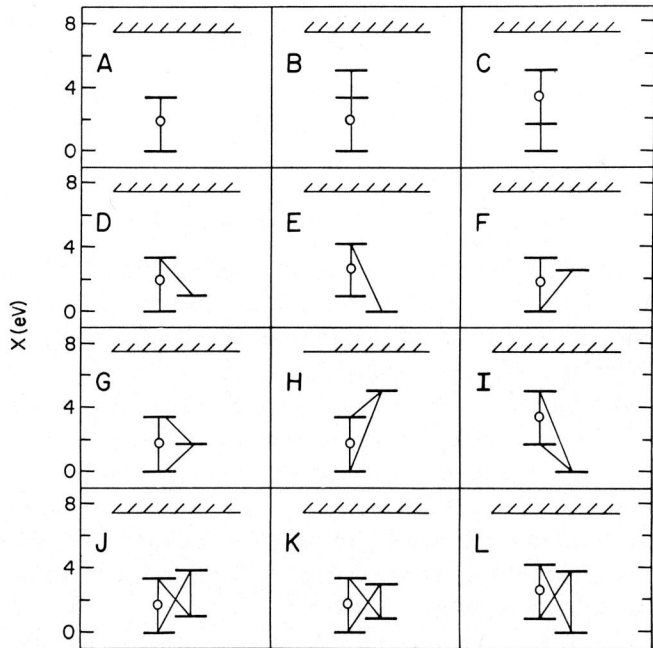

Fig. IV-15. Illustration of twelve different configurations for a line of the same wavelength and equivalent width.

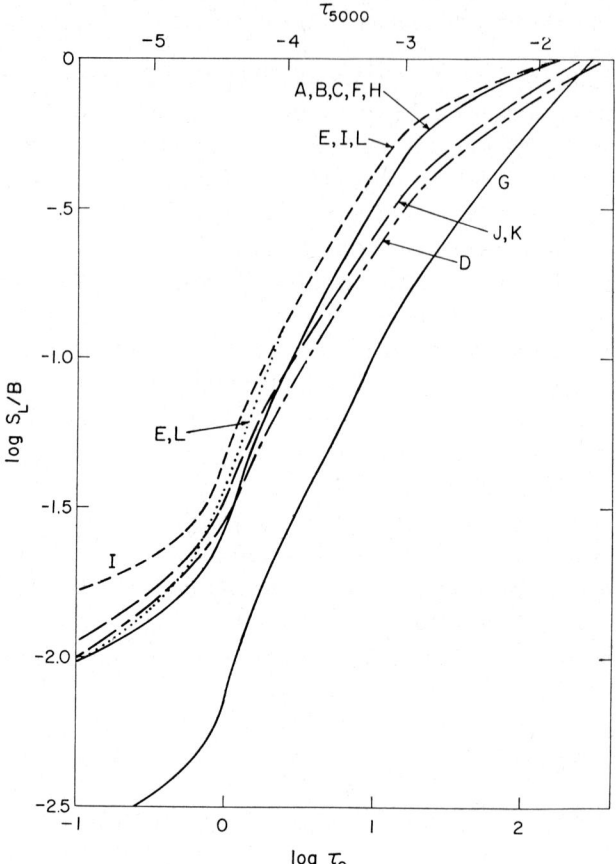

Fig. IV-16. Solutions for the twelve cases in Figure IV-15.

$(0.11 \pi a_0^2)$ and collisional ionization cross-sections of πa_0^2. The continuum radiation temperatures are 5200° for all levels with excitation energies greater than 2 eV and 4500° for all lower lying levels. Note that in some cases the reference line is a resonance line and in other cases a subordinate line.

Source functions for the twelve cases in Figure IV-15 are shown in Figure IV-16. They fall into six distinct classes whose members are: (1) A, B, C, F and H; (2) D; (3) E and L; (4) G; (5) I; and (6) J and K. Note, however, that cases (2) and (6) are very similar and that, as noted earlier, case (4) (G) is the most strikingly unique. In grouping the lines into six classes, we have overlooked small differences in S_L between some members of a given class. These differences, in all cases, are less important than the differences between classes.

For convenience in discussing the different cases let us label the lower and upper levels of the reference transition as 1 and 2, respectively in each case. In cases B through I the third level is labeled level 3 and in cases J, K and L the extra upper level is labeled level 4. We then note the following results: Level 3 changes S_{21} when level 3 lies near level 1 in all cases except C. When level 3 is metastable its influence on S_{21} at large values of τ_0 is strongest when level 3 lies above level 1 and near $\tau_0 = 3$ it is strongest when level 3 lies below level 1. The strongest effect of level 3 is felt when the 3-1 and 3-2 transitions are both permitted, and is particularly strong when level 3 lies between levels 1 and 2 with permitted transitions to each level. Level 4 has relatively little effect upon S_{21} as may be seen by comparing cases J and K with case D and case L with case E. The reason for the small influence of level 4 lies in the fact that the 2-4 transition is forbidden in each case considered. If this were not true, level 4 could more readily influence the 2-1 source function by changing the population of level 2 differentially with respect to level 1.

The reader should note, also, that we are considering only the case of strong lines where optical pumping effects are unimportant. For weaker lines, optical pumping becomes important and the influence of the added energy levels becomes much more important as illustrated in the preceding discussion.

12. Discussion

We have, in this chapter, tried to consider enough simple cases of interlocking to illustrate the general nature of their effects. Specifically, we have shown that added energy levels and added spectral lines may, in some cases, seriously alter the behavior of the source function for a particular spectral line. In some cases the source function moves nearer to the Planck function but in equally as many cases it moves further from the Planck function.

We have seen, also, that we quite often find rather close equality between source functions for different lines, sometimes in the sense that b_j/b_i is nearly the same for the different lines and sometimes in the sense that T_{ji} is nearly the same for different lines near $\tau_0 = 1$. Neither case, however, is associated with a tendency for the source function to approach the Planck function. Thus, the empirical demonstration that two lines, or even a group of lines, have a common source function in no way implies LTE. Eleven of the cases considered in Figure IV-16, for example, give source functions that differ by less than a factor two near $\tau = 1$ even though the source function is a factor 30 less than the Planck function.

The few interlocking cases considered in this chapter represent only a small fraction of those that actually exist in most atoms. Thus we cannot claim to have considered all of the important effects. Nor can we guarantee even that we have considered the most important effects.

It seems clear from the examples in this chapter that in most spectral lines interlocking effects cannot be ignored. Thus, multilevel atomic models are necessary in the majority of cases. In certain specific cases, notably for resonance lines of some ionized metals where both of the energy levels involved are far removed from other energy levels, the two-level approach may give satisfactory results. Examples where this seems to be true are found in the resonance lines of Mg II. The H and K lines of

Ca II show some influence from the metastable 3d levels as illustrated by the results in Section 6 and case D of Section 11. The two-level approach will, in this case, give the general behavior of the source function but will fail to give the proper detailed behavior. In still other cases, such as the Na D lines, the two-level approach may give rather accurate results for the source function as a function of τ_0, but may fail badly in relating τ_0 to τ_c, i.e., in giving the proper values for r_0.

It appears therefore that the two-level atomic model really works well only in rare cases and that, as a general rule, a multilevel atomic model must be used. Exactly which energy levels must be used in the atomic model can be picked, at this time, only by trial and error. As a practical way to proceed one can start with a two-level model then add extra transitions one at a time until both the source functions for the line in question and τ_0 for that line are unaffected by the addition of extra lines.

The most important considerations in adding extra energy levels and spectral lines are, on the one hand, their proximity to and their coupling with the energy levels of the line in question and, on the other hand, their possible influence on the ionization equilibrium. Apart from these general guidelines, however, no specific rules can be given at this time. Each case will differ somewhat depending upon the specific values of the atomic parameters associated with the coupled levels.

References

Athay, R. G. and Canfield, R. C.: 1969, *Astrophys. J.* **156**, 695.
Athay, R. G., Avrett, E. H., Beebe, H. A., Johnson, H. R., Poland, A. I., and Cuny, Y.: 1968, *Resonance Lines in Astrophysics*, National Center for Atmospheric Research, Boulder.
Auer, L. H. and Mihalas, D.: 1969, *Astrophys. J.* **158**, 641.
Avrett, E. H.: 1968, *Resonance Lines in Astrophysics*, National Center for Atmospheric Research, Boulder.
Avrett, E. H. and Kalkofen, W.: 1968, *J. Quant. Spectr. Radiative Transfer* **8**, 219.
Chapman, G. D. and Krauss, L.: 1966, *Canadian J. Phys.* **44**, 753.
Cuny, Y.: 1967, *Ann. Astrophys.* **30**, 143.
Finn, G. D. and Jefferies, J. T.: 1968, *J. Quant. Spectr. Radiative Transfer* **8**, 1705.
Finn, G. D. and Jefferies, J. T.: 1969, *J. Quant. Spectr. Radiative Transfer* **9**, 469.
Jefferies, J. T.: 1960, *Astrophys. J.* **132**, 775.
Jefferies, J. T.: 1968, *Spectral Line Formation*, Blaisdell, Waltham.
Johnson, H. R.: 1964, *Ann. Astrophys.* **27**, 695.
Kalkofen, W.: 1968, *Resonance Lines in Astrophysics*, National Center for Atmospheric Research, Boulder.
Kalkofen, W.: 1970, *Extended Atmosphere Stars*, NBS Special Publication **332**, 120.
Rybicki, G.: 1970, *Atlas Symposium No. 3*, Interdisciplinary Symposium on Applications of Transport Theory, Oxford.
Skumanich, A. and Domenico, B. A.: 1970, *Atlas Symposium No. 3*, Interdisciplinary Symposium on Applications of Transport Theory, Oxford.
Thomas, R. N.: 1957, *Astrophys. J.* **125**, 260.
Waddell, J. H., III: 1963, *Astrophys. J.* **138**, 1147.

CHAPTER V

LINE PROFILES

1. The Eddington-Barbier Relation

The specific intensity at frequency v coming from a position μ on a stellar disk is given by

$$I_v(\mu) = \int_0^\infty S_v \, e^{-\tau_v/\mu} \, d\tau_v/\mu. \tag{V-1}$$

For S_v of the form

$$S_v = a_v + b_v \tau_v, \tag{V-2}$$

Equation (V-1) gives

$$I_v = a_v + b_v \mu \tag{V-3}$$

so that

$$I_v = S_v(\tau_v = \mu). \tag{V-4}$$

Equation (V-4) is the familiar Eddington-Barbier relation widely used in solar applications for inversion of the integral in Equation (V-1).

If the dependence of S_v upon optical depth is not linear, Equation (V-4) is not strictly valid. However, for

$$S_v = a_v + b_v \tau_v^n, \quad n \geqslant 1, \tag{V-5}$$

Equation (V-1) gives

$$I_v = a_v + n! \, b_v \mu^n, \tag{V-6}$$

and for

$$S_v = a_v + b_v \tau_v^{1/2} \tag{V-7}$$

Equation (V-1) gives

$$I_v = a_v + \frac{\sqrt{\pi}}{2} b_v \mu^{1/2}. \tag{V-8}$$

Equations (V-5) and (V-6) give, for $n=2$,

$$I_v = S_v(\tau_v = \sqrt{2}\mu), \tag{V-9}$$

and Equations (V-7) and (V-8) give

$$I_v = S_v\left(\tau_v = \frac{\pi}{4}\mu\right). \tag{V-10}$$

Neither of these latter two relations departs very much from the simple Eddington-Barbier relation and we seem justified in concluding that the Eddington-Barbier relation is reasonably correct in most circumstances.

There are, of course, particular sets of circumstances under which the Eddington-Barbier relation gives poor results. It was noted in Chapter III (see Figure III-7) that when the Doppler width increases rapidly with height in a strong line and for frequencies near the edge of the Doppler core where ϕ_ν varies rapidly with ν, τ_ν varies extremely slowly either with τ_c or with τ_0. Under such circumstances we may have S_ν approximately of the form given by Equation (V-2) at $\nu = \nu_0$ but more nearly of the form given by Equation (V-5), where n is some relatively large number, at values of ν near the edge of the Doppler core. The relationship between I_ν and S_ν for different values of n is given in Table V-1. Since the effective value of n may vary with fre-

TABLE V-1

Ratio of τ_ν/μ for $I_\nu = S_\nu$ when S_ν is of the form given by Equation (V-5)

n	2	4	6	8	10
τ_ν/μ	1.4	2.2	3.0	3.8	4.5

quency within the Doppler core and may become relatively large at certain frequencies, the Eddington-Barbier relation must be used cautiously and with the realization that it may be much less reliable near the edge of the Doppler core than either near line center or in the line wings where ϕ_ν varies more slowly with frequency.

It should be remembered that the Eddington-Barbier relation is an integrated effect and that the statement $I_\nu = S_\nu$ at $\tau_\nu = \mu$ does not mean that one 'sees' only the phenomena near $\tau_\nu = \mu$ nor even that one 'sees' preferentially the phenomena near $\tau_\nu = \mu$. The integrand in Equation (V-1) multiplied by equal elements in $d\tau_\nu$ (which we call the contribution function) for

$$S_\nu = a_\nu \left(1 + \frac{b_\nu}{a_\nu} \tau_\nu^n\right),$$

is shown in Figure V-1 for $\mu = 1$ for the four cases $n = \frac{1}{2}$, 1, 2 and 3 at $b_\nu/a_\nu = 1.5$ and for the two additional cases $b_\nu/a_\nu = 3.0$ and 4.5 for $n = 1$. Three positions are marked on each curve: a vertical bar for the depth at which $I_\nu = S_\nu$, a circle for the depth of maximum contribution and a cross for the depth above which 50% of I_ν is contributed. In the following, we refer to these three depths as $I_\nu(S_\nu)$, $I_\nu(\text{max})$, and $I_\nu(\frac{1}{2})$.

For $n = \frac{1}{2}$, $I_\nu(\text{max})$ occurs at a such smaller τ than either $I_\nu(S_\nu)$ or $I_\nu(\frac{1}{2})$. For $n = 1$, $I_\nu(\text{max})$ and $I_\nu(S_\nu)$ coincide when b_ν/a_ν is very large, but for $b_\nu/a_\nu \lesssim 3$, $I_\nu(\text{max})$ remains at substantially smaller τ than $I_\nu(S_\nu)$. The point $I_\nu(\text{max})$ moves deeper into the atmosphere as n increases and as b_ν/a_ν increases. $I_\nu(\text{max})$ and $I_\nu(S_\nu)$ are at the same depth for $n = 1$ at very large b_ν/a_ν, for $n = 2$ at $b_\nu/a_\nu = 1.2$, and for $n = 3$ at $b_\nu/a_\nu = 0.25$.

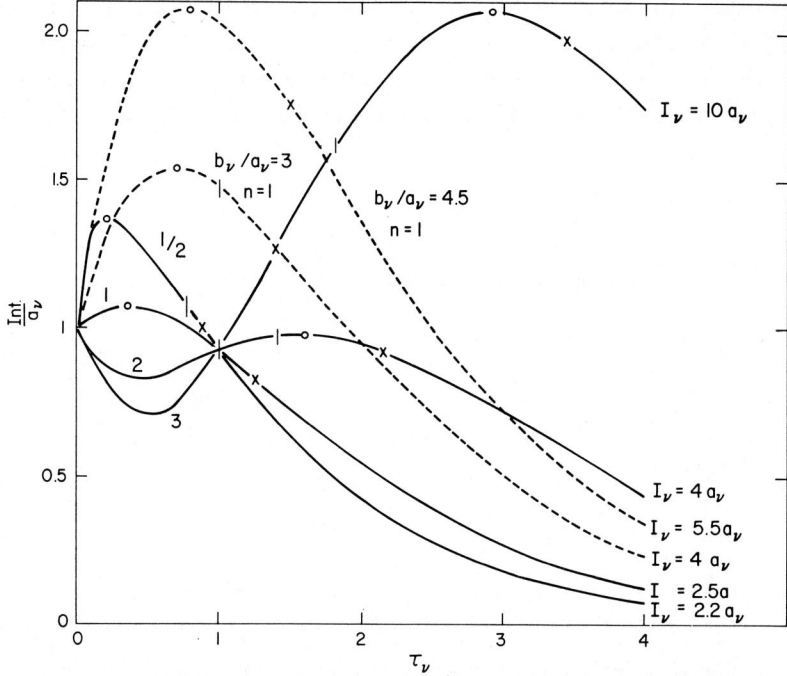

Fig. V-1. Illustration of the depths of maximum visibility (0), depths where $I_\nu(0) = S_\nu(\tau_\nu)$ (|) and depths at which half of $I_\nu(0)$ is formed (x). The solid curves are for $\beta = 1.5$ and for the indicated values of n, where $B = a_\nu + b_\nu \tau_\nu^n$, $\beta = b_\nu/a_\nu$. Dashed curves are shown for $\beta = 3$ and 4.5, and $n = 1$. The plots show the contribution to $I_\nu(0)$ from equal increments in optical depth.

Note that the contribution functions in Figure V-1 are actually quite broad. The function is most sharply peaked when n is small and becomes increasingly broad as n increases. For a given n, the contribution function has about the same relative width independently of b_ν/a_ν. Larger values of b_ν/a_ν suppress somewhat the relative contributions for τ less than unity and increase somewhat the relative contributions for τ greater than unity. This effect is small compared to that of changing n, however.

Because of the broadness of the contribution function one cannot always assign a property of the atmosphere inferred from a line profile to the depth $\tau_\nu = \mu$ with safety. An observed brightness fluctuation, for example, could arise from a change in S_ν or in τ_ν anywhere between $0 \lesssim \tau_\nu \lesssim 4$ for $n \approx 1$. The Eddington-Barbier relation is useful, with certain restrictions, for a crude inversion of the integral in Equation (V-1) to obtain $S_\nu(\tau_\nu)$. In other applications, such as the assignment of an 'observed' structural feature or an 'observed' velocity to a particular depth in the atmosphere, considerable caution should be exercised.

Failure of the Eddington-Barbier relation may be particularly pronounced when gradients in the Doppler width are present. As shown in Figure III-7, a gradient in the Doppler width may have a very marked influence on the relationship between S_ν and

τ_ν at wavelengths near the edge of the Doppler core. This effect will be discussed in Section 3 of this chapter.

In the case of stars where we observe only the net flux H_ν the Eddington-Barbier relation takes on a somewhat different form. For this case

$$H_\nu = \tfrac{1}{2}\int_{\tau_\nu}^{\infty} S_\nu E_2(t - \tau_\nu)\,dt - \tfrac{1}{2}\int_{0}^{\tau_\nu} S_\nu E_2(\tau_\nu - t)\,dt, \qquad \text{(V-11)}$$

and for S_ν given by Equation (V-7) we obtain

$$H_\nu = \tfrac{1}{4}(a_\nu + \tfrac{2}{3}b_\nu), \quad n = 1, \qquad \text{(V-12)}$$

$$H_\nu = \tfrac{1}{4}(a_\nu + b_\nu), \quad n = 2, \qquad \text{(V-13)}$$

and

$$H_\nu = \tfrac{1}{4}(a_\nu + \tfrac{1.2}{5}b_\nu), \quad n = 3. \qquad \text{(V-14)}$$

Thus,

$$4H_\nu = S_\nu(\tau_\nu = \tfrac{2}{3}), \quad n = 1, \qquad \text{(V-15)}$$

$$4H_\nu = S_\nu(\tau_\nu = 1), \quad n = 2, \qquad \text{(V-16)}$$

and

$$4H_\nu = S_\nu(\tau_\nu = 1.34), \quad n = 3. \qquad \text{(V-17)}$$

Apart from the scaling factor of 4 these results are quite similar to those for $I_\nu(\mu)$. The weighted values of τ_ν for $4H_\nu = S_\nu$ are less than the value of τ_ν for $I_\nu = S_\nu$ by an additive amount of about $-\tfrac{1}{3}$.

The preceding comments about the effects of variable Doppler width and the differences between $I_\nu(\max)$, $I_\nu(S_\nu)$ and $I_\nu(\tfrac{1}{2})$ illustrated in Figure V-1 apply equally as well to the corresponding quantities $H_\nu(\max)$, $H_\nu(S_\nu)$ and $H_\nu(\tfrac{1}{2})$ with a similar reduction in τ_ν by about $-\tfrac{1}{3}$.

The subsequent discussion in this chapter is devoted to $I_\nu(\mu)$. However, the reader should have little difficulty in visualizing the corresponding effects in H_ν.

2. Depth Dependence of S_ν Near Line Center

Near the center of a strong line we may ignore the contribution of B to S_ν and the contribution of τ_c to τ_ν. Also, depth variations in the shape of ϕ_ν are not usually of major importance in altering the relationship between S_L and τ_ν.

It has been shown in preceding chapters that in the vicinity of $\tau_0 = 1$, S_L/B is generally a decreasing function of τ_0. Since in solar type stars B is itself a decreasing function of τ_0 at all values of $\tau_{5000} \gtrsim 10^{-4}$, it follows that whenever $\tau_0 = 1$ lies at $\tau_{5000} > 10^{-4}$ S_L is a more strongly decreasing function of τ_0 than is either S_L/B or B. At $\tau_{5000} < 10^{-4}$ in the solar case B increases with decreasing τ_0. However, as the results in Figures IV-3a and IV-3B show, even under this circumstance S_L may decrease steadily outwards.

In a number of the illustrations in Chapter IV S_L/B increases outwards in the region just below the temperature minimum, i.e., at optical depth near $\tau_{5000} = 10^{-2}$. Even in the most extreme cases, however, S_L decreases steadily outwards, albeit in some cases the gradient in S_L is small. Of all the examples considered in Chapters III and IV, not a single one results in S_L increasing outwards near $\tau_0 = 1$.

The strong tendency for S_L to decrease outwards in the vicinity of $\tau_0 = 1$ for strong lines carries certain implications regarding the behavior of I_v near line center. We expect, for example, that I_v will increase away from line center, i.e., the core of the line will not be flat bottomed, and we expect I_v to darken toward the limb at least out to $\mu = 0.1$. These general properties are shown by all of the strong Fraunhofer lines. They result purely from the effect of the boundary and even a relatively strong chromospheric effect such as is found in the K_3 region of the Ca II K-line does not obliterate these tendencies.

Within the Doppler core of the line S_L/B is proportional to $\tau_0^{1/2}$ if B is either constant or linear in τ_c and if optical pumping effects are unimportant. For these restricted cases, then, S_L is of the form

$$S_L = \tau_0^{1/2} (a_v + b_v \tau_c)_{\tau_0 = 1}, \qquad (V-18)$$

and the τ_0 dependence of S_L lies between the approximate limits $\tau_0^{1/2}$ and $\tau_0^{3/2}$. When $\tau_0 = 1$ lies in the chromosphere where B increases outwards S_L varies more slowly with τ_0. The cases shown in Figures IV-3a and IV-3b are typical. Near $\tau_0 = 1$ in the stronger lines of the multiplets shown S_L varies approximately as $\tau_0^{1/4}$, and near $\tau_0 = 1$ with weaker lines of the multiplets S_L varies approximately as $\tau_0^{1/3}$. These results, together with those in Figure V-1 suggest that the Eddington-Barbier relation is reasonably valid but that both $I_v(\max)$ and $I_v(S_v)$ are located at τ_0 significantly less than unity.

3. Frequency Dependence of S_v

Although S_L and B are each assumed to be independent of frequency throughout the spectral band of the line, the mixture of S_L and B that forms S_v,

$$S_v = \frac{\phi_v}{\phi_v + r_0} S_L + \frac{r_0}{\phi_v + r_0} B, \qquad (V-19)$$

is not independent of frequency. If, for example, we consider the case $r_0 = 10^{-3}$, $S_L/B = 10^{-1}$ and $\phi_v = \exp -(\Delta v/\Delta v_D)^2$, we obtain the values of S_v/B given in row 2 in Table V-2. Substantial frequency dependence is already evident at $\Delta v/\Delta v_D = 1.5$. For the more extreme case of $r_0 = 10^{-6}$ and $S_L/B = 10^{-2}$, which is typical of the region $\tau_0 = 1$ in the Na D lines (Figure III-3a), the frequency dependence becomes evident near $(\Delta v/\Delta v_D) = 2.5$.

It is clear from the results in Table V-2 that any empirical analyses of line profiles based upon the assumption of frequency independence must be restricted to wavelengths very near line center. It is equally clear that empirical analyses of line profiles designed

TABLE V-2

Frequency Dependence of S_ν for $\phi_\nu = \exp -(\Delta\nu/\Delta\nu_D)^2$

	$\Delta\nu/\Delta\nu_D$	0	1	1.5	2	2.5	3
$r_0 = 10^{-3}$ $S_L/B = 10^{-1}$	S_ν/B	0.101	0.102	0.108	0.147	0.407	0.904
$r_0 = 10^{-6}$ $S_L/B = 10^{-2}$	S_ν/B	0.0100	0.0100	0.0100	0.0101	0.0105	0.0180

to detect frequency dependence in S_ν should not have too much trouble detecting it. If center-limb data are used in such analyses, additional allowance must be made for the possibility of anisotropy in the Doppler broadening.

4. Mapping of S_ν into I_ν

Because of the approximate validity of the Eddington-Barbier relation, it is commonly assumed that for strong lines the profile of the line, I_ν, is a direct reflection of $S_\nu(\tau_\nu)$, each point in the profile being given by the simple assignment of $I_\nu = S_\nu(\tau_\nu = 1)$. There is reasonable justification for this assumption near line center and in the line wings. In both of these cases τ_ν is not markedly sensitive to depth variations in ϕ_ν and there is little chance of seriously modifying the Eddington-Barbier relation.

The situation is quite different in the outer parts of the Doppler core, however. In these parts of the profile ϕ_ν is a very strong function of $\Delta\nu/\Delta\nu_D$ and consequently τ_ν is very sensitive to depth dependence of $\Delta\nu_D$. We illustrated this effect schematically in Figure III-7 for a discontinuous change in $\Delta\nu_D$. To understand the effect further, we write Equation (V-1) in the form

$$I_\nu = \int_0^\infty S_\nu \exp - \left[\int_0^{\tau_0} (\phi_\nu + r_0) \frac{dt_0}{\mu} \right] (\phi_\nu + r_0) \frac{d\tau_0}{\mu}. \tag{V-20}$$

Consider the case where $\phi_\nu \gg r_0$. The optical depth τ_ν is then given by $\int_0^{\tau_0} \phi_\nu \, dt_0$. If ϕ_ν is constant with depth, τ_ν has the same relationship to τ_0 throughout the atmosphere, and S_ν is essentially the same function of τ_ν at all frequencies. The only factor that mars this constancy of the dependence of S_ν upon τ_ν is the differential mixing of S_L and B with frequency and the fact that S_L and B will generally have a different dependence upon τ_0. However, this effect is usually not of major importance. It may give rise to some differences in the depth dependence of S_ν at different points in the profile and consequently to some differential behavior of I_ν with respect to $S_\nu(\tau_\nu = 1)$. Other factors influencing the optical depth dependence of S_ν tend to be more important, however.

Let us suppose that within the solar atmosphere there is a gradient in Δv_D such as would result from the mean broadening velocities given in column 2 of Table V-3. The resultant values of ϕ_ν for fixed values of Δv located at $\Delta v/\Delta v_D(100) = 0, 1, 2, 3, 4, 5$ and 7.5 are given in columns 3 to 9 of Table V-3. The Voigt parameters in ϕ_ν have

TABLE V-3

ϕ_ν at different values of $\Delta v/\Delta v_D$, where Δv_D is evaluated at 100 km and $a = \sqrt{\pi} \times 10^{-3}$

Height (km)	$\bar{v}$ km s^{-1}	ϕ_0	$\phi_1(1.5)$	$\phi_2(1.5)$	$\phi_3(1.5)$	$\phi_4(1.5)$	$\phi_5(1.5)$	$\phi_{7.5}(1.5)$
800	8	1	0.965	0.869	0.729	0.570	0.415	0.138
750	7.5	1	0.961	0.852	0.670	0.527	0.368	0.105
700	7	1	0.955	0.832	0.661	0.480	0.317	0.0755
650	6.5	1	0.948	0.808	0.620	0.427	0.264	0.0500
600	6	1	0.939	0.779	0.570	0.368	0.210	0.0297
550	5.5	1	0.928	0.743	0.512	0.304	0.156	0.0152
500	5	1	0.914	0.698	0.445	0.237	0.105	0.00638
450	4.5	1	0.895	0.641	0.369	0.169	0.0622	0.00198
400	4	1	0.869	0.570	0.282	0.105	0.0297	0.000415
350	3.5	1	0.832	0.480	0.191	0.0530	0.0100	0.0000742
300	3	1	0.779	0.368	0.105	0.0184	0.00201	0.0000357
250	2.5	1	0.700	0.237	0.0394	0.00325	0.000183	0.0000298
200	2	1	0.570	0.105	0.00647	0.000204	0.0000542	0.0000238
150	1.5	1	0.368	0.018	0.000233	0.0000625	0.0000402	0.0000177
100	1.5	1	0.368	0.018	0.000233	0.0000625	0.0000400	0.0000177

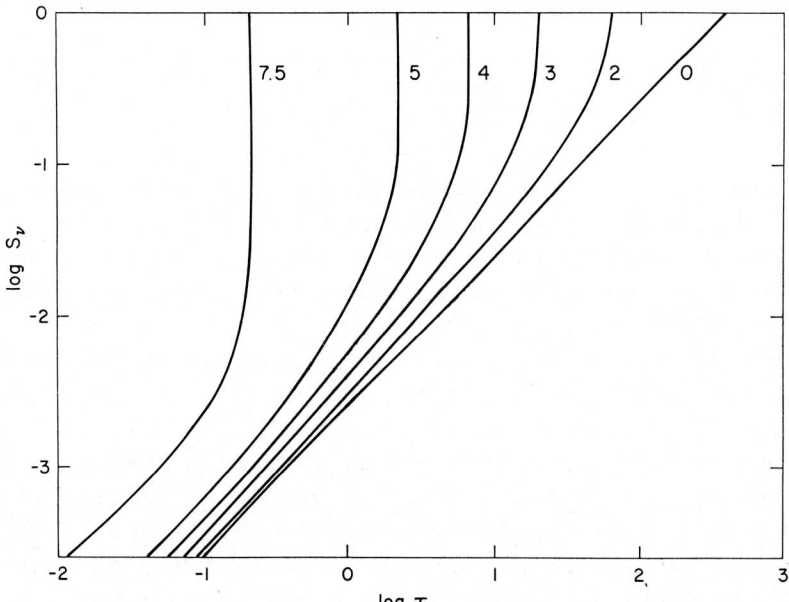

Fig. V-2. The variation of S_ν with τ_ν for Doppler width variations given in Table V-3. The labels on the curves are the values of $\Delta v/\Delta v_D$.

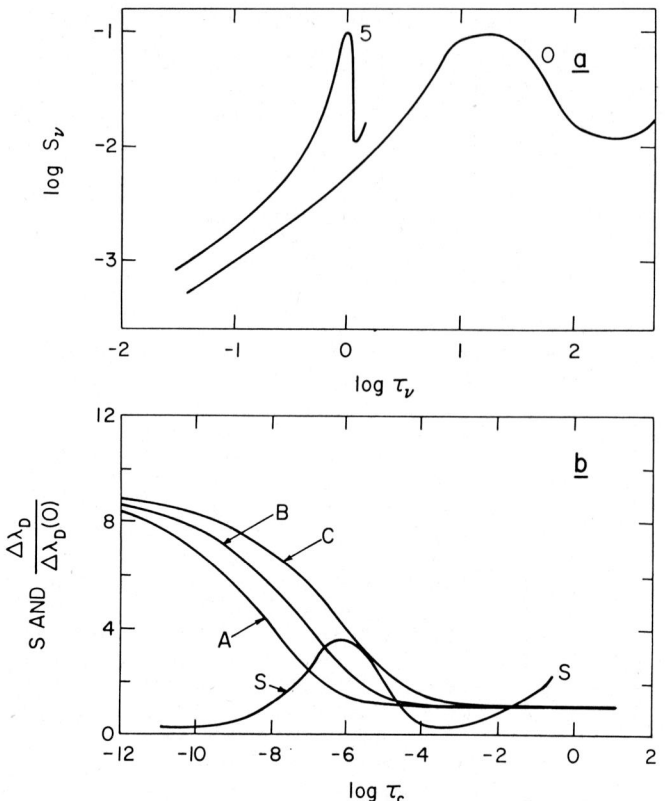

Fig. V-3. (a) – S_ν vs. τ_ν at $\Delta\nu/\Delta\nu_D = 0$ and 5 for the Doppler width variations in Table V-3. (b) – Variation of S_L with depth and three variations of $\Delta\lambda_D/\Delta\lambda_D(0)$ with depth for the profiles illustrated in Figure V-4.

been taken as $\underline{a} = \sqrt{\pi} \times 10^{-3}$ and $\Delta\nu_D(100)$ has been evaluated at the bottom line in Table V-3. Because of the increase of the broadening velocity, $\bar{v}$, with height, the Doppler width at 750 km in Table V-3 is five times as large as it is on the bottom line of the table. Hence a wavelength displacement at $\Delta\nu = 7.5\,\Delta\nu_D(100)$ coincides with $\Delta\nu = 1.5\,\Delta\nu_D(750)$.

Figure V-2 illustrates the effect of the velocity gradient in Table V-3 upon $S_\nu(\tau_\nu)$. At the position marked by $\tau_\nu = 1$, which is arbitrary, of course, S_ν is linear in τ_0. However, at $\Delta\nu = 5\,\Delta\nu_D(100)$ S_ν varies approximately as the square of τ_ν, and at $\Delta\nu \approx 6\,\Delta\nu_D(100)$ S_ν rises almost discontinuously. The Eddington-Barbier relation will be reasonably good, for the case illustrated in Figure V-2, out to $\Delta\nu \approx 4\,\Delta\nu_D(100)$. It will fail markedly at $\Delta\nu = 5\,\Delta\nu_D(100)$ and become much worse at $\Delta\nu = 6\,\Delta\nu_D(100)$. Far enough in the wings that the Voigt term dominates over the Doppler core at all depths (approximately $\Delta\nu \geqslant 15\,\Delta\nu_D(100)$) S_ν will again become linear in τ_ν and the Eddington-Barbier relation will again be valid.

Compression of the τ_ν scale due to a gradient in $\Delta\nu_D$ may produce still different types of failure of the Eddington-Barbier relation than those due simply to a change in the power-law of $S_v(\tau_v)$. We show in Figure V-3a a source function that passes through a maximum near $\tau_0 = 20$. The width of the maximum at $S_v = S_v(\text{max})/2$ extends from approximately $\tau_0 = 6$ to $\tau_0 = 45$. It is assumed that at larger values of τ_v, S_v increases monotonically to B and that B exceeds $S_v(\text{max})$. In other words, S_v mimics the curves in Figures III-4a and III-4b.

If ϕ_v were constant with depth, S_v would map directly into the line profile I_v to produce an emission peak with an intensity given to good accuracy by $I_v(\text{max}) = S_v(\text{max}) = 10^{-1}$. The maximum would occur at $\Delta v \approx 1.7 \Delta v_D$, and would be similar in appearance to the K_2 peak in the Ca II K-line. The second curve in Figure V-3a shows the same source function plotted against τ_v at $\Delta v = 5 \Delta v_D(100)$ from Table V-3. At the points $S_v = S_v(\text{max})/2$ the width of the maximum now extends only from $\tau_v = 0.8$ to $\tau_v = 1.1$. Thus, the maximum in S_v appears, at this wavelength, to be optically thin. As a result I_v will show a peak intensity that still resembles that observed at K_2 but which

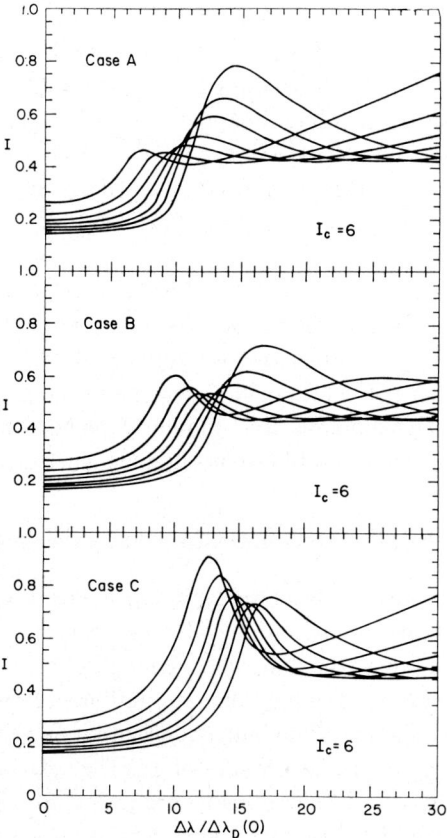

Fig. V-4. Profiles computed for the cases illustrated in Figure V-3b (Athay and Skumanich, 1968b; courtesy *Solar Physics*, Reidel, Dordrecht).

is only a fraction of $S_\nu(\text{max})$. At no point in the profile will $I_\nu = S_\nu(\text{max})$ except in the line wings where S_ν has again increased to a value near $S_\nu(\text{max})$.

Note also that the minimum in S_ν near $\tau_0 = 250$ is compressed at $\tau_\nu(s)$ into a shallow region of optical thickness less than unity. Thus, the mapping of S_ν into I_ν will show a value of $I_\nu(\text{min})$ that lies considerably above $S_\nu(\text{min})$.

To recover the true run of $S_\nu(\tau_\nu)$ from the profile of I_ν resulting from $S_\nu(\tau_\nu)$ in Figure V-3a and ϕ_ν in Table V-3 by inversion of Equation (V-1) would require that $\Delta\nu_D$ be known as a function of depth. Otherwise, one might well infer completely erroneous values of $S_\nu(\tau_\nu)$. Conversely, if one assumed that I_ν were related to S_ν via the Eddington-Barbier relation the inferred values of $\Delta\nu_D$ would necessarily be erroneous.

The preceding two illustrations consider only one particular type of variation of ϕ_ν with depth. They serve to illustrate, however, that I_ν is a sensitive function of the depth variations in ϕ_ν as well as the depth variation of $S_\nu(\tau_0)$. It is particularly important in empirical analyses of line profiles to make note of this. Most such analyses are based upon the *a priori* assumption that ϕ_ν is constant with depth even though the results quite often show that ϕ_ν is in fact depth dependent.

Athay and Skumanich (1968a) used the phenomena represented in Figure V-3 to explain the solar center-limb behavior of the K_3, K_2 and K_1 components of the K-line of Ca II. Their results for three different depth variations of $\Delta\nu_D$ and the single depth variation of S_ν shown in Figure V-3b are given in Figure V-4 Cases A, B, and C. The K_2 peak can be made to limb-darken or limb-brighten depending upon the location of the gradient in $\Delta\nu_D$. The K_1 minimum either limb-darkens or remains constant. K_3 always shows limb-darkening.

For a constant $\Delta\nu_D$ the source function in Figure V-3b would produce a K_2 peak intensity approximately seven times the K_3 intensity. The profiles in Figures V-4, Cases A, B and C show K_2/K_3 ratios ranging from less than 2:1 to just under 5:1. As might be expected, the largest values of K_2 occur when the strong gradient in $\Delta\nu_D$ is farthest removed from the maximum in S_ν. Similarly the minimum value of K_1 occurs in cases A and B where there is little or no gradient in $\Delta\nu_D$ through the region of $S_\nu(\text{min})$.

5. Microturbulence, Macroturbulence, and Differential Motion

There is little reason to suppose that stellar atmospheres are quiescent. In fact, a wealth of evidence indicates that motions of a complex nature are present in the atmospheres of most stars. For the Sun, where enough photons are present to permit observations with high spectral and high time resolution and where individual surface features can be resolved, there is abundant evidence that non-thermal motions play a major role in determining the characteristics of ϕ_ν. The solar lines are broadened by so-called 'microturbulence' and 'macroturbulence'; they are often asymmetrical and they are typically displaced from point to point on the surface of the Sun. The latter two phenomena are time dependent, the time dependence of the last giving rise to the wiggly-line phenomena.

It is now clearly recognized that much of the motion in the solar atmosphere is of a periodic wave character, and that the wave motion accounts at least partially for the non-thermal line broadening. However, it still appears that other types of motion are present and that, in fact, most of the energy in the non-thermal motions resides in modes whose character is largely unknown. It may be that as observations and analyses of the wave motion improve we will eventually discover that most of the energy resides in such motion. Meanwhile, we continue to describe the line broadening motions as being due to 'microturbulence' and 'macroturbulence.' The true nature of such motions, however, is left largely in the realm of the mystical. The reference to turbulence means only that the motions are assumed to be approximately random in character, when integrated along any particular line of sight. Micro and macro refer to the mean 'eddy' size relative to the photon mean free path length. The scale is assumed to be microscopic if the eddy size is much less than $\Delta\tau=1$ and macroscopic if the scale is large compared to $\Delta\tau=1$. For a detailed discussion of micro- and macroturbulence, the reader is referred to Huang and Struve (1960). Only a brief discussion will be given here.

In spite of the wealth of data available for the Sun, our knowledge of even the most basic properties of the micro- and macroturbulence is in a state of almost total confusion. We have not been able as yet, for example, to decide whether within the photosphere the microturbulence is isotropic and increasing outwards or anisotropic and decreasing outwards. Either model seems to fit the data equally well. In the case of macroturbulence, we are perhaps even less certain of the characteristics of motion.

Microturbulence is defined such that ϕ_ν is broadened on a microscopic basis exactly as though the motion were thermal. Since the equivalent width of a spectral line is proportional to the Doppler width of ϕ_ν, microturbulence, if present, increases the equivalent widths of the lines. Macroturbulence, on the other hand, produces local displacements of the profiles without affecting the equivalent widths. The wiggly line phenomena produced by oscillatory motions in the solar atmosphere is primarily macroscopic in scale as are such phenomena in stellar spectra as rotation and height gradients in Δv_D. The latter produces an apparent macroturbulence because of the integration over μ and the fact that the values of Δv_D are a function of μ.

The simple distinction of two scales of motion (micro and macro) is a gross oversimplification, of course. A given type of motion may produce effects upon ϕ_ν and upon I_ν that lie between the two extremes. The problem is complex and the complexity accounts for much of the confusion.

In some sense the intensity at a monochromatic point in a line profile is not independent of the intensity at adjacent points. I_ν must be single valued, for example, and this means that one point in a profile cannot be displaced very far in frequency without simultaneously displacing adjacent points that lie on the side of the profile toward which the displacement occurs. The entire Doppler core of a line generally extends only about three Doppler widths from line center. Obviously, the center of the line cannot be displaced by three Doppler widths without some displacement occurring at the edge of the core. This must happen even if the only part of the atmosphere that is in motion is the region around $\tau_0 \lesssim 1$. There may be no motion whatsoever at optical depths where the edges of the Doppler core are formed. Nevertheless, if the center of the core has a large Doppler displacement the edges of the core must show some fraction of the displacement in the same direction. This tendency for a certain amount of coherency in Doppler displacement often gives the illusion of macroturbulence through a large range of depths whereas in reality only a limited region of the atmosphere may be in motion. This effect is illustrated in Figure V-5 where we show three Hα profiles

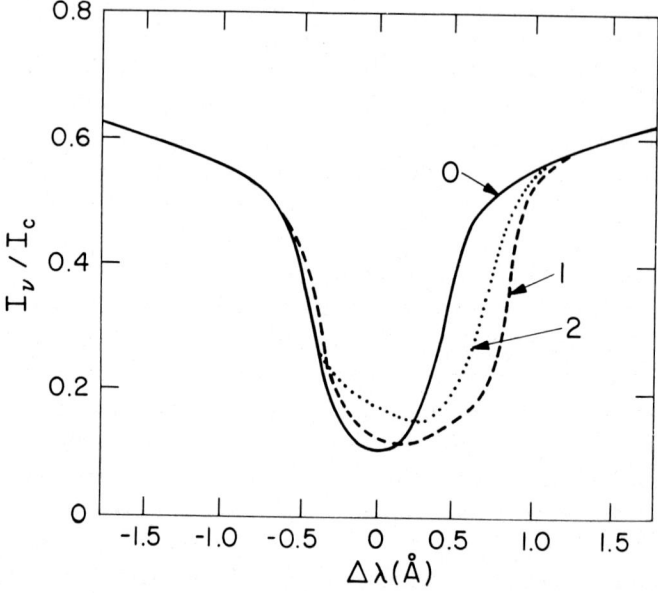

Fig. V-5. Three Hα profiles computed for

0 – Stationary atmospheres
1 – Downward motion for $1 \ll \tau_0 \ll 100$
2 – Downward motion for $\tau_0 \lesssim 1$

computed for cases labeled as: 0 – chromosphere at rest, 1 – chromosphere moving between heights of 500 km and 1000 km above the limb, and 2 – chromosphere moving at heights exceeding 1500 km above the limb. The motion is downward at a velocity such that the center of ϕ_ν is shifted $+0.5$ Å. Line center is formed ($\tau_0 = 1$) at about 1200 km in the chromosphere. For case 1, the portion of the chromosphere in motion extends from the regions of formation ($\tau_\nu = 1$) where $I_\nu/I_c \approx 0.4$ to $I_\nu/I_c \approx 0.15$. For case 2, however, the motion is confined to regions for $\tau_0 < 1$. In spite of these restrictions on the motion, the profiles show velocity displacement well into the red wing of the line to where $I_\nu/I_c \approx 0.55$ and $\Delta\lambda \approx 1.2$ Å. At this wavelength the radiation originates ($\tau_\nu = 1$) primarily in the photosphere.

Even in case 2, where none of the line forming layer (as usually conceived from the Eddington-Barbier relation) is in motion, an observer would erroneously infer from the line displacement that the entire chromosphere below about 1200 km as well as the upper photosphere is moving downward at an appreciable velocity. Note also that the equivalent width of the line is increased as a result of the motion. It would appear to the observer, therefore, that the additional broadening of the line was due to 'microturbulence' when, in fact, the motion is both systematic and of large scale.

It is logical to ask, at this point, whether an observer could, by any more or less direct analytical method, use the profiles in Figure V-5 to infer the true amplitude of the velocity and the actual layer of the atmosphere that is in motion. Clearly, the standard technique of using the center of symmetry of the profile will give completely erroneous results. The velocities obtained will be too small and at the wrong depths. Also, the difference profiles obtained by subtracting the pairs 0 and 1 or 0 and 2 give maximum difference at $\Delta\lambda > 0.5$ on the red side of the line and $|\Delta\lambda| < 0.5$ on the violet side. The averages of the two difference maxima on the two sides of the profile are $\Delta\lambda = 0.35$ Å (case 2) and $\Delta\lambda = 0.55$ Å (case 1). This method therefore gives a better measure of the velocity amplitude, in this example, but gives no indication as to what part of the atmosphere is in motion. There appears to be no direct method for arriving at the conclusion in either case 1 or case 2 that only part of the atmosphere is in motion.

Weaker spectral lines show effects very similar to those in Figure 4-5 (Athay, 1970a). In the case of weak lines, however, the depth of the line forming layer becomes more nearly of the order of the atmospheric density scale height than in the case of strong lines and there is less likelihood of strong gradients in velocity within the line forming layer. Velocity gradients are present in the Sun, however, and the gradients themselves inevitably influence the deductions made from profile analysis. Just how serious such influences are is not presently known. They may well be partially responsible for the failure of solar astronomers to reach any common agreement concerning the amplitude of the microturbulent and macroturbulent motions.

The examples shown in Figure V-5 illustrate the basic complexity of the effects of differential motion on line profiles. Empirical analyses aimed at separating microturbulence and macroturbulence are very likely indeed to arrive at erroneous conclusions unless it can be demonstrated unambiguously that such conclusions are unique. The demonstration of uniqueness implies that large quantities of data be satisfied and there seems little hope of demonstrating the uniqueness of conclusions based upon only a few line profiles.

The influences on S_L of a gradient in microturbulence and of differential motion in an atmosphere were discussed in Sections 5 and 6 of Chapter III. We noted there that gradients in microturbulence may produce quite substantial effects upon S_L but that differential motion did not have a strong effect upon S_L.

Differential motion in an atmosphere may produce a variety of effects upon line profiles. Profiles computed by Hummer and Rybicki (1968a, b) for a slab atmosphere expanding differentially with the optical distance from the slab center according to the relation

$$\mu = \mu_0 + \tau\mu_1 \tag{V-21}$$

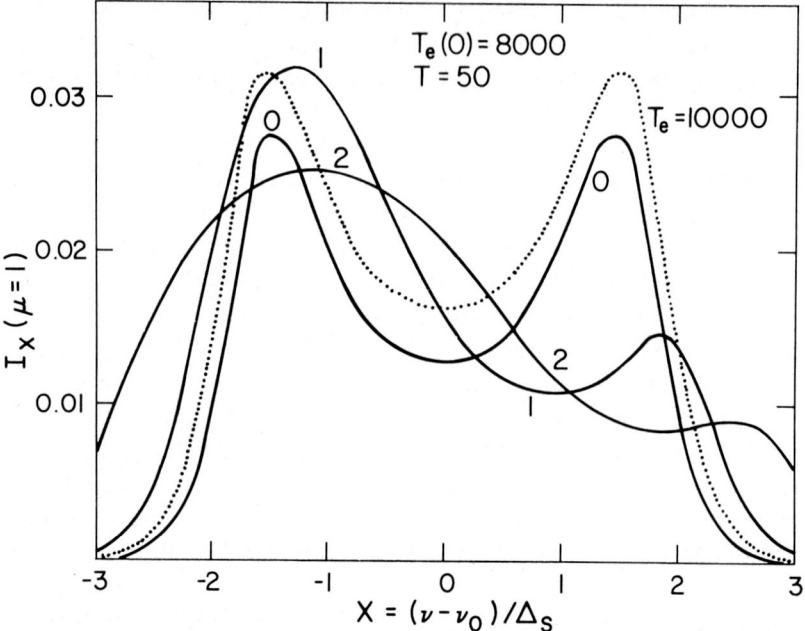

Fig. V-6. Profiles for a differentially moving atmosphere computed by Hummer and Rybicki (1968b) (see text for discussion) (Hummer and Rybicki, 1968b; courtesy *Resonance Lines in Astrophysics*, National Center for Atmospheric Research, Boulder).

are shown in Figure V-6. u_1 is chosen to give $u=0$ at $\tau=T/2$, where T is the total thickness of the slab. All velocities are measured in units of the Doppler width. Two profiles are shown in Figure (V-6) for the stationary case. For the solid curve T_e increases from 8000° at the slab boundaries to a maximum of 10 000° at the slab center. For the dotted curve, T_e is constant. Profiles are shown also for $u_0=1$ and $u_0=2$. These profiles are strongly asymmetric and show an apparent red shift. Note, however, that the central reversal separating the emission peaks is violet shifted. The apparent red shift results from the velocity gradient and its effect upon τ_v rather than from a recessional velocity of the slab. The same type of shift is apparent even when u is of the form

$$\mu = 0, \quad T/2 \le \tau \le T,$$
$$\mu = \mu_0 + \tau\mu_1, \quad \tau \le T/2,$$
(V-22)

i.e., when all motion is toward the observer.

Motion of the type indicated by Equation (V-21) produces a marked distortion of the τ_v grid. On the side of the profile toward which ϕ_v is shifted the distortion of $S_v(\tau_v)$ is similar to that illustrated in Figure V-2, i.e., the τ_v scale is expanded. On the opposite side of the profile, however, an opposite effect is present. The τ_v scale is compressed, reducing the slope of the $S_v(\tau_v)$ curves. Also, at equal distances from line center the surface $\tau_v=1$ lies much deeper in the atmosphere on the side of the profile away from the shift in ϕ_ν than it does on the side in the direction of the shift in ϕ_ν. The two effects combine to produce the apparently red shifted profiles in Figure V-6.

Similar calculations made by Kalkofen (1970) for a semi-infinite atmosphere expanding differentially with height give violet shifted absorption lines similar to the P-Cygni type. The profiles have emission wings; the red emission peak being stronger as in Figure V-6.

Figure V-7b illustrates the effect on the K_2 peaks and the K_3 core of the solar Ca II line of downward motion in different layers of the solar atmosphere (Athay, 1970b). The motion is confined to restricted layers in each of the cases shown. For the different models the motion is primarily in the regions: A – 0–700 km, B – 500–1000 km, C – 1000–1500 km and D > 1500 km. Vertical dashed lines in Figure V-7b indicate the approximate position of the center of ϕ_ν in the moving layer. The relative location of the maximum in S_L is shown in Figure V-7a together with the curve of systematic velocity for model A and the curve of mean microscopic velocity, $\bar{v}$, due to combined thermal and microturbulent motions. The gradient in $\bar{v}$ was chosen to produce a partial reduction in the K_2 intensity as discussed earlier in Section 4 of this chapter. Most of the motion is in the K_1 region for model A, in the K_2 region for model B, and in the K_3 region for models C and D.

It is of interest that downward motion in the K_3 region enhances the violet K_2 peak and depresses the red K_2 peak. Just the opposite effect results from downward motion in the K_1 region. Motion in the K_2 region, model B, however, seems to have little effect upon K_2 other than a small red shift of both K_2 and K_3. A similar red shift occurs in models C and D, where the motion is in K_3.

Enhancement and suppression of the K_2 features for models C and D can be understood in terms of the expansion and compression of the τ_ν scale on opposite sides of

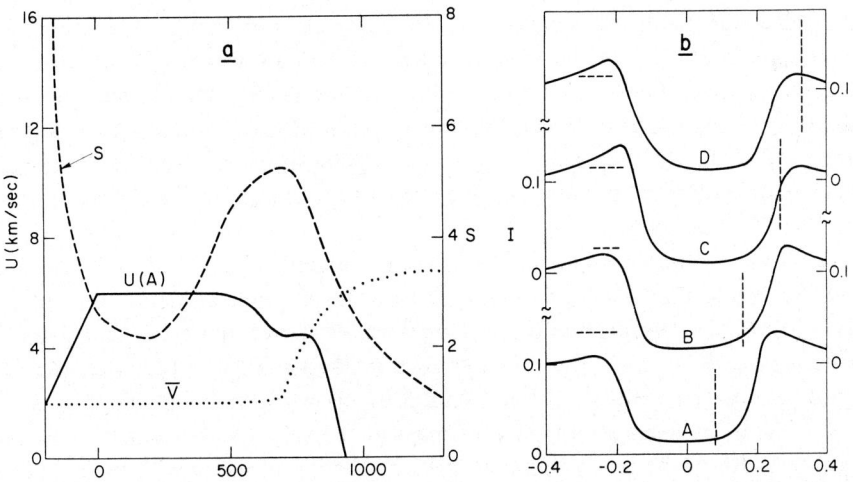

Fig. V-7. The effect of differential motion on the K_2 and K_3 features in the Ca II K-line. The velocity and source function models are shown in part a and the resultant line profiles are shown in part b.

the line. Expansion of the τ_v scale on the red side of the line tends both to make the maximum in S_L optically thin (as illustrated in Figure V-3a) and to increase the effective gradient in $\bar{v}$, whereas compression of τ_v on the violet side reduces the effective gradient in $\bar{v}$ and broadens the maximum in S_L. Opposite compression and expansion effects are present in model A.

The red shift in K_2 and K_3 produced by the downward motion in models C and D is less than 20% of the shift in ϕ_v, again exhibiting the fact that when gradients in ϕ_v are present the superficial evidences of motion may lead to the faulty conclusion that the motion is of much lower amplitude than is actually the case.

6. Line Cores

The profile shape factor ϕ_y changes from a Gaussian form near line center to a dispersion form in the line wings. Since the Gaussian function in ϕ_y decreases from 2×10^{-3} at $y=2.5$ to 5×10^{-6} at $y=3.5$ the transition of ϕ_y from Gaussian to dispersion form will occur near $y=3$ for a wide range of values of the Voigt parameter a.

To illustrate this point more explicitly, we again write ϕ_y in the form

$$\phi_y = e^{-y^2} + \frac{a}{\sqrt{\pi} y^2}, \quad y \geq 1, \tag{V-23}$$

as discussed in Chapter II, Section 5. For a strong iron line at 4000 Å formed in the high photosphere, v is approximately 2 km s^{-1} and v is 1.3×10^{15}. Thus, $\Delta v_D \approx 10^{10}$ s^{-1}. If the damping constant is of the order of 10^8 s^{-1}, which is typical for a strong line, then a is of the order of 10^{-3}. The transition from the Gaussian core to the dispersion wing in ϕ_y will occur, therefore, at $y \approx 3$ where $\phi_y \approx 10^{-4}$.

It can be argued in a somewhat general way that lines in which τ_0 exceed τ_c by considerably more than a factor of 10^4, i.e. lines for which $r_0 \ll 10^{-4}$, will exhibit well developed wings. Such lines should usually show clear evidence of two distinctive regions in the profiles, viz., the Doppler core and the damping wings. Because of the radical change in ϕ_y between the two regions, the profile character should be notably different in the two regions. The change of slope in $d \log (S/B)/d \log \tau_0$ evidenced in Chapter III at the depth $\tau_0 \approx a^{-1}$ will tend to enhance the differences in the profile characteristics still further. However, the effect of the change in the character of ϕ_y is much the dominant effect.

At $v=2$ km s^{-1} and $\lambda=4000$ Å, $\Delta \lambda_D$ is approximately 0.027 Å. If the Doppler core extends to $3 \Delta \lambda_D$, the transition from core to wing in a typical metal line will occur at $\Delta \lambda = \pm 0.08$ Å. An inspection of the *Utrecht Atlas* shows that many of the strongest solar iron lines near $\lambda 4000$ do in fact show a marked change in the character of the line profiles near $\Delta \lambda \approx 0.08$ Å. At $\Delta \lambda \lesssim 0.08$ Å $dI_\lambda/d\lambda$ is markedly greater than at $\Delta \lambda \gtrsim 0.08$ Å. This 'bell-shape' is typical of most of the very strong solar lines. Furthermore, the width of the inner region of large $dI_\lambda/d\lambda$ is consistent with the assumption that the Doppler core extends to $\pm 3 \Delta \lambda_D$ (cf. Athay, 1961). The simple 'rule-of-thumb' that the wings of the bell-shaped profiles start at $\Delta \lambda \approx \pm 3 \Delta \lambda_D$ provides a quick esti-

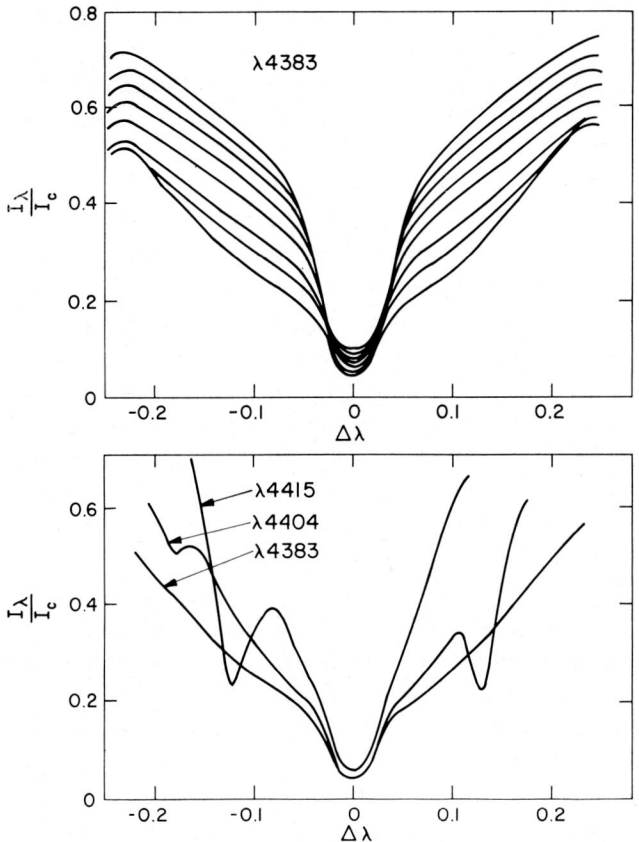

Fig. V-8. Observed profiles of $\lambda 4383$ of Fe I at $\mu = 1, \frac{1}{2}, \frac{1}{3}, \frac{1}{4}, \frac{1}{5}, \frac{1}{6}, \frac{1}{8}$ and $\frac{1}{10}$ and of three lines of multiplet 41 of Fe I at $\mu = 1$ (Athay *et al.*, 1971; courtesy *Solar Physics*, Reidel, Dordrecht).

mate of the 'average' value of $\Delta\lambda_D$ in the Doppler core region. An excellent example of such a line is provided by $\lambda 4383$ of Fe I, which is illustrated in Figure V-8a from Athay, *et al.* (1971). The shoulders of this line occur at $\mu = 1$ at approximately ± 0.070 Å corresponding to $\Delta\lambda_D = 0.070/3 = 0.023$ Å, or to a mean broadening velocity of 1.6 km s^{-1} in the Doppler core. At $\mu = 0.12$ the core of $\lambda 4383$ has broadened to $\Delta\lambda = \pm 0.12$ Å, corresponding to $\Delta\lambda_D = 0.04$ Å and to a mean broadening velocity of 2.8 km s^{-1}.

In most strong solar lines, the Doppler cores are relatively smooth with I_λ increasing monotonically with $\Delta\lambda$. The cores are concave in shape (towards I_c) because of the rapid variation in ϕ_y across the core.

7. Line Wings

Solar Fraunhofer lines with well developed wings show an interesting reversal from a convex curvature to a concave curvature as the line wing becomes very strong (Athay et al., 1971). The concave character is restricted to some 45 or so of the strongest lines between $\lambda 3720$ and $\lambda 8700$ and is restricted to the inner portions of the line wings. Most of the lines showing this concavity are Fe I lines although lines of Al I, Mg I, Ca I, Si I and Ca II also exhibit it. The concavity is not evident, however, in either the hydrogen lines or the Na D lines, which are of a strength where the concavity appears in other lines.

The H and K lines are concave out to about ± 5 Å and ± 6 Å, respectively. In other notable cases of concavity, such as Fe I, $\lambda 3720$ and Mg I 5184, however, the concavity extends to less than 1 Å from line center. There is a distinct tendency for the concavity to be more pronounced in the violet end of the spectrum and to be absent in the far red end of the spectrum, as witnessed by almost no evidence of concavity in $\lambda 8542$ of Ca II.

Figure V-8 (Athay et al., 1971) exhibits three lines of multiplet 41 of Fe I, the strongest of which shows strong concavity but the weakest of which shows only very weak concavity. Corliss and Tech (1968) give relative gf values for the lines of approximately $1:2.5:4.4$. It appears therefore that this change in character of the line wings is quite sharply defined with line strength.

Figure V-8 illustrates the behavior of I_y/I_c as a function of μ for the $\lambda 4383$ line. The concavity in the wing persists out to $\mu \approx 0.5$ and is clearly absent at $\mu = 0.33$. Thus, the concavity is sharply defined in μ is well as in line strength. It is strongest at large μ, in strong lines and at short wavelengths.

In the case of the H and K lines of Ca II which show the K_2 and H_2 reversals, the concave nature of the wings is easily understood. Since the source function passes through a minimum in the regions of the atmosphere where K_1 forms, there must be a corresponding minimum in I_λ; this leads quite logically to a concave structure. The explanation is clearly not so simple for the other cases where no K_2-like reversal is present.

To investigate the possible causes of the concave structure in line wings, we make several simplifying assumptions. Since the wings form at $\tau_0 > 10^4$, we assume that S_L is thermalized to B, and, hence, that S_ν is independent of frequency. We assume, also, that r_0 and y are independent of depth and that locally S_y varies with τ_0 according to the relation

$$S_y = S_1 \tau_0^n, \tag{V-24}$$

where n remains to be determined and where S_1 is an implicit function of y.

A line wing is concave when

$$(d^2 I_y/dy^2) > 0 \tag{V-25}$$

and convex when

$$d^2 I_y/dy^2 < 0. \tag{V-25a}$$

We seek to relate I_y to S_y and τ_y to y. Hence, we adopt the Eddington-Barbier relation

$I_y = S$ at $\tau_y = 1$ and write

$$\tau_y = \tau_0 (r_0 + \phi_y) = \tau_0 \left(r_0 + \frac{a}{\pi^{1/2} y^2} \right), \quad \text{(V-26)}$$

and

$$I_y = \frac{S_1}{\left(r_0 + \frac{a}{\pi^{1/2} y^2} \right)^n}. \quad \text{(V-27)}$$

We then find

$$\frac{d^2 I_y}{dy^2} = S_1 \left[\frac{n(n+1)}{(r_0 + \phi_y)^{n+2}} \frac{4\phi_y^2}{y^2} - \frac{n}{(r_0 + \phi_y)^{n+1}} \frac{6\phi_y}{y^2} \right], \quad \text{(V-28)}$$

from which it follows that for concave profiles

$$\frac{(n+1)\phi_y}{r_0 + \phi_y} > \tfrac{3}{2}, \quad \text{(V-29)}$$

and for convex profiles

$$\frac{(n+1)\phi_y}{r_0 + \phi_y} < \tfrac{3}{2}. \quad \text{(V-30)}$$

It is readily apparent from the inequalities (Equations (V-29) and (V-30)) why the concavity occurs preferentially in the inner wings in strong lines, near the center of the disk, and in the violet end of the spectrum. The value of n is larger in the violet and for larger τ_c (i.e., larger μ) and $\phi_y/(r_0 + \phi_y)$ is larger for strong lines (small r_0) and for the near wings (smaller values of y).

As noted earlier, we expect ϕ_y to be of the order of 10^{-4} to 10^{-5} in the near wings of the lines. Thus, if $r_0 \ll 10^{-5}$, $\phi_y/(r_0 + \phi_y) \approx 1$ and we expect concavity for $n > \tfrac{1}{2}$. Photospheric models give $n > \tfrac{1}{2}$ at $\lambda 4000$ for $\tau_{5000} \gtrsim 0.03$ and at $\lambda 5000$ for $\tau_{5000} \gtrsim 0.1$. At $\lambda 8000$ $n < \tfrac{1}{2}$ through all but the deep photospheric layers. It is not surprising therefore that strong lines in the violet end of the spectrum show concave wings whereas those in the far red do not.

In the violet end of the spectrum n is of order unity in the photosphere for $\tau_{5000} > 0.2$. At smaller values of τ_{5000} or larger wavelengths, the effective value of n decreases. For $r_0 = \tfrac{1}{3} \phi_y$, the concavity condition is $n > 1$, a condition which we do not expect in the upper photosphere. Thus, the extent of the concavity into the line wings is limited in the violet approximately by the condition $\phi_y > \tfrac{1}{3} r_0$. At 1 Å from line center in a line at $\lambda 4000$ y is of the order of 40. For $a = 10^{-3}$, the concavity will extend to ± 1 Å only if $r_0 < 10^{-6}$.

The absence of concave wings in the Na D lines is partly a result of the decreased likelihood for concavity toward the red end of the spectrum and partly a result of the fact that r_0 and ϕ_y are of comparable magnitude in the near wings. According to Athay and Canfield (1969) the shoulder of the D_2 line forms near $\tau_c = 10^{-1}$, which means that in the near wings $r_0 > 10^{-1} \phi_y$. By contrast the shoulder of Mg b_1 forms near

$\tau_c = 10^{-2}$ (Athay and Canfield, 1969) and in the near wing, which is concave, $r_0 < 10^{-1} \phi_y$.

The lack of concavity in the wings of the infrared triplet of Ca II undoubtedly arises from the wavelength effect. Other explanations must be found for the lack of concavity in the wings of the hydrogen lines, however, none of which shows concave wings even though Hβ and Hγ are very strong and in the blue end of the spectrum. Again, the explanation is found in the relative sizes of r_0 and ϕ_y at the depths where the wings form. In the cooler layers of the upper photosphere there are relatively few hydrogen atoms excited to the second quantum level. This means that r_0 is relatively large in these layers and that the wings of the hydrogen lines tend to form deeper in the photosphere where r_0 and ϕ_y are of comparable magnitude. That this is in fact the proper explanation is confirmed by the fact that the Hβ wings start at a relative intensity of about 60% of the continuum whereas the concave wings in $\lambda 5184$ of Mg I and $\lambda 4227$ of Ca I start at relative intensities of approximately 20 and 15%, respectively.

A second interesting feature found in the wings of strong lines with concave form (Figure V-8) is the cross-over in the curves at $\mu = 1$ and $\mu = \frac{2}{3}$ at $I_y/I_c \approx 0.5$ (Athay et al. 1971). This same tendency is evident in most strong lines although it is often obscured by blends with other lines. It seems always to occur near $I_y/I_c \approx 0.5$. The cross-over means that $(dI_y/I_c)/d\mu = 0$ at some intermediate value of μ.

If we write S in the form

$$S_y = a_y + b_y \tau_c \tag{V-31}$$

and

$$\tau_y = \tau_c \left(1 + \frac{\phi_y}{r_0}\right) \tag{V-32}$$

we obtain, after dropping the subscript on a and b,

$$S_y = a + \frac{r_0}{r_0 + \phi_y} b\tau_y. \tag{V-33}$$

Integration of Equation (V-1) with $r_0/(r_0 + \phi_y) = \text{const.}$ then gives

$$\frac{I_y}{I_c}(\mu) = \frac{a + \dfrac{r_0}{r_0 + \phi_y} b\mu}{a + b\mu}. \tag{V-34}$$

Upon differentiation of Equation (V-34) with respect to μ, we find

$$\frac{dI_y/I_c}{d\mu} = \frac{b \dfrac{r_0}{r_0 + \phi_y}}{a + b\mu} - \frac{b\left(a + \dfrac{r_0}{r_0 + \phi_y} b\mu\right)}{(a + b\mu)^2}$$

or

$$\frac{dI_y/I_c}{d\mu} = \frac{ab^2}{I_c}\left(\frac{r_0}{r_0 + \phi_y} - 1\right). \tag{V-35}$$

The right hand side of Equation (V-35) is always negative when a and b are positive. Hence, the cross-over of the I_y/I_c curves cannot be explained with the simple form we have picked for $S_y(\tau_c)$ and for $r_0/(r_0+\phi_y)=$ constant. Since we are here dealing with an effect of height (μ) dependence in I_y, a depth dependence in the quantity $r_0/(r_0+\phi_y)$ can be of considerable importance. Athay *et al.* (1971) have shown that, in fact, it is the height dependence of $r_0/(r_0+\phi_y)$ rather than higher order terms in $S_y(\tau_c)$ that produces the cross-over in $S_y(\tau_c)$. Nevertheless, the general behavior of the curvature in the wings appears to be accounted for crudely by the preceding simplified arguments.

8. Center-Limb Effects

Most solar Fraunhofer lines show marked center-limb effects in both the absolute values of I_v at a given frequency and the profiles of I_v as a function of frequency. Such effects may be partially understood in terms of the angular dependence of I_v emitted by a plane parallel atmosphere. Two influences in the solar atmosphere, however, prohibit the complete analogy with the plane parallel atmosphere. Near the extreme limb, $\mu < 0.1$, curvature effects become important and must be taken into account. Furthermore, the solar atmosphere is not spherically symmetric, let alone plane parallel.

Distinctive point-to-point fluctuations in I_v are observed at all depths in the photosphere and chromosphere. Such features are indicative of irregular variations of temperature, density and velocity fields over the solar surface. As a result surfaces of constant S_v or constant τ_v are roughened by 'hills' and 'valleys'. The angular projections of I from a roughened spherical layer may be markedly different from those from a smooth spherical layer with the same 'average' S_v and τ_v. Again, the roughening effect upon I_v becomes more severe near the extreme limb.

Although we must keep in mind that sphericity and rugosity may both be important in center-limb variations of I_v, it is instructive to consider the effects in terms of a plane parallel atmosphere.

Center-limb effects upon the core regions of the Ca II K-line were discussed somewhat in Section 4 of this chapter where we were illustrating the effects of gradients in $\varDelta v_D$. It is observed in the K-line that, relative to the central disk continuum intensity $I_c(0)$, I_v darkens towards the limb at all wavelengths. This is typical of all strong solar lines. However, the K line of Ca II presents a special case in that the K_2 maxima and K_1 minima are associated with a maximum and a minimum, respectively, in S_v. Most center-limb effects can be qualitatively explained in terms of a simple application of the Eddington-Barbier relation, which is not true in the case of the K line. There we found it necessary to resort to relatively strong gradients in $\varDelta v_D$ to produce marked deviations from the Eddington-Barbier relation when we retained the assumption of plane parallelism. Beebe (1971) has offered an alternative explanation of these effects in terms of a two-component atmosphere with the colder components preferentially obscuring the hotter component near the limb. This provides an example where a height gradient in one quantity appears, superficially at least, to be equivalent to a particular type of rugosity.

To illustrate the case where neither deviations from the Eddington-Barbier relation nor rugosity are required, qualitatively at least, consider the center-limb profiles of the strong Fe I line at $\lambda 4383$ shown normalized to the local value of $I_c(\mu)$ in Figure V-8. Within the line, I_ν darkens to the limb relative to $I_c(1)$ and brightens to the limb relative to $I_c(\mu)$, which is typical for all strong lines. The results in Figure V-8 are consistent with the conclusion that limb-darkening occurs at all wavelengths within the line but is most pronounced in the continuum and least pronounced in the line core.

Certain features of the profiles in Figures V-8 are of interest in connection with the preceding discussion. The 'fluting' (or broadening) of the line core seen near the limb means clearly that the average value of Δv_D in the Doppler core regions increases to the limb. This may be interpreted either as an increase of Δv_D with height or as an anisotropy in Δv_D such that the horizontal component of Δv_D exceeds the vertical component.

The fluting at the limb is common to all solar lines and has been a subject of much discussion. Nevertheless, we are still uncertain as to which of the alternatives is the more correct one. In the case of the Ca II H and K lines (and the restriction of plane parallelism) it is necessary to invoke an appreciable increase in Δv_D with height in the chromosphere in order to explain both the K_2/K_3 ratio at $\mu = 1$ and the center-limb effects in K_2 and K_1 (Athay and Skumanich, 1968a, 1968b). The increase of Δv_D in the chromosphere is further evidenced by the width of the O I line in the chromosphere (Athay and Canfield, 1970). Thus, in the strong solar lines it seems evident that at least part of the fluting of the core is due to an increase of Δv_D with height. Whether this is a sufficient explanation for other lines as well remains to be seen.

Center-limb data can be used in a variety of ways to derive information about line source functions. Some of the uses of center-limb data are discussed in following sections of this chapter. A particular application of such data used by Waddell (1962, 1963) is of interest at this point.

Waddell (1962) noted that if two lines have a common source function, but have different values of τ_0, they should give identical profiles at values of μ given by

$$\frac{\mu(1)}{\mu(2)} = \frac{\tau_\nu(1)}{\tau_\nu(2)}. \tag{V-36}$$

This conclusion follows directly from Equation (V-1) since the assumptions are that for the two lines both S_ν and τ_ν/μ are identical (see the discussion of Equations (I-11) to (I-15)).

Waddell assumed that for a multiplet τ_ν is directly proportional to $\tilde{\omega}f$, and, hence, that profiles of strong lines with common values of S_ν should be congruent at

$$\frac{\mu(1)}{\mu(2)} = \frac{(\tilde{\omega}f)_1}{(\tilde{\omega}f)_2}. \tag{V-37}$$

However, the stricter condition is that $\tau_\nu \propto \tilde{\omega} f (\phi_\nu + r_0)$. Even if the line is strong enough that we may neglect r_0, the congruency condition is

$$\frac{\mu(1)}{\mu(2)} = \frac{(\tilde{\omega}f)_1 \, \phi_\nu(1)}{(\tilde{\omega}f)_2 \, \phi_\nu(2)}. \tag{V-38}$$

This reduces to Equation (V-26) only if $\phi_\nu(1) \equiv \phi_\nu(2)$, i.e., only if $\varDelta v_D$ is independent of μ.

Comparison of two or more profiles following the prescription of Equation (V-37) will shed information on the equality of S_1 to S_2 and $\phi_\nu(1)$ to $\phi_\nu(2)$. If the profiles are congruent, either both equalities are satisfied or $\phi_\nu(1)$ and $\phi_\nu(2)$ differ in just such a way as to compensate for differences in S_1 and S_2. The latter condition, obviously, could arise only by chance and must, therefore, occur only rarely, if at all.

If profiles compared according to Equation (V-37) are not congruent, three possibilities exist: 1) $S_1 \neq S_2$, 2) $S_\nu(1) \neq S_\nu(2)$, or 3) $\phi_\nu(1) \neq \phi_\nu(2)$. The second condition implies that the inequality arises from a failure of the condition $\phi_\nu S_L \gg r_0 B$.

Congruency tests of profiles have been carried out for only very limited cases, partly due to the fact that observations must be made with extreme care to insure that all

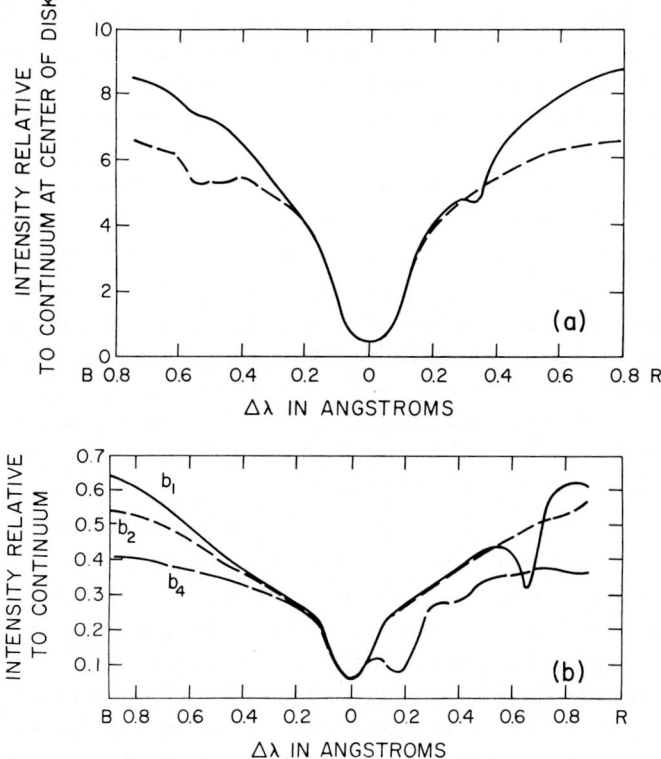

Fig. V-9. Superposition of profiles of (a) – the Na D and (b) – Mg b lines at equal values of $\tilde{\omega}f/\mu$ (Waddell, III, 1962, 1963; courtesy *Astrophysical Journal*, copyright 1962, 1963 by the University of Chicago. All rights reserved).

effects of instrumental resolution, scattered light, and solar inhomogenieties have been eliminated. Also, the relative values of $\tilde{\omega} f$ must be accurately known.

Superposition of the profiles of the Na D lines and Mg b lines (Waddell, 1962, 1963) are shown in Figures V-9a and V-9b. Both sets of profiles are closely congruent throughout the Doppler core, with the exception of the blend in the b_4 line. In the wings the congruency fails. Clearly, this is due to the fact that r_0 is no longer small compared to ϕ_v, i.e., continuum opacity is becoming important in the wings. Note that in the region $0.1 < \Delta\lambda < 0.4$ where the b_1 and b_2 profiles both tend to be slightly concave, suggesting that ϕ_λ is still large compared to r_0, the two profiles are still closely congruent. Wherever the wings are convex the congruency fails as expected.

9. Profile Synthesis

A large amount of work has been done in attempts to synthesize observed line profiles with computed profiles. The majority of such work has been based upon the LTE assumption and will not be dealt with here. Most of the work based on kinetic equilibrium has been applied to solar problems, mainly because the solar data are more precise and contain center-limb effects which aid the analysis.

An excellent review of work on the solar H and K lines is given by Linsky and Avrett (1970). All of the recent work using kinetic equilibrium has concentrated on the K_3, K_2 and inner K_1 regions, specifically the region of K_1 just beyond K_2 where the profile passes through a second minimum (first minimum at K_3). We shall refer to this minimum region in K_1 simply as K_1.

Jefferies and Thomas (1960) were the first to use the kinetic equilibrium approach, as outlined in this book, to investigate the K_2 emission reversals. However, earlier work by Miyamoto (1953a, b, 1954a, b) followed a rather similar approach. Jefferies and Thomas succeeded in showing that the K_2 K_3, and K_1 components could be explained, qualitatively at least, by a chromospheric temperature rise preceded by a temperature minimum as is illustrated in Chapter III. Subsequent extensions of this work by Dumont (1967a, b), Athay and Skumanich (1968a, b, c, d), Linsky (1968), Beebe and Johnson (1968), Athay (1970b), and Beebe (1971) have considerably refined the synthesis of the H and K profile. Nevertheless, much still remains to be done before completely satisfactory explanations of the profiles are achieved.

There appears to be no essential difficulty in explaining the general form of the K_2 and K_3 components in the profile, apart from the asymmetries and the point to point variations on the solar disk. The only recent attempt to account for the asymmetries is that illustrated in Figure V-7 (Athay, 1970b). Studies on spatial variations in the profiles are in progress by A. Skumanich.

No satisfactory explanation of the K_1 feature has as yet been given. The basic difficulty with this portion of the profile is that the intensity falls too low to be accounted for by current solar models. At K_1 the radiation temperature is approximately 4200 K at $\mu = 1$ and 4100 K at $\mu = 0.1$. The lowest accepted value of the solar temperature minimum, however, is about 4300 K and all computations of the H and K source functions have given S_L very close to B in the K_1 region, i.e., K_1 is formed below the thermalization depth. Efforts to extend the thermalization length by including multi-level effects from the ^2D states (Dumont, 1967a, b; Linsky, 1968) or by including effects of frequency coherency (Athay and Skumanich, 1968d) have helped to some extent, but they fall short of the required goal. The same difficulties are present in the corresponding lines of MgII (Athay and Skumanich, 1968b). In this latter case, however, the difficulty is enhanced by the greater sensitivity of the Planck function to small temperature changes. The difference in intensity at 4300K and 4100K is nearly a factor of 2 and cannot be attributed to observational errors.

A proper answer to the difficulties encountered in the K_1 region of the profile may lie in the inhomogeneous nature of the solar atmosphere. It may be, for example, that local regions of reduced temperature minimum exist. Beebe and Johnson (1968) and Beebe (1970) have used this latter approach to synthesize the K_1 profile. A second plausible alternative is that the temperature minimum, generally, is lower than it is currently believed to be.

The mechanism proposed by Athay and Skumanich (1968a) for explaining the limb darkening in K_1 is of interest from the qualitative point of view but fails quantitatively for the reason that it raises $I_\nu(K_1)$ above B_{min} for large values of μ and allows $I_\nu(K_1)$ to approach B_{min} at small μ. Since we are already faced with the difficulty that $I_\nu(K_1) < B_{min}$ at $\mu = 1$, this appears to be a step in the wrong direction. Evidence supporting a rapid rise in $\Delta\lambda_D$ in the chromosphere and hence the type of effect proposed by Athay and Skumanich is very strong. Thus, the rise in $\Delta\lambda_D$ very likely contributes to the limb darkening even though it enhances the difficulty of fitting the absolute intensity at K_1.

Attempts to synthesize the solar Na D profiles have been made by Johnson (1964), Chamaraux (1967), Finn *et al.* (1967), and Athay and Canfield (1969). The latter authors obtained successful fits to both the Na D and Mg b profiles, including their center-limb variations and with a model atmosphere that fits the K_2 and K_3 profiles reasonably well. They later attempted to fit the resonance lines of OI at $\lambda 1304$ and $\lambda 1302$ with the same model atmosphere. Although the general features of the OI profiles and their equivalent widths were satisfactorily fitted the detailed shape of the line profile was not satisfactory. The model atmosphere used is similar to the one used in Chapter IV.

It is of interest to note that Athay and Canfield were not able to fit the Na D and Mg b profiles with an isotropic microturbulence in the photosphere. They were forced to use anisotropic motions in the photosphere (below the temperature minimum) with the horizontal component exceeding the vertical component by about 60%. However, the amplitude of the motions was not large and the primary contribution to $\Delta\lambda_D$ came from the thermal motions, which, of course, are isotropic. The use of anisotropic motions seemingly contradicts the conclusions from the profile superpositions shown in Figure V-9. This point has not been resolved.

Efforts to fit the solar hydrogen lines have not been notably successful. Although the central intensities and wing intensities of the Balmer-α line appear to be readily accounted for the Doppler cores are systematically too narrow in the computed profiles (Cuny, 1967; Schoolman, 1969). The same difficulties are encountered in Lyman-α and Lyman-β, namely the computed lines are too narrow (Morton and Widing, 1961; Avrett, 1965; Hearn, 1967b; Cuny, 1968). There is no difficulty, however, in explaining the equivalent widths of the Lyman lines if no attempt is made to fit the profiles (Athay, 1965; Hearn, 1967a).

While considerable work has been done to synthesize solar line profiles, it cannot be claimed that any one line is completely and satisfactorily explained. Nevertheless, the kinetic equilibrium approach has scored major achievements in bringing much closer harmony between the computed and observed profiles with model atmospheres derived from independent data even though the computations to date have used only simplified atomic models and simplified model atmospheres. The theory certainly holds promise for improved analysis of the solar lines and for the extraction of a wealth of data concerning the solar atmosphere.

Attempts to synthesize line profiles for some early type stellar models have been made by Auer and Mihalas (1969a, b, 1970) and Mihalas and Auer (1970a, b). The problem of synthesizing stellar lines is somewhat different from that of synthesizing solar lines. In the latter case we know a good deal about the model atmosphere in the line forming layers independently of the analysis of the lines themselves. In stars other than the Sun virtually all that is known about the higher layers of the atmosphere must be derived from the line data. In addition the lines themselves are the most important absorbers and emitters of radiation in the higher layers and the lines therefore largely determine the temperature structure. These effects will be elaborated in Chapter VII. Here we note only that the determination of the model atmosphere and the spectral line profile synthesis are closely coupled problems and should be treated simultaneously. One can do this properly, of course, only if the energy supply is known.

For the Sun we know that mechanical energy dissipation is the dominant energy source in the chromosphere and corona and that many of the stronger Fraunhofer lines are formed partially in the regions of mechanical energy dissipation. This effect is taken into account in the Sun through our knowledge of the temperature structure derived from continuum data both on the disk and beyond the limb at eclipse. Without taking these effects into account, i.e., if we used a pure radiative equilibrium solar model, we would be unable to explain the cores of the strong Fraunhofer lines. The predicted hydrogen lines would be much too weak, and other lines, such as Na D, Mg b, and the H and K lines would have much too low central intensities.

Virtually nothing is known of mechanical energy dissipation in the early type stars so one proceeds by assuming pure radiative equilibrium. Also, since the early type stars have relatively few strong lines in their spectra it is feasible to include all or most of the strong lines in the radiative equilibrium equations. Thus, it is possible in this case to solve the radiative transfer equations, with radiative equilibrium imposed, simultaneously for the lines and the continuum. The solution provides both the model atmosphere and the predicted spectrum, including line profiles.

Using this approach, Auer and Mihalas (1969a, b, 1970) have investigated the profiles of Balmer-α, β and γ, Paschen-α and β and Brackett-α for stars with effective temperatures of 12 500 to 15 000 K. They successfully match the observed profiles for stars with these effective temperatures. LTE computations, by comparison, fail badly in these same cases.

Mihalas and Auer (1970b) have extended the computations to include hotter stars of O and B type ($25\,000° \leqslant \tau_{\text{eff}} \leqslant 50\,000°$) and to include lines of He I and He II. For He I and He II, however, they impose LTE in all levels except the ground states. In still later work (in press) they have dropped the LTE restriction. The computed lines again agree well with observations and represent large departures from the LTE situation. Equivalent widths of the Balmer lines of hydrogen increase by factors of three when the LTE restriction is dropped.

The apparent success of the kinetic equilibrium approach in the early type stars furnishes dramatic proof that the method offers a great improvement over the LTE approach. In solar type stars the improvement is less dramatic but nevertheless pronounced.

10. A Standard Set of Data

The remainder of this chapter is devoted to the discussion of certain techniques employed in the analysis of line profiles. Each technique is based upon assumptions regarding the nature of $S_\nu(\tau, v)$ and $\Delta\lambda_D(\tau)$. It is sometimes difficult when working with actual observational data to know whether a given method of analysis is providing satisfactory answers. Even when discrepancies arise it is not always clear why

TABLE V-4

Adopted values of ξ, v_{ther}, $\bar{v}$, $\Delta\lambda_D$ and a for standard line profiles

Height (km)	ξ	v_{ther} (km s^{-1})	$\bar{v}$	$\Delta\lambda_D$ (mÅ)	a	τ_c
1980	9.5	2.54	9.83	148		
1965	8.8	2.41	9.12	137		
1870	8.3	2.30	8.61	129		
1650	7.5	2.20	7.82	117	4.58 − 4	
1350	6.5	2.11	6.83	103	5.24 − 4	
1200	5.3	2.07	5.68	84.9	6.30 − 4	
1000	4.5	2.03	4.94	74.0	7.30 − 4	
700	3.4	1.97	3.93	58.9	9.42 − 4	5.5 − 6
500	2.5	1.89	3.13	46.9	1.51 − 3	3.0 − 5
450	1.9	1.82	2.63	39.4	2.27 − 3	5.3 − 5
400	1.3	1.76	2.18	32.7	3.87 − 3	1.2 − 4
300	0.5	1.69	1.76	26.4	9.43 − 3	6.5 − 4
180	0.7	1.71	1.85	27.8	1.20 − 2	3.4 − 3
0	1.1	1.74	2.06	30.9	2.45 − 2	2.1 − 2
− 100	1.5	1.80	2.34	35.1	5.16 − 2	6.7 − 2
− 170	1.7	1.89	2.54	38.3	8.79 − 2	0.19
− 240	1.9	2.01	2.77	41.5	1.34 − 1	0.50
− 340	2.1	2.15	3.00	45.0	2.65 − 1	1.84

TABLE V-5

Source function data, $\tau_{02} = 2\tau_{01}$

Height (km)	Case A τ_{02}	S_1	S_2	r_{02}	Case B τ_{02}	S_1	S_2	r_{02}	B
1650	3.88 − 4	1.15 − 6	1.04 − 6	3.19 − 5	1.04 − 3	1.65 − 6	1.34 − 6	1.30 − 5	5.50 − 5
1350	3.04 − 3	1.05 − 6	9.45 − 6	3.11 − 5	6.21 − 3	1.44 − 6	1.18 − 6	1.73 − 5	3.83 − 5
1200	8.16 − 3	9.22 − 7	8.31 − 7	2.72 − 5	1.49 − 2	1.18 − 6	1.00 − 6	1.65 − 5	3.19 − 5
1000	3.29 − 2	8.54 − 7	7.81 − 7	2.49 − 5	5.09 − 2	1.04 − 6	9.23 − 7	1.82 − 5	2.62 − 5
700	3.02 − 1	9.00 − 7	8.66 − 7	2.15 − 5	3.41 − 1	9.76 − 7	9.96 − 7	2.29 − 5	1.93 − 5
500	1.44	1.26 − 6	1.29 − 6	1.81 − 5	1.69	1.11 − 6	1.35 − 6	2.04 − 5	1.16 − 5
450	3.60	1.48 − 6	1.53 − 6	1.70 − 5	3.40	1.26 − 6	1.60 − 6	1.63 − 5	7.24 − 6
400	8.91	1.95 − 6	1.97 − 6	1.57 − 5	8.22	1.58 − 6	2.00 − 6	1.86 − 5	4.81 − 6
300	5.34 + 1	2.94 − 6	2.94 − 6	1.56 − 5	3.44 + 1	2.39 − 6	2.72 − 6	3.96 − 5	2.77 − 6
180	2.76 + 2	3.66 − 6	3.66 − 6	1.60 − 5	1.08 + 2	3.28 − 6	3.39 − 6	5.45 − 5	5.28 − 6
0	1.58 + 3	4.47 − 6	4.47 − 6	1.73 − 5	3.22 + 2	4.64 − 6	4.52 − 6	1.36 − 4	4.14 − 6
−100	4.83 + 3	6.62 − 6	6.62 − 6	1.86 − 5	5.67 + 2	6.84 − 6	6.76 − 6	3.31 − 4	6.52 − 6
−170	1.30 + 4	1.21 − 5	1.21 − 5	1.94 − 5	9.00 + 2	1.21 − 6	1.20 − 5	5.54 − 4	1.21 − 5
−240	2.44 + 4	2.30 − 5	2.30 − 5	5.26 − 5	1.45 + 3	2.27 − 5	2.28 − 5	8.32 − 4	2.30 − 5
−340	4.00 + 4	4.55 − 5	4.55 − 5	1.87 − 4	2.80 + 3	4.53 − 5	4.54 − 5	1.52 − 3	4.55 − 5

TABLE V-6

$I_\nu(\Delta\lambda, \mu)$, case A

$\Delta\lambda$ (mÅ) \ μ	1	0.8	0.64	0.5	0.4	0.32	0.25	0.2	0.16	0.125
					Line 1					
0	0.0540	0.0515	0.0493	0.0472	0.0456	0.0441	0.0428	0.0417	0.0408	0.0401
22.7	0.0572	0.0541	0.0514	0.0489	0.0469	0.0451	0.0436	0.0423	0.0414	0.0405
45.4	0.0771	0.0707	0.0647	0.0590	0.0546	0.0510	0.0478	0.0456	0.0438	0.0423
268.1	0.139	0.131	0.122	0.112	0.103	0.0931	0.0823	0.0730	0.0646	0.0565
90.8	0.189	0.181	0.173	0.164	0.157	0.149	0.141	0.132	0.123	0.113
136	0.231	0.221	0.212	0.203	0.195	0.188	0.181	0.176	0.171	0.166
182	0.260	0.247	0.235	0.224	0.215	0.206	0.198	0.191	0.185	0.179
363	0.370	0.343	0.319	0.296	0.278	0.262	0.247	0.234	0.223	0.212
545	0.482	0.436	0.398	0.362	0.334	0.310	0.287	0.268	0.252	0.237
726	0.589	0.526	0.472	0.422	0.384	0.351	0.320	0.297	0.276	0.256
1135	0.768	0.684	0.607	0.533	0.475	0.425	0.379	0.344	0.314	0.286
2270	0.938	0.855	0.765	0.670	0.591	0.522	0.456	0.405	0.362	0.322
4086	0.985	0.909	0.821	0.721	0.636	0.560	0.487	0.430	0.381	0.336
7037	0.999	0.926	0.839	0.738	0.652	0.574	0.498	0.439	0.388	0.342
$\Delta\lambda_c$[a]	7700	8400	9000	—						
					Line 2					
0	0.0467	0.0448	0.0431	0.0414	0.0401	0.0391	0.0381	0.0374	0.0369	0.0365
22.7	0.0484	0.0462	0.0442	0.0424	0.0410	0.0398	0.0387	0.0378	0.0372	0.0367
45.4	0.0589	0.0543	0.0506	0.0471	0.0446	0.0426	0.0408	0.0397	0.0385	0.0376
68.1	0.112	0.103	0.0928	0.0819	0.0725	0.0638	0.0557	0.0498	0.0453	0.0419
90.8	0.165	0.157	0.149	0.141	0.132	0.123	0.113	0.103	0.0923	0.0808
136	0.204	0.196	0.189	0.183	0.177	0.172	0.167	0.162	0.158	0.153
182	0.226	0.217	0.209	0.200	0.193	0.187	0.181	0.176	0.172	0.168
363	0.306	0.287	0.271	0.255	0.242	0.231	0.220	0.211	0.202	0.194
545	0.383	0.354	0.329	0.305	0.285	0.268	0.252	0.239	0.227	0.215
726	0.462	0.420	0.384	0.351	0.325	0.302	0.280	0.263	0.248	0.233
1135	0.630	0.561	0.502	0.446	0.403	0.367	0.333	0.307	0.284	0.263
2270	0.874	0.787	0.700	0.612	0.541	0.480	0.422	0.378	0.341	0.306
4086	0.964	0.884	0.795	0.697	0.615	0.542	0.472	0.418	0.372	0.329
7037	0.992	0.917	0.829	0.729	0.644	0.567	0.492	0.434	0.385	0.339
$\Delta\lambda_c$[a]	9000									

[a] $\Delta\lambda_c$ gives the value of $\Delta\lambda$ at which the line merges with the continuum.

TABLE V-7
$I(\Delta\lambda, \mu)$, case B

$\Delta\lambda$ (mÅ) \ μ	1	0.8	0.64	0.5	0.4	0.32	0.25	0.2	0.16	0.125
					Line 1					
0	0.0504	0.0492	0.0481	0.0472	0.0406	0.0462	0.0458	0.0457	0.0457	0.0457
22.7	0.0525	0.0508	0.0493	0.0481	0.0472	0.0465	0.0461	0.0459	0.0458	0.0457
45.4	0.0707	0.0648	0.0598	0.0554	0.0523	0.0500	0.0482	0.0472	0.0466	0.0462
68.1	0.168	0.153	0.138	0.123	0.110	0.0973	0.0849	0.0747	0.0661	0.0584
90.8	0.326	0.294	0.266	0.238	0.216	0.196	0.175	0.158	0.143	0.126
136	0.509	0.455	0.407	0.361	0.326	0.295	0.267	0.245	0.226	0.208
182	0.626	0.558	0.497	0.438	0.392	0.352	0.315	0.286	0.262	0.239
363	0.875	0.784	0.697	0.610	0.540	0.478	0.419	0.373	0.335	0.299
545	0.951	0.863	0.772	0.676	0.597	0.527	0.459	0.407	0.362	0.321
726	0.977	0.894	0.803	0.704	0.622	0.548	0.477	0.421	0.374	0.330
1135	0.995	0.918	0.829	0.729	0.644	0.567	0.492	0.434	0.384	0.338
2270	1.00	0.931	0.844	0.743	0.656	0.577	0.501	0.441	0.390	0.343
4086		0.934	0.847	0.746	0.659	0.580	0.503	0.443	0.392	0.344
2037		0.935	0.848	0.747	0.660	0.581	0.504	0.443	0.392	0.344
$\Delta\lambda_c$[a]	2270	5500	6000	6000	6000	7000	7000	7500	4500	3000
					Line 2					
0	0.0518	0.0500	0.0484	0.0469	0.0458	0.0449	0.0441	0.0436	0.0432	0.0431
22.7	0.0535	0.0514	0.0496	0.0478	0.0465	0.0454	0.0445	0.0439	0.0435	0.0432
45.4	0.0638	0.0593	0.0557	0.0523	0.0500	0.0481	0.0464	0.0453	0.0445	0.0438
68.1	0.126	0.114	0.102	0.0896	0.0793	0.0702	0.0617	0.0557	0.0512	0.0477
90.8	0.243	0.220	0.199	0.179	0.161	0.145	0.129	0.115	0.101	0.0878
136	0.386	0.347	0.314	0.282	0.258	0.237	0.217	0.202	0.188	0.175
182	0.488	0.436	0.391	0.348	0.315	0.287	0.260	0.240	0.223	0.207
363	0.764	0.680	0.605	0.530	0.471	0.419	0.370	0.332	0.300	0.271
545	0.890	0.799	0.710	0.621	0.550	0.486	0.426	0.379	0.339	0.302
726	0.943	0.855	0.763	0.668	0.590	0.521	0.454	0.402	0.358	0.318
1135	0.983	0.902	0.811	0.712	0.628	0.554	0.481	0.425	0.377	0.332
2270	1.001	0.927	0.839	0.738	0.652	0.574	0.498	0.439	0.388	0.341
4086		0.932	0.845	0.744	0.658	0.579	0.502	0.442	0.391	0.344
2037		0.934	0.847	0.746	0.659	0.580	0.503	0.443	0.392	0.344
$\Delta\lambda_c$[a]	2770	5500	5500	7000	5000	5500	5000	4500	5000	4000

[a] $\Delta\lambda_c$ gives the value of $\Delta\lambda$ at which the line merges with the continum.

they have arisen; nor is it clear whether two conflicting results are both in error or whether one or the other is correct. Similarly, two seemingly independent methods giving the same answer may both be incorrect.

To partially alleviate the difficulties of interpretation enumerated in the preceding paragraph, we will apply each of the analytical methods, where possible, to a 'standard set of data'. The standard data consist of computer generated line profiles for two upper-level doublets at $\lambda 4500$: one doublet for which the two source functions are strongly coupled so that $S_1 = S_2$ to a good approximation and a second doublet for which the source functions are weakly coupled so that $S_1 \neq S_2$. Values of ε, δ and $\Delta\lambda_D$ are the same for the two cases. The lines used are strong and have well developed wings.

The model atmosphere from which the line profiles are generated is given in Table IV-1 with the exception that a microturbulent velocity, ξ, has been added. Values of ξ, v_{thermal} and $\bar{v}$ are given in Table V-4 together with $\Delta\lambda_D$. Source functions and optical depths are given in Table V-5 and the line profile data are given in Tables V-6 and V-7. The ratio of optical depths in the two lines is 2:1. We designate the weaker line by subscript 1 and the stronger line by subscript 2. Case A is the strongly coupled case. In both case A and case B $\tau_{02} = 1$ occurs near 550 km in the chromosphere.

Selected sets of the standard profiles are plotted in Figures V-10a and V-10b for cases A and B. The sets chosen are: line 1, $\mu = 1, 0.5, 0.25$ and 0.125, and line 2, $\mu = 0.5$. Note the fluting of the line core at small values of μ and the change in curvature of the line wings in Figure V-10a from slightly concave at $\mu = 1$ to convex at small μ.

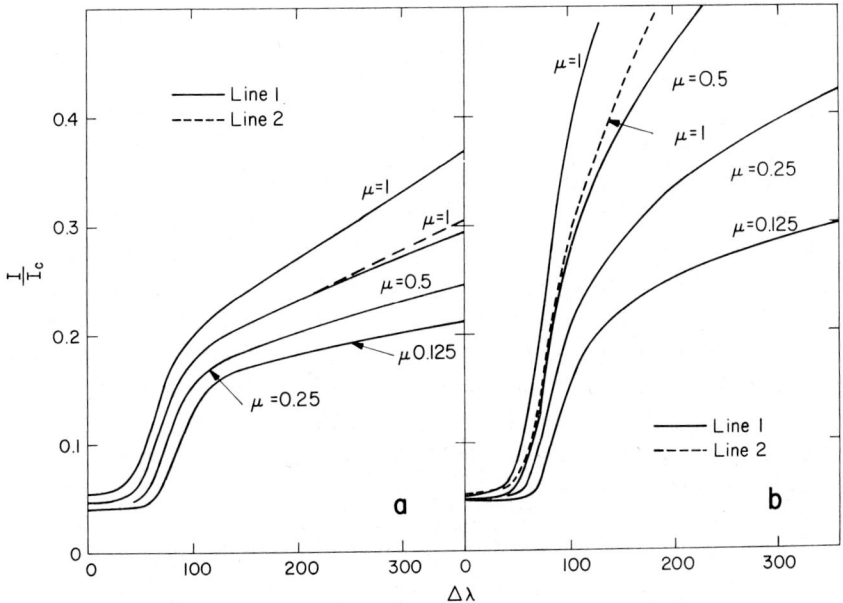

Fig. V-10. Profiles of standard doublet lines with (a) – strong coupling and (b) – weak coupling.

These changes follow the pattern discussed in the preceding section. Note also that for case A the profile of line 2 at μ is essentially identical to that of line 1 at $\mu/2$ out to approximately $\Delta\lambda = 200$ mÅ. For case B where S_1 and S_2 differ by about 40% over much of the atmosphere the corresponding superposition is not very good.

The change from case A to case B was made by changing two parameters: the collision rate between the upper levels of the doublet and the ionization potential. The latter parameter influences the doublet coupling via the continuum, and was increased for case B in order to reduce the coupling. Case A represents more nearly a neutral metal with strong coupling and case B represents more nearly an ionized metal with weak coupling. This change is reflected in Table V-5 most notably in the changes in r_0 and τ_0 at low chromospheric and photospheric levels ($h<300$ km). The wings of the line profiles are markedly different in cases A and B as a result of the changes in r_0. The damping parameters and Doppler widths, hence the ϕ_ν's, are identical in both cases A and B.

11. Evaluation of $\Delta\lambda_D$ from Cores of Strong Lines – One μ Position

One of the important pieces of information contained in line profiles is the Doppler width of the absorption coefficient. Under certain circumstances the Doppler width of ϕ_ν can be derived from the observed profiles without recourse to an elaborate analysis giving both S_ν and τ_ν as functions of τ_0. One such method, albeit a rather crude one, has already been mentioned, viz., the identification of the shoulders of the strong bell-shaped profiles with the edge of the Doppler core at approximately $\Delta\lambda_S = 3 \Delta\lambda_D$. Simply by locating the shoulder position, $\Delta\lambda_S$, one obtains a quick estimate of $\Delta\lambda_D$. The value of $\Delta\lambda_D$ so obtained represents an average value taken over the entire Doppler core of the line. Since the Doppler core is formed in the layers given approximately by $1 \leqslant \tau_0 \leqslant 10^4$, height gradients in $\Delta\lambda_D$ will appreciably affect the average value obtained by this method. Thus, the method is crude, but it provides a quick, convenient estimate of $\Delta\lambda_D$.

Better methods for evaluating $\Delta\lambda_D$ for strong lines are available providing appropriate data are available. We consider next the method proposed by Goldberg (1958).

It follows from Equation (V-1) that two wavelengths λ_1 and λ_2 will have equal values of I_{λ_1} and I_{λ_2} if, at all depths, $S_{\lambda_1} \equiv S_{\lambda_2}$ and $\tau_{\lambda_1}/\mu_1 \equiv \tau_{\lambda_2}/\mu_1$. In particular, if we choose $\mu_1 = \mu_2$ and choose λ_1 and λ_2 such that $I_{\lambda_1} = I_{\lambda_2}$ in two lines of a closely spaced multiplet where we have reason to believe that the two line source functions are equal and if the lines are strong enough that $S_\lambda = S_L$ then the condition $I_{\lambda_1} = I_{\lambda_2}$ implies that $\tau_{\lambda_1} = \tau_{\lambda_2}$. The latter equality may be used to derive the wavelength dependence of τ_λ if the relative values are known for τ_{01} and τ_{02}.

For each line in a multiplet

$$\tau\lambda = \int_0^{\tau_0} d\tau_0 (\phi_\lambda + r_0) \tag{V-39}$$

and

$$S_\lambda = \frac{\phi_\lambda S_L}{\phi_\lambda + r_0} + \frac{r_0 B}{\phi_\lambda + r_0}. \tag{V-40}$$

Since r_0 differs for each line in the multiplet whereas ϕ_λ is the same for each line, assuming that the lines are close together, the assumption that $S_{\lambda_1} \equiv S_{\lambda_2}$ generally requires both that $S_\lambda = S_L$ and that S_L itself be independent of wavelength. The latter condition has been assumed in all of the preceding discussion, the former is valid only if $\phi_\lambda S_L \gg r_0 B$ and $\phi_\lambda \gg r_0$. The only other condition for which we can expect $S_{\lambda_1} = S_{\lambda_2}$ is for the trivial case $S = B$.

The restriction $\phi_\lambda S_L \gg r_0 B$ is a rather severe one. Suppose, for example, that a line with $r_0 = 10^{-5}$ has $S_L/B = 10^{-2}$. We then find that $\phi_\lambda S_L \gg r_0 B$ requires $\phi_\lambda \gg 10^{-3}$, or that $\phi_\lambda > 10^{-2}$, say. This, in turn, requires that $\Delta\lambda < 2.2\, \Delta\lambda_D$. Few Fraunhofer lines have $r_0 \ll 10^{-5}$ and few of the strong lines have $S_L/B \gg 10^{-2}$ near line center. Thus, it will be almost universally true that the condition $\phi_\lambda S_L \gg r_0 B$ will fail in the outer limits of the Doppler core.

Consider next the equality of $\tau_{\lambda 1}$ and $\tau_{\lambda 2}$. If we set $\phi_\lambda \gg r_0$, Equation (V-39) gives $\tau_\lambda = \int_0^{\tau_0} \phi_\lambda d\tau_0$. We assume that in the case of multiplets we may write $d\tau_0 \propto (\tilde{\omega} f)\, dh$ so that $\tau_\lambda \propto (\tilde{\omega} f) \int_\infty^h \phi_\lambda\, dh$. We then have

$$\frac{\tau_{\lambda 1}}{\tau_{\lambda 2}} = \frac{(\tilde{\omega} f)_1 \int_\infty^h \phi_{y1}\, dh}{(\tilde{\omega} f)_2 \int_\infty^h \phi_{y2}\, dh}. \tag{V-41}$$

In the case where $\Delta\lambda_D$ is a function of depth the quantity $\int_\infty^h \phi_{y1}\, dh / \int_\infty^h \phi_{y2}\, dh$ is necessarily a function of depth since y_1 necessarily differs from y_2 and the depth dependence of ϕ_{y1} necessarily differs from that of ϕ_{y2}. This does not mean, of course, that when $\Delta\lambda_D$ varies with depth points of equal intensity will not be found in the two profiles. It means rather that at the points of equal intensity $\tau_{\lambda 1}$ will not equal $\tau_{\lambda 2}$ at more than one depth in the atmosphere.

Assume, for the moment, that $\Delta\lambda_D$ is constant with depth and that $S_{\lambda 1} = S_{\lambda 2}$ and $I_{\lambda 1} = I_{\lambda 2}$ so that Equation (V-41) may be rewritten

$$\frac{\phi_{y2}}{\phi_{y1}} = \frac{(\tilde{\omega} f)_1}{(\tilde{\omega} f)_2}. \tag{V-42}$$

More specifically, if we choose the weaker line as line 1 and choose $y_1 = 0$, we have

$$e^{-y_2^2} = \frac{(\tilde{\omega} f)_1}{(\tilde{\omega} f)_2}. \tag{V-43}$$

For $(\tilde{\omega} f)_1/(\tilde{\omega} f)_2 = \frac{1}{2}$ Equation (V-43) gives $\Delta\lambda_2 = 0.83\, \Delta\lambda_D$ and for $(\tilde{\omega} f)_1/(\tilde{\omega} f)_2 = \frac{1}{3}$ it gives $\Delta\lambda_2 = 1.05\, \Delta\lambda_D$. Usually $(\tilde{\omega} f)_1/(\tilde{\omega} f)_2$ is known for multiplets and the values

TABLE V-8

Values of $\Delta\lambda_D$ and $\Delta\lambda_D/\lambda$ using Goldberg's method
(Unno, 1959)

Element	λ	$\Delta\lambda_1$ (mÅ)	$\Delta\lambda_2$ (mÅ)	$\Delta\lambda_D$ (mÅ)	$\Delta\lambda_D/\lambda$ $\times 10^6$
Ca II	3933.7	14.0	61.0	66.0	16.7
	3968.5	48.0	82.5	80.5	20.3
		68.0	102.5	91.9	23.2
Mg I	3832.3	0	32.3	45.7	11.9
	3838.3	49.3	60.2	48.7	12.7
		62.5	75.0	58.5	15.3
Fe I	4383.6	0	26.9	33.1	7.54
	4407.8	45.2	57.1	42.9	9.77
Fe I	4202.0	0	32.5	31.7	7.49
	4271.8	38.7	52.6	28.9	6.83
Ca I	6102.7	0	38.6	31.2	5.09
	6162.2	40.8	68.5	43.4	7.07
		63.8	96.5	57.0	9.30
Ca I	6102.7	0	31.8	31.0	5.07
	6122.2	40.8	56.5	37.2	6.09
		63.8	81.3	48.1	7.87
Fe I	4461.7	0	22.6	18.3	4.08
	4489.8	29.8	43.1	25.2	5.63
		39.7	55.3	31.1	6.94
Fe I	6408.0	0	31.7	32.0	4.99
	6411.7	43.6	58.3	39.0	6.08
		62.0	83.5	56.3	8.78

are such that $\Delta\lambda_2 - \Delta\lambda_1$ is of the order of $\Delta\lambda_D$ when $\Delta\lambda_1$ is small. Thus, the method is very useful for determining $\Delta\lambda_D$ provided the double restrictions $S_{\lambda 1} = S_{\lambda 2}$ and $\Delta\lambda_D$ constant with depth are satisfied and provided, of course, that accurate line profiles are available. This method for evaluating $\Delta\lambda_D$ was proposed in its general form by Goldberg (1958) but was used earlier in a modified form by de Jager (1952) and by Athay and Thomas (1958). It has been used quite extensively with solar lines (cf., Unno, 1959a, b).

Table (V-8) contains a summary of results obtained by Unno (1959b) for strong solar lines using Goldberg's method. The line pairs are ordered according to the height at which they form with stronger (higher) lines at the top of the table and weaker (lower) lines at the bottom. For each pair of lines there are two or three pairs of wavelengths $\Delta\lambda_1$ and $\Delta\lambda_2$. These are ranked from top to bottom in the same sense, running from pairs of $\Delta\lambda$ formed at greater heights to pairs formed at lower heights. The quantity $\Delta\lambda_D/\lambda$ is proportional to the broadening velocity and shows a clear tendency to decrease with decreasing height of line formation in going from one pair of lines to another. This led Unno (1959b) to the conclusion, which is probably correct, that the broadening velocity decreases with decreasing height. However, note that for each pair of lines, except one, the gradient of $\Delta\lambda_D/\lambda$ is just the reverse of the general trend, i.e., $\Delta\lambda_D$ increases away from line center. Other authors have found similarly discordant results.

The strong tendency for $\Delta\lambda_D$, as determined by the Goldberg method, to increase away from line center even in opposition to the actual gradient in $\Delta\lambda_D$ can be attributed to a failure of the two basic assumptions $S_{\lambda 1} = S_{\lambda 2}$ and $\Delta\lambda_D$ constant. In the edges of the Doppler core $r_0 B$ contributes to S_λ. The contribution of $r_0 B$ will be strongest for line 1 (larger r_0). This will preferentially increase $S_{\lambda 1}$ and $I_{\lambda 1}$ and, in turn, will increase the value of $\Delta\lambda_2$ required for $I_{\lambda 1} = I_{\lambda 2}$. Furthermore, the relative increase in $\Delta\lambda_2$ will become larger as $\Delta\lambda_1$ is increased. Thus, a contribution to S_ν from $r_0 B$ will lead to systematic increase of the inferred value of $\Delta\lambda_D$ with increasing $\Delta\lambda_1$. Because ϕ_ν is a rapid function of frequency near $\Delta\lambda/\Delta\lambda_D = 2-3$, this apparent increase of $\Delta\lambda_D$ may easily override the true gradient in $\Delta\lambda_D$.

The gradient in $\Delta\lambda_D$ evident from comparison of different line pairs in Table V-4, in this case, has an effect similar to that of the contribution of $r_0 B$. Since $\Delta\lambda_D$ increases with height, the value of $\int_\infty^h \phi_{y_2} \, dh$ is increased relative to $\int_\infty^h \phi_{y_1} \, dh (y_2 > y_1)$. Thus, to obtain a given ratio of $\int_\infty^h \phi_{y_2} \, dh / \int_\infty^h \phi_{y_1} \, dh$, which is imposed by the condition $I_{y_1} = I_{y_2}$, the value of y_2 is increased. This effect, also, will become progressively worse with increasing $\Delta\lambda_1$. Thus, both the contribution of $r_0 B$ and the outward increase of $\Delta\lambda_D$ conspire to make $\Delta\lambda_D$ increase away from line center. Similar effects have been noted by Roddier (1965), Banos (1968), and by Dunn and Olson (1971).

In the weaker lines in Table V-4 the influence of $r_0 B$ is probably the more important contribution to the increase of $\Delta\lambda_D$ with increasing $\Delta\lambda_1$ since in these lines r_0 is relatively large and $\Delta\lambda_2$ often exceeds $2\Delta\lambda_D$. In the stronger lines (MgI and CaII), however, r_0 is small and $\Delta\lambda_2 < 2\Delta\lambda_D$. On the other hand, the height gradient in $\Delta\lambda_D$ is stronger for these lines and this probably accounts for the increase of $\Delta\lambda_D$ with $\Delta\lambda_1$.

Analysis of the standard lines in Tables V-6 and V-7 by the Goldberg method gives the results in Table V-9, which are similar to those in Table V-8, in that for $\Delta\lambda_1 > 30$ mÅ they show an increase of $\Delta\lambda_D$ away from line center even though $\Delta\lambda_D$ is decreasing

TABLE V-9

Estimate of $\Delta\lambda_D$ for standard lines using Goldberg's method

Case A			Case B		
$\Delta\lambda_1$	$\Delta\lambda_2$	$\Delta\lambda_D$	$\Delta\lambda_1$	$\Delta\lambda_2$	$\Delta\lambda_D$
0	39	47	0	–	–
15	40	41	15	–	–
30	45	39	30	30	0
45	56	40	45	51	29
60	69	41	60	69	41
75	89	57	75	85	48
90	112	80	90	112	80

At $\tau_{01} = 1$ true $\Delta\lambda_D = 44$

with depth. The value of $\Delta\lambda_D$ obtained for $\Delta\lambda_1 = 0$ in case A is in excellent agreement with the value of 47 mÅ given by the model atmosphere at $\tau_{01} = 1$. The results for case B are not good, even at line center, as is to be expected since the source functions are not equal.

In both cases A and B and for both lines of the multiplet the shoulders of the profiles occur at $\Delta\lambda_S = 90$ mÅ at $\mu = 1$ and at $\Delta\lambda_S = 110$ mÅ at $\mu = 0.125$. Using the approximation that $\Delta\lambda_S = 3\ \overline{\Delta\lambda_D}$, we find $\overline{\Delta\lambda_D} = 30$ mÅ at $\mu = 1$ and $\overline{\Delta\lambda_D} = 37$ mÅ at $\mu = 0.125$. As may be seen from Table V-4 these average values are reasonably consistent with the known values of $\Delta\lambda_D$ between the approximate heights of 500 km to 0 where the core forms at $\mu = 1$ and between 700 km and 180 km where the core forms at $\mu = 0.125$.

There is nothing intrinsic in the Goldberg method that restricts its use to the center of the solar disk. In the case of stars where we observe only H, the method will still work provided the restrictions $S_{\lambda 1} = S_{\lambda 2}$ and $\Delta\lambda_D = $ const. are satisfied. In this case, however, $\Delta\lambda_D$ must be constant with μ as well as with height, i.e., all broadening motions must be isotropic. Although, to the author's knowledge, this method has not been used in stellar work, it could perhaps give useful results if restricted to the innermost cores of strong multiplets.

12. Evaluation of $\Delta\lambda_D$ from Cores of Strong Lines – Center-Limb Data

The preceding estimates of $\Delta\lambda_D$ require profiles at one value of μ only and they make no attempt to invert the integral Equation (V-1) to obtain $S_\nu(\tau_\nu)$. Center-limb data, in principle, contain more information than data at a single value of μ and therefore permit more detailed, and presumably more reliable, analyses. In particular, it is possible using center-limb data to invert Equation (V-1) to obtain both $S_\nu(\tau_\nu)$ and the form of ϕ_ν. The uniqueness of such inversions has been the subject of much discussion and will not be explicitly dealt with here. We rely instead upon analysis of the standard data to indicate the self-consistency of a particular method. This, of course, does not test the uniqueness of the inversion since we do not attempt to prove that atmospheric models of a different character cannot produce data quite similar to the standard data.

In the case of multiplet lines that are strongly coupled we have shown that the profiles provide essentially equivalent sets of data in the line cores. This has the consequence that there is not much additional information to be gained by a combined analysis of two lines. The second line might be used to effectively extend the μ scale of the other line, but beyond this type of extension the information is redundant. In the following, we analyze just the stronger line of the doublet for case A. We could, in principle, extend the μ scale to $\mu = 2$ by equating the data for the weaker line of the multiplet at $\mu = 1$ to the stronger line at $\mu = 2$. Instead we analyze just the data for the range $\mu = 1$ to $\mu = 0.125$. At the end of the analysis we investigate the effect of extending the data to $\mu = 2$.

A particularly simple method of inverting the integral in Equation (V-1) is to assume some explicit form for $S_\lambda(\tau_\lambda)$ that permits a direct integration. A form commonly used, because of its simplicity, is

$$S_\lambda = a_\lambda + b_\lambda \tau_\lambda + (c_\lambda/2)\,\tau_\lambda^2, \qquad (\text{V-44})$$

which gives

$$I_\lambda = a_\lambda + b_\lambda \mu + c_\lambda \mu^2. \qquad (\text{V-45})$$

Curtis and Jefferies (1965, 1967) and Jefferies (1968) suggest that a more appropriate form is

$$S_\lambda = \alpha_\lambda + \beta_\lambda \log \tau_\lambda + \gamma_\lambda (\log \tau_\lambda)^2. \qquad (\text{V-46})$$

This gives

$$I_\lambda = a_\lambda + b_\lambda \log \mu + c_\lambda (\log \mu)^2, \qquad (\text{V-47})$$

where

$$\alpha_\lambda = a_\lambda + \kappa \gamma b_\lambda - \kappa^2 F c_\lambda, \\ \beta_\lambda = b_\lambda + 2\kappa \gamma c_\lambda, \qquad (\text{V-48})$$

and

$$\gamma_\lambda = c_\lambda,$$

with $\kappa = \log e = 0.4343$, $\gamma = 0.5772$ (Euler's constant) and $F = \gamma^2 + \pi^2/6 = 1.9781$. A third form commonly used is

$$S_\lambda = a_\lambda + b_\lambda \tau + c_\lambda E_2(\tau). \qquad (\text{V-49})$$

$E_2(\tau)$ is the second exponential integral. Equation (V-49) gives

$$I_\lambda = a_\lambda + b_\lambda \mu + c_\lambda \{1 - \mu \ln(1 + \mu^{-1})\}. \qquad (\text{V-50})$$

Although various arguments have been given in favor of one particular form for $S_\lambda(\tau_\lambda)$ or another these arguments are generally based upon some *a priori* expectation about the form of $B(\tau_\lambda)$. Because S differs substantially from B we expect that $S_\lambda(\tau_\lambda)$ will differ from $B(\tau_\lambda)$. Furthermore, when the Doppler width varies with depth the ratio τ_λ/τ_0 may vary markedly with depth. Thus, $S_\lambda(\tau_\lambda)$ may have a markedly different form from $S_\lambda(\tau_0)$. It is not evident *a priori*, therefore, which form for $S_\nu(\tau_\nu)$ will give the most accurate inversion of the integral equation when the expansion coefficients a_λ, b_λ and c_λ are evaluated using a restricted set of data.

Using a set of center-limb data, $I_\lambda(\mu)$, one may evaluate the coefficients a_λ, b_λ, and c_λ, which are then sufficient to specify $S_\lambda(\tau_\lambda)$ from the equation appropriate to the particular form chosen to represent $I_\lambda(\mu)$. Such a procedure will represent $I_\lambda(\mu)$ between certain limits on μ, say $0.1 \leqslant \mu \leqslant 1$. It follows from the Eddington-Barbier relation that the resultant $S_\lambda(\tau_\lambda)$ is most valid in the range $0.1 \leqslant \tau_\lambda \leqslant 1$. Outside this range of τ_λ there is no guarantee whatsoever that $S_\lambda(\tau_\lambda)$ bears any meaningful relationship to the actual run of S_λ in the atmosphere. Even within the range $0.1 \leqslant \tau_\lambda \leqslant 1$ the specific form of S_λ may be seriously influenced by the form assumed for $I_\lambda(\mu)$. We illustrate this effect in Figure V-11a where we have plotted $\log S_\lambda$ vs. $\log \tau_\lambda$ for three wavelengths within the Doppler core of line 2, Case A, in the standard sets of

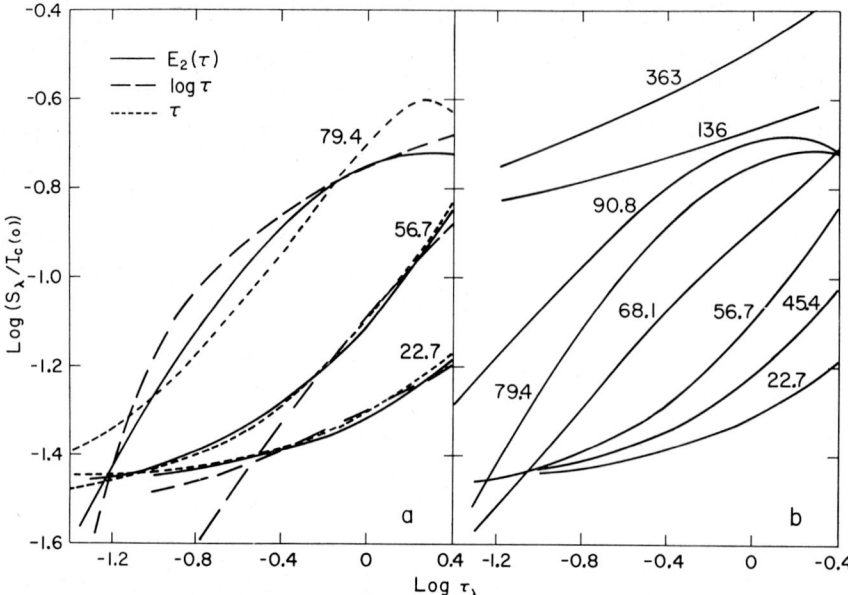

Fig. V-11. Source function segments derived from standard center-limb data (a) – for three different expansions of $S_\nu(\tau_\nu)$ and (b) – for the E_2 expansion (b).

data. The three cases shown at each wavelength are obtained by least squares fits to Equations (V-45), (V-47), and (V-50). The assumed forms for $I_\lambda(\mu)$ fit the data well in all cases (the maximum fitting error for any one data point is less than 4% of the value of I_λ at that point). At 79.4 mÅ the three representations for $S_\lambda(\tau_\lambda)$ give values that differ by as much as 60% within the range $0.1 \leqslant \tau_\lambda \leqslant 1$. Outside this range the differences become markedly larger. At 56.7 mÅ the $\log \tau_\lambda$ expansion gives a value of S_λ at $\tau_\lambda = 0.1$ that differs from the values given by the other two expansions by more than 100%. This divergence of the results of the $\log \tau_\lambda$ expansion from the results of the τ_λ and $E_2(\tau_\lambda)$ expansions at small μ is evident at several wavelengths and evidently results from the fact that $|\log \tau_\lambda|$ increases rapidly as τ_λ decreases below 0.1.

Sections of $S_\lambda(\tau_\lambda)$ between $\log \tau \lambda$ equal to 0.4 and -1.2 are shown in Figure V-11b for several different values of $\Delta \lambda$ for the $E_2(\tau)$ expansion. The fitting coefficients are given in Table V-10 for all three representations of $S_\lambda(\tau_\lambda)$. The amount of limb-darkening at a given wavelength is given approximately by

$$1 + b_\lambda/a_\lambda, \tag{V-51}$$

where b_λ and a_λ are the values for the τ_λ expansion (Equation (V-44)). Note from Table V-10 that b_λ/a_λ has abnormally large values at values of $\Delta\lambda = 68.1$–90.8. This is the wavelength region in the steep part of the line core (Figure V-10), and it is the region in which the height gradient in $\Delta\lambda_D$ seriously modifies the ratio τ_λ/τ_0 as a function of τ_0. The increased limb-darkening at these wavelengths is reflected in $S_\lambda(\tau_\lambda)$

TABLE V-10
Coefficients for $I_\lambda(\mu)$ equations for line 2, Case A

			Equation (V-45)					
$\Delta\lambda$	a_λ	b_λ	c_λ	b_λ/a_λ	$\Delta\lambda$	a_λ	b_λ	c_λ	b_λ/a_λ
0	0.0345	0.0152	−0.0030	0.44	136	0.144	0.0982	−0.039	0.68
22.7	0.0345	0.0176	−0.0037	0.51	182	0.156	0.108	−0.038	0.79
34.0	0.0344	0.0208	−0.0037	0.61	363	0.173	0.198	−0.066	1.2
45.4	0.0343	0.0266	−0.0020	0.79	545	0.184	0.287	−0.089	1.6
56.7	0.0319	0.0509	−0.0039	1.6	726	0.192	0.368	−0.100	1.9
68.1	0.0242	0.142	−0.054	5.9	1135	0.198	0.562	−0.133	2.8
79.4	0.0316	0.226	−0.116	7.2	2270	0.187	1.01	−0.329	5.4
90.8	0.0615	0.219	−0.119	3.6	4086	0.179	1.29	−0.506	7.2
					7037	0.177	1.39	−0.583	7.9

Equation (V-47)				Equation (V-50)			
$\Delta\lambda$	a_λ	b_λ	c_λ	$\Delta\lambda$	a_λ	b_λ	c_λ
22.7	0.0484	0.0227	0.0109	22.7	0.0405	0.0102	−0.0072
34.0	0.0514	0.0287	0.0143	34.0	0.0403	0.0135	−0.0070
45.4	0.0586	0.0440	0.0234	45.4	0.0375	0.0226	−0.0038
56.7	0.0785	0.0859	0.0471	56.7	0.0360	0.0445	−0.0045
68.1	0.114	0.122	0.0453	68.1	0.108	0.0362	−0.100
79.4	0.144	0.099	−0.0246	79.4	0.228	−0.0122	−0.238
90.8	0.165	0.072	−0.0230	90.8	0.269	−0.028	−0.252
				136	0.211	0.018	−0.082
				182	0.218	0.031	−0.075
				363	0.286	0.062	−0.137
				726	0.363	0.162	−0.206
				1135	0.420	0.293	−0.268
				2270	0.718	0.358	−0.637

by the much steeper slopes, $d \log S_\lambda / d \log \tau_\lambda$, shown in Figure V-11b. It is the steeper variation in $S_\lambda(\tau_\lambda)$, in turn, that leads to the relatively larger spread in the values of S_λ between the different representations of $S_\lambda(\tau_\lambda)$ exhibited in Figure (V-11a). Near line center and in the near line wings where $d \log S_\lambda / d \log \tau_\lambda$ is smaller the different expansions of $S_\lambda(\tau_\lambda)$ give relatively good agreement.

It must be remembered that we are working here with nearly perfect 'data' for a perfectly plane parallel atmosphere. Even so, it is plainly evident from Figure V-11a, that we cannot assume that the reconstruction of $S_\lambda(\tau_\lambda)$ is free from error. There are errors, and they are substantial at some wavelengths. It follows, of course, that in comparing the section of the $S_\lambda(\tau_\lambda)$ curve at one wavelength, λ_1, against the section for another wavelength, λ_2, we will encounter differential errors. This point will be illustrated in more detail in the subsequent discussion.

In any real photometric data there are errors in I_λ and in both $\Delta\lambda$ and μ. The errors in I_λ and μ, in particular, can be quite important. Also, real atmospheres are not plane parallel. Temperature and density inhomogeneities as well as curvature effects introduce differential effects at different μ positions. These differential effects with μ as well as the errors in I_λ, $\Delta\lambda$ and μ produce errors in the data that reflect as still larger errors in $S_\lambda(\tau_\lambda)$. Since even perfect data lead to substantial errors in $S_\lambda(\tau_\lambda)$, real data must inevitably lead to even larger errors in $S_\lambda(\tau_\lambda)$.

As an illustration of the difficulties introduced by our failure to recover $S_\lambda(\tau_\lambda)$ accurately, let us attempt to derive the Doppler widths, $\Delta\lambda_D$, from the data in the line core. Such analyses have been carried out by several authors, notably by White (1963) for the solar hydrogen Balmer lines and by Curtis and Jefferies (1965, 1967; see also Jefferies, 1968) for the solar sodium D lines.

In the core of strong lines the ratio of $\tau_{\lambda 1}$ to $\tau_{\lambda 2}$ is given by Equation (V-41). For a given line, of course, $(\omega f)_1 \equiv (\omega f)_2$. If $\Delta\lambda_D$ were constant with depth and if $S_\lambda(\tau_\lambda)$ were accurately recovered from the data, the individual sections of $S_\lambda(\tau_\lambda)$ shown in Figure (V-11b) for $\Delta\lambda \leqslant 90.8$ should lie parallel to the curve $S(\tau_0)$ and should be displaced only in the τ_0 direction. Since the individual curves are not parallel to each other, they obviously do not conform to this simple interpretation, mostly because $\Delta\lambda_D$ is not constant with depth. By ignoring this rather obvious lack of parallelism, we assume $\Delta\lambda_D$ to be constant and set

$$\ln\frac{\tau_{\lambda 1}}{\tau_{\lambda 2}} = \frac{\Delta\lambda_2^2 - \Delta\lambda_1^2}{\Delta\lambda_D^2}. \tag{V-52}$$

This permits an immediate derivation of $\Delta\lambda_D$, with the results shown in Figure (V-12a). The mean results show a gradient similar to the true gradient but give values of

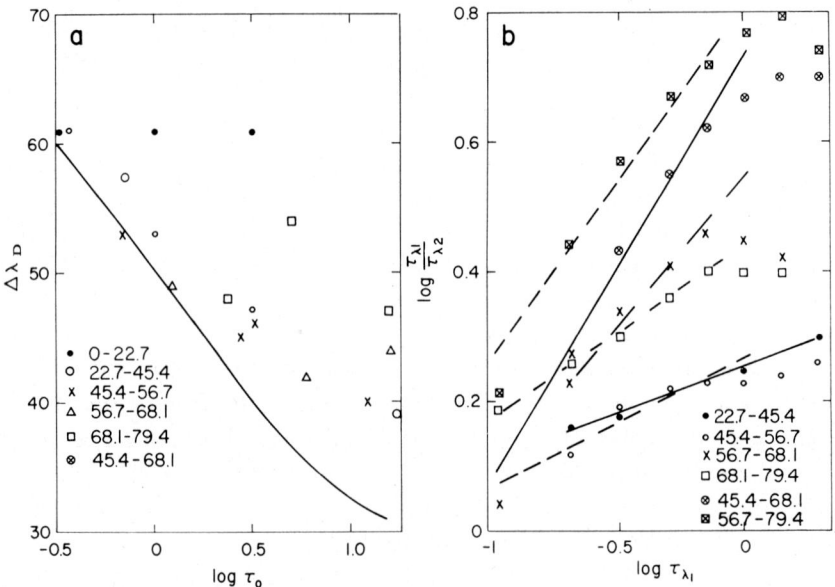

Fig. V-12. (a) – Initial estimates of $\Delta\lambda_D$, and (b) – illustration of Equation (V-52) for different wavelength pairs. The solid curve in Part (a) gives the true run of $\Delta\lambda_D$.

$\Delta \lambda_D$ that are 10 to 20% too large. A second iteration using the first estimate of $\Delta \lambda_D(\tau_0)$ to evaluate the integrals in Equation (V-41) would probably give somewhat better results.

Instead of following the iterative procedure, which was employed by White (1963) and by Curtis and Jefferies (1965, 1967), we choose instead to assume a specific form for $\Delta \lambda_D(\tau_0)$ with unspecified constants and to use the data to evaluate the constants.

We write $y = Dy(1)$, where

$$D^2 = 1 + \alpha \ln \tau_0 \tag{V-53}$$

and where $y(1)$ is the value of y at $\tau_0 = 1$. Since $y(1)$ is defined at a fixed value of τ_0, we have $y_2^2(1) = (\Delta \lambda_1^2 / \Delta \lambda_2^2) y_1^2(1) = l^2 y_1^2(1)$. For convenience of notation we write $y_1^2(1)$ in the following as y^2. For any given pair of wavelengths l^2 is known and y^2 and α are to be evaluated from the $S_\lambda(\tau_\lambda)$ curves.

It follows from Equations (V-41) and (V-53) that

$$\tau_{\lambda 1} = e^{-y^2} \frac{\tau_0^{1-\alpha y^2}}{1 - \alpha y^2}, \quad \alpha y^2 < 1, \tag{V-54}$$

and

$$\tau_{\lambda 2} = e^{-l^2 y^2} \frac{\tau_0^{1-\alpha l^2 y^2}}{1 - \alpha l^2 y^2}, \quad \alpha l^2 y^2 < 1. \tag{V-55}$$

We combine these latter two equations to obtain

$$\ln \frac{\tau_{\lambda 1}}{\tau_{\lambda 2}} = \ln \frac{l^2(1-m) + m - 1}{l^2 + m - 1} + (m-1) \ln \frac{l^2 - 1}{l^2 + m - 1}$$
$$+ (l^2 + m - 1) y^2 + m \ln \tau_{\lambda 1} \tag{V-56}$$

with

$$m = \frac{(l^2 - 1) \alpha y^2}{1 - \alpha y^2}. \tag{V-57}$$

Equation (V-56) defines a straight line for the two variables $\ln(\tau_{\lambda 1}/\tau_{\lambda 2})$ and $\ln \tau_{\lambda 1}$ with slope m and an intercept, $\mathscr{I}$, at $\ln \tau_{\lambda 1} = 0$ given by the first three terms on the right hand side. The two unknowns in Equation (V-56) are m and y^2. We determine both m and $\mathscr{I}$ by plotting $\log \tau_{\lambda 1}/\tau_{\lambda 2}$ vs. $\log \tau_{\lambda 1}$ and drawing the best straight line through the points represented by $0.1 \leqslant \tau_{\lambda 1} \leqslant 1$. From m and $\mathscr{I}$ we obtain y^2 and we combine m and y^2 in Equation (V-57) to obtain α. The construction of the plots to determine m and $\mathscr{I}$ are shown for the $E_2(\tau_\lambda)$ expansion in Figure V-12b together with hand-drawn straight lines. No attempt has been made to obtain least square fits. Data for the τ_λ expansion of $S_\lambda(\tau_\lambda)$ are similar in appearance to those in Figure V-12b but for the $\log \tau_\lambda$ expansion the lines have markedly greater slopes. In both the τ and $\log \tau$ expansions the wavelength pair 68.1–79.4 gives *negative* values of m, which are contradictory with all the remaining values of m.

Values of $\Delta \lambda_D(1)$ and α obtained from the different wavelength pairs and the different expansions of $S_\lambda(\tau_\lambda)$ are given in Table V-11. Of the three expansions, the best

TABLE V-11
Solutions for $\Delta\lambda_D$ ($\tau_0 = 1$) and α

$S_\nu(\tau)$		22.7–45.4	45.4–56.7	56.7–68.1	68.1–79.4	45.4–68.1	56.7–79.4
τ	$\Delta\lambda_D(1)$	47	43	39	–	39	46
	α	0.22	0.26	0.21	–	0.23	0
$\log\tau$	$\Delta\lambda_D(1)$	41	40	41	–	42	39
	α	0.29	0.33	0.09	–	0.24	0.17
$E_2(\tau)$	$\Delta\lambda_D(1)$	47	42	37	45	36	40
	α	0.19	0.22	0.22	0.18	0.22	0.19

agreement is between the τ_λ and $E_2(\tau_\lambda)$ cases, although there is a somewhat surprising degree of agreement among most of the results. The average value of $\Delta\lambda_D(1)$ in Table V-11 is 40 mÅ and the average value of α is 0.20. By comparison the actual values of $\Delta\lambda_D$ are represented quite well for $\tau_0 > 1$ by $\Delta\lambda_D(1) = 50$ mÅ and $\alpha = 0.55$. Thus, we have underestimated both $\Delta\lambda_D(1)$ and α. The average values of $\Delta\lambda_D$ obtained in this way intersect the true value of $\Delta\lambda_D$ at $\tau_0 \approx 0.7$.

The internal consistency among the results in Table V-11 tempts us to conclude that we have, in fact, successfully evaluated $\Delta\lambda_D(1)$ and α with reasonable accuracy. We know, however, that this is not true and we see demonstrated once again the fallacy of arguments based only upon internal consistency.

If the results in Table V-11 were, in fact, valid, we should be able to reconstruct them in different ways. For example each of the pairs of values of $\Delta\lambda_D(1)$ and α together with their appropriate values of y^2, m and l^2 can be used in Equation (V-54) to compute the value of τ_0 at which $\tau_{\lambda 1} = 1$ as well as the value of $\Delta\lambda_D$ at this value of τ_0. These values of $\Delta\lambda_D$ and τ_0 should then define a curve of $\Delta\lambda_D$ versus τ_0 that is consistent with the results in Table V-11. We show a plot of such values in Figure V-13 together with the true run of $\Delta\lambda_D$ with τ_0 and the run given by $\Delta\lambda_D(1) = 40$ mÅ and $\alpha = 0.20$. It is of interest to note the following results: the τ_λ expansion gives nearly the true value for $\Delta\lambda_D(1)$ but with a mean value of α in rough agreement with the results in Table V-11; the $\log\tau_\lambda$ expansion gives both $\Delta\lambda_D(1)$ and α in fair agreement with the results in Table V-11; and the $E_2(\tau_\lambda)$ expansion gives values for both $\Delta\lambda_D(1)$ and α that are in reasonably good agreement with the true values. Thus, the $\log\tau_\lambda$ expansion has the greatest internal consistency but the $E_2(\tau_\lambda)$ expansion gives the most reliable results. Perhaps the reason that the $\log\tau_\lambda$ expansion has the highest internal consistency is to be found in the fact that the μ values in the standard data are equally spaced in $\log\mu$ which tends to make $\log\mu$ the natural basis for a series expansion.

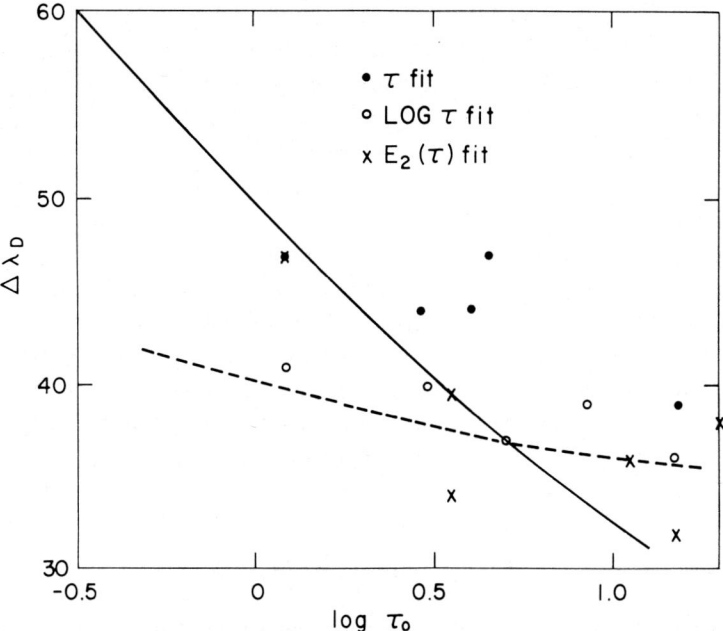

Fig. V-13. Illustration of final attempt to evaluate $\Delta\lambda_D(\tau_0)$ (see text for comments).

Data points that are equally spaced in $\log\mu$ between $0.1 \leq \mu \leq 1$, to a close approximation, are equally spaced in both $E_2(\mu)$ and $\mu \ln(1+\mu^{-1})$. Thus, equal spacing in $\log\mu$ is favorable for the $E_2(\tau_v)$ term in Equation (V-49).

Although the preceding results seem to suggest some definite trends, it must be remembered that we have considered only a single case. A different set of data may lead to quite different conclusions about the relative validity of a given method of analysis. The preceding analyses, when bolstered by a knowledge of the true values of $\Delta\lambda_D(\tau_0)$ clearly suggest that the $E_2(\tau_\lambda)$ expansion is the preferred one. Without the *a priori* knowledge of $\Delta\lambda_D(\tau_0)$, however, the internal consistency of the $\log\tau_\lambda$ expansion would lead one to accept it as the preferred expansion and, consequently, to adopt quite erroneous values for $\Delta\lambda_D(1)$ and α. This is a somewhat disturbing conclusion, but one which, nevertheless, is realistic.

Attempts to recover $S(\tau_0)$ from the data in Figure V-11b are directly coupled, of course, to the determination of $\Delta\lambda_D$. The process of deriving $\Delta\lambda_D$ used in the preceding discussion corresponds to a parallel displacement of the individual curves in Figure V-11b laterally to construct the best single curve of $S_\lambda(\tau_0)$. When gradients in $\Delta\lambda_D$ are present the individual sections of $S_\lambda(\tau_0)$ should be displaced different amounts for different values of S_λ, i.e., the parallelism is not retained in the translation. Unless $\Delta\lambda_D$ is properly recovered the curve of $S_\lambda(\tau_0)$ constructed by the above process will not agree with the true curve. An underestimate for $\Delta\lambda_D$ will result in an overestimate for τ_0/τ_λ and an underestimate for $S_\lambda(\tau_0)/S_\lambda(1)$. Conversely, an overestimate for $\Delta\lambda_D$ will result in an underestimate for τ_0/τ_λ and an overestimate for $S_\lambda(\tau_0)/S_\lambda(1)$. Since the construction of $S_\lambda(\tau_0)$ is a piece-wise process requiring larger and larger shifts in τ_λ, the errors in $\Delta\lambda_D$ are cumulative and produce increasingly serious errors in S_λ as τ_0 increases.

The preceding analysis to obtain $\Delta\lambda_D$ was repeated for the range $0.125 \leqslant \mu \leqslant 2$ using the weaker line to give values of $I_\lambda(\mu)$ for μ between 1 and 2. The results are very similar to those for the case $0.125 \leqslant \mu \leqslant 1$. Alternative expansions of $S_\lambda(\tau_\lambda)$ still give values of $S_\lambda(\tau_\lambda)$ that disagree as badly as in Figure V-11a. Also, the scatter in $\Delta\lambda_D(\tau_0)$ is about the same as in Figure V-13 with the same general results. Thus, the improvement in extending the data to $\mu = 2$ appears to be minimal.

In the Doppler cores of strong lines we expect to have $S_L \ll B$ and $r_0 \ll \phi_y$. It is not likely therefore that we can obtain useful information about either B or r_0 through an empirical analysis of the core region. To obtain information on B and r_0 we analyze next the line wings where we expect $S_L \approx B$ and where r_0 and ϕ_y are more nearly of the same order.

13. Analysis of Line Wings

Spectral lines with strongly developed wings provide additional opportunities for the extraction of useful information. Sufficiently far from line center to be clearly in the line wings we expect τ_0/τ_λ to be of the order of 10^4 or greater. It is very likely, in the solar case at least, that at $\tau_0 \gg 10^4$ S_L is very close to B, in which case $S_\lambda = B$ and S_λ is again independent of frequency. The frequency independence allows us to again make use of the displacement between the $S_\lambda(\tau_\lambda)$ curves in a simple manner. Because the source function is independent of frequency, equal values of S_λ for different λ must refer to a single depth in the atmosphere.

In the line wings we may write τ_λ at wavelength λ_1 as

$$\tau_{\lambda 1} = \tau_{c1} + \tau_1 \tag{V-58}$$

where τ_{c1} and τ_1 are the continuum and line optical depths at λ_1. At equal values of $S_{\lambda 1}$ and $S_{\lambda 2}$ we must have $\tau_{c1} = \tau_{c2}$. Thus, at three wavelengths λ_1, λ_2 and λ_3 and for $S_{\lambda 1} = S_{\lambda 2} = S_{\lambda 3}$ we may write

$$\frac{\tau_{\lambda 1}}{\tau_{\lambda 2}} = \frac{1 + \tau_1/\tau_c}{1 + \tau_2/\tau_c} \tag{V-59}$$

and

$$\frac{\tau_{\lambda 2}}{\tau_{\lambda 3}} = \frac{1 + \tau_2/\tau_c}{1 + \tau_3/\tau_c}. \tag{V-60}$$

We suppose now that τ_1, τ_2 and τ_3 are related by the conditions

$$\tau_1/\tau_2 = (\Delta\lambda_2/\Delta\lambda_1)^2 \sigma = l_{21}^2 \sigma \tag{V-61}$$

and

$$\tau_2/\tau_3 = (\Delta\lambda_3/\Delta\lambda_2)^2 \sigma = l_{32}^2 \sigma. \tag{V-62}$$

These latter two equations together with Equation (V-59) and (V-60) yield

$$\frac{\tau_2}{\tau_c} = \frac{1 - \tau_{\lambda 1}/\tau_{\lambda 2}}{\dfrac{\tau_{\lambda 1}}{\tau_{\lambda 2}} - l_{21}^2 \sigma} \tag{V-63}$$

and

$$\frac{\tau_2}{\tau_c} = \frac{1 - \dfrac{\tau_{\lambda 2}}{\tau_{\lambda 3}}}{\dfrac{\tau_{\lambda 2}}{\tau_{\lambda 3}} - \dfrac{1}{l_{32}^2 \sigma}}. \tag{V-64}$$

In the line wings we expect σ to be of order unity. The quantities $\tau_{\lambda 1}/\tau_{\lambda 2}$ and $\tau_{\lambda 2}/\tau_{\lambda 3}$ are obtained from the $S_\lambda(\tau_\lambda)$ curves for the condition $S_{\lambda 1} = S_{\lambda 2} = S_{\lambda 3}$. Plots of $S_\lambda(\tau_\lambda)$ for the $E_2(\tau_\lambda)$ expansion are shown in Figure V-14 for line 2, case A. By comparison to the data for the Doppler core, the wing data are relatively well behaved.

By reading values of $\tau_{\lambda 1}/\tau_{\lambda 2}$ and $\tau_{\lambda 2}/\tau_{\lambda 3}$ from the curves in Figure V-14 we may solve Equations (V-63) and (V-64) for both σ and τ_2/τ_c. The results will tell us whether σ is significantly different from unity, the expected value in the wings. Also, from the ratio τ_2/τ_c we may determine $\tau_{\lambda 2}/\tau_c$ and from the known curve of $S_{\lambda 2}(\tau_{\lambda 2})$ we

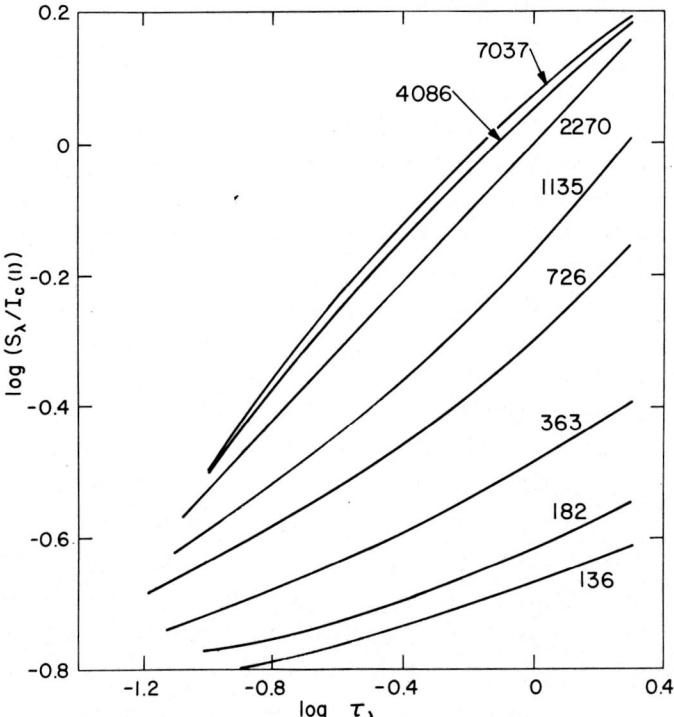

Fig. V-14. Variation of S_λ with τ_λ in the line wings.

may construct a curve giving $S_\lambda(\tau_c)$. Or, since we expect $S_\lambda = B$, we obtain $B(\tau_c)$.

The preceding method for determining $(\tau_{\lambda 2}/\tau_c)$ and $B(\tau_c)$ differs from the method used by Curtis and Jefferies (1965, 1967) and Jefferies (1968). We have simplified the situation by assuming $S_\lambda = B$. In the Curtis-Jefferies method it is not necessary to make this latter assumption. Instead, it is assumed that two lines of a multiplet have equal values of S at a given depth and that they have equal values of ϕ_λ at a given $\Delta\lambda$.

For each line Curtis and Jefferies then choose two wavelengths and a reference optical depth, say, $\tau_{\lambda ij}$, where i designated the wavelength and j the line. For each value of $S_{\lambda ij}$ and its associated $\tau_{\lambda ij}$, we write

$$\tau_{\lambda ij} = \tau_{cij} + \tau_{ij}, \tag{V-65}$$

and

$$S_{\lambda ij} = \frac{r_{\lambda ij} B_{ij}}{1 + r_{\lambda ij}} + \frac{S_{ij}}{1 + r_{\lambda ij}}, \tag{V-66}$$

where

$$r_{\lambda ij} = r_{0ij}/\phi_{\lambda ij}. \tag{V-67}$$

We may write as an alternative for $\tau_{\lambda ij}$

$$\tau_{\lambda ij} = \tau_{cij} + \int_0^{\tau_{cij}} r_{\lambda ij}^{-1} \, d\tau_c. \tag{V-68}$$

Both r_{0ij} and $\phi_{\lambda ij}$ will normally vary with depth. In the line wings $\phi_{\lambda ij}$ contains the damping parameter, which, in the case of collision damping, varies with depth, and the Doppler width, which is depth dependent also. Because of the depth variation of $r_{\lambda ij}$, we cannot safely assume that $\tau_{cij}/\tau_{ij} = r_{\lambda ij}$.

At a given geometrical depth in the atmosphere the $S_{\lambda ij}$ and $\tau_{\lambda ij}$ are related by

$$\begin{aligned}
\tau_{\lambda 11} &= \tau_{c11} + \tau_{11}, \\
\tau_{\lambda 12} &= \tau_{c11} + g\tau_{11}, \\
\tau_{\lambda 21} &= \tau_{c11} + (1/l_\sigma^2) \tau_{11}, \\
\tau_{\lambda 22} &= \tau_{c11} + (g/l_\sigma^2) \tau_{11},
\end{aligned} \tag{V-69}$$

and

$$\begin{aligned}
S_{\lambda 11} &= \frac{r_{\lambda 11} B_{11}}{1 + r_{\lambda 11}} + \frac{S_{11}}{1 + r_{\lambda 11}}, \\
S_{\lambda 12} &= \frac{g r_{\lambda 11} B_{11}}{1 + g r_{\lambda 11}} + \frac{S_{11}}{1 + g r_{\lambda 11}}, \\
S_{\lambda 21} &= \frac{r_{\lambda 11} B_{11}}{l^2 \sigma + r_{\lambda 11}} + \frac{l^2 \sigma S_{11}}{l^2 \sigma + r_{\lambda 11}}, \\
S_{\lambda 22} &= \frac{g r_{\lambda 11} B_{11}}{l^2 \sigma + g r_{\lambda 11}} + \frac{l^2 \sigma S_{11}}{l^2 \sigma + g r_{\lambda 11}},
\end{aligned} \tag{V-70}$$

where $g = (\omega f)_2/(\omega f)_1$ and where we have again set

$$\phi_{\lambda 1}/\phi_{\lambda 2} = (\Delta\lambda_2/\Delta\lambda_1)^2 \sigma = l^2 \sigma.$$

The subscripts on l are dropped in Equations (V-69) and (V-70) for convenience.

The unknowns in the set of Equations (V-70) are σ, $r_{\lambda 11}$, B_{11} and S_{11}, and the four equations are sufficient to determine the four unknowns. However, Equations (V-70) are subject to the restriction that the values of $S_{\lambda ij}$ be known at a single value of τ_{c11}. There is no way *a priori* to pick such values of $S_{\lambda ij}$. In this method, therefore, it is necessary to iterate the set of Equations (V-69) with the set (V-70). We may, for example, set

$$r_{\lambda 11}^{-1} = r_{\lambda 11}^{-1}(1) \tau_{c11}^n, \qquad \text{(V-71)}$$

in which case

$$\tau_{11}/\tau_{c11} = r_{\lambda 11}^{-1}(1) \frac{\tau_{c11}^n}{n+1}. \qquad \text{(V-72)}$$

Next, choose a reference value of $\tau_{\lambda 11}$ and a set of values for $r_{\lambda 11}$, σ, and n. The value of τ_{c11} is then determined by the first of Equations (V-69). From this set of parameters compute $\tau_{\lambda 12}$, $\tau_{\lambda 21}$ and $\tau_{\lambda 22}$ and find the corresponding values of $S_{\lambda 11}$, $S_{\lambda 12}$, $S_{\lambda 21}$, $S_{\lambda 22}$. Then solve the set of Equations (V-70) for S_{11}, B_{11} and for new values of σ, and $r_{\lambda 11}$. Use the new $r_{\lambda 11}$ in Equation (V-72) with the same n to determine a new τ_{c11} and use this new τ_{c11} together with the new σ and $r_{\lambda 11}$ to obtain a new set of $\tau_{\lambda ij}$ from Equations (V-69). Continue this process until convergence is achieved for the initial values of n.

This process may be repeated for different initial values of $\tau_{\lambda 11}$ and, after a number of values of $\tau_{\lambda 11}$ have been solved, the results will give $r_{\lambda 11}$ as a function of τ_{c11}. Thus, we may now derive a new value for n and repeat the iterative process on Equations (V-69) and (V-70) until we again obtain a new value for n. Hopefully, the double iterative process will converge to give finally a reliable set of values for B_{11}, S_{11}, $r_{\lambda 11}$, σ, each as functions of τ_c.

Curtis and Jefferies (1965, 1967) (see also Jefferies, 1968) have applied a version of this type of iterative analysis to the solar Na D lines with rather striking success. Rather than reproduce this rather complicated analysis here, we consider the much simpler approach using the single line analysis with Equations (V-63) and (V-64). This method has the advantage of simplicity, but the disadvantage of being somewhat more restrictive than the Curtis-Jefferies method. It is not directly applicable to those parts of the line profile where S_λ is frequency dependent. On the other hand, it is possible to test the results by applying the method separately to each line. If S_L differs from B at the depths in the atmosphere considered, S_λ is necessarily frequency dependent and, in addition, is necessarily different for the two lines of the multiplet. Hence, if the two desired curves for $S_\lambda(\tau_c)$ for the two lines differ the method should be abandoned. If they agree, however, it may be concluded that S_λ is indeed close to B. Since we expect the condition $S_L = B$ to be a common feature in line wings,

TABLE V-12

Evaluation of σ in line wings

$\tau_{\lambda 3}$ \ Wavelengths	4086, 2270, 1135	2270, 1135, 726	1135, 726, 365	726, 363, 182
2	1.1	1.05	0.97	1.2
1	1.2	1.1	1.05	1.1
0.5	1.02	1.04	1.15	1.1

the restriction of $S_\lambda = B$ is not overly severe and, as indicated, it may be tested using the second line of the multiplet.

There is, incidentally, nothing in the Curtis-Jefferies method that requires the use of more than one spectral line. The method can be used equally well with a single line by using four different wavelengths, provided the analysis is restricted to the line wings where $\sigma = $ const. The only modification necessary is to replace g with l_{31}^2 and l_{41}^2 in appropriate places. When this method is applied to a single line the wavelength separation must be kept small enough to ensure that all four segments of the $S_\lambda(\tau_\lambda)$ curves overlap sufficiently in the range $\tau_\lambda = \mu_{\text{obs}}$. By using the doublet method the wavelength separation of $\Delta\lambda_1$ and $\Delta\lambda_2$ may be kept larger.

We proceed now to evaluate σ, $\tau_2(\tau_c)$ and $B(\tau_c)$ using the single line method with the explicit assumption that $S_\lambda = S_L = B$. The method is straightforward and requires only graph paper and desk calculations. Values of σ obtained from Equations (V-63) and (V-64) and the $S_\lambda(\lambda_\lambda)$ plots in Figure V-14 are given in Table V-12. As is readily apparent, σ is close to unity as expected.

Jefferies (1968) notes that in the far wings of the line where $\phi_\lambda/r_0 \ll 1$ and, hence, $\tau_1/\tau_c \ll 1$, we may write

$$\ln\frac{\tau_{\lambda 1}}{\tau_{\lambda 2}} = \ln\frac{1 + \tau_1/\tau_c}{1 + \tau_2/\tau_c} = \frac{\tau_1}{\tau_2} = l_{21}^2 \sigma. \tag{V-73}$$

The wavelength pair $\Delta\lambda = 4086$ and $\Delta\lambda = 2270$ give $\sigma = 1.02$ using Equation (V-73).

TABLE V-13

Evaluation of τ_2/τ_c and τ_c in line wings

$\Delta\lambda_1$	4086		2270		2270		1135		726	
$\Delta\lambda_2$	2270		1135		726		363		182	
$\tau_{\lambda 2}$	τ_2/τ_c	τ_c	τ_2/τ_c	τ_c	τ_2/τ_c	τ_c	τ_2/τ_c	τ_c	τ_2/τ_c	τ_c
2	0.39	1.44	2.0	0.67	5.3	0.32	12	0.15	36	0.054
1	0.67	0.60	2.2	0.31	4.3	0.19	10	0.091	19	0.050
0.5	0.77	0.28	2.1	0.16	3.5	0.11	7.2	0.061	13	0.036
0.2	0.52	0.13	1.4	0.08	2.5	0.057	4.9	0.034	7	0.025
$\frac{1}{2}(2270/\Delta\lambda_2)^2$	0.5		2.0		4.9		19		78	

At larger $\Delta\lambda$ we are too near the continuum for good results and at smaller $\Delta\lambda$, ϕ_λ/r_0 is not small compared to unity.

The values of σ in Table V-12 together with the observed values of $\tau_{\lambda 1}/\tau_{\lambda 2}$ yield the values of τ_2/τ_c and $\tau_{\lambda 2}/\tau_c$ in Table V-13. The resultant values of $B(\tau_c)$ are shown in Figure V-15 as crosses. Excellent agreement is obtained with the actual run of $B(\tau_c)$ (solid curve). Also shown in Figure V-15 are a few values of $B(\tau_c)$ obtained from the other line of the doublet (circles) using the same analytical technique. The two lines are in good agreement, which provides a check on the assumption that $S_\lambda = B$.

We have assumed in the analysis that τ_2/τ_c is independent of depth. A test of the validity of this assumption is provided by a comparison of the quantity $\frac{1}{2}(2270/\Delta\lambda_2)^2$, given in the last row of the table, with the entries in the rows above. If τ_2/τ_c were constant with depth the entries for the four values of $\tau_{\lambda 2}$ should have the same ratios for different $\Delta\lambda_2$ as the entries in the last row. This follows from the fact that σ is unity. In the row for $\tau_{\lambda 2} = 0.2$, successive pairs of values for τ_2/τ_c have ratios of 2.7, 1.8, 2.0, and 1.4 whereas those in the last row have ratios of 4.0, 2.4, 4.0 and 4.0.

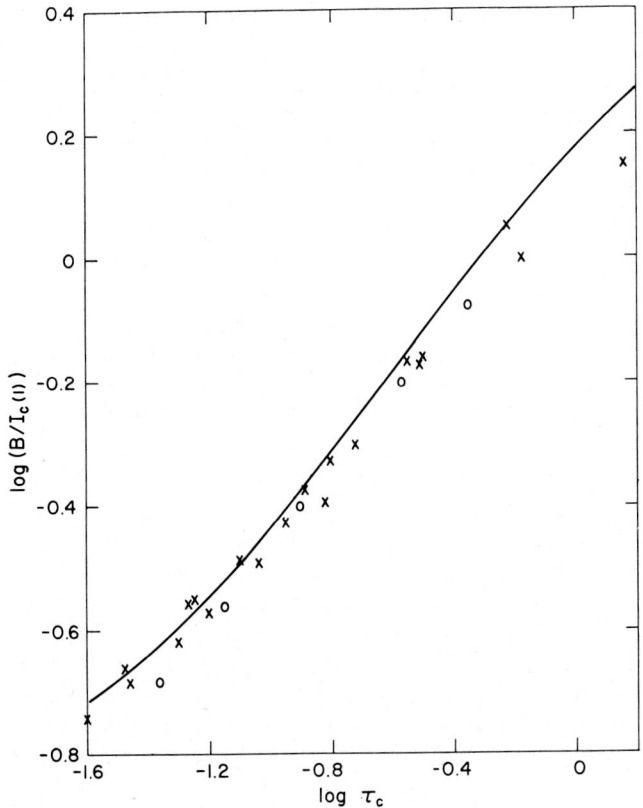

Fig. V-15. Comparison of derived values of B (points) to the actual values (curve).

In the full range between $\Delta\lambda_2 = 2270$ and 182, the ratio of the τ_2/τ_c values is only 13 at $\tau_{\lambda 2} = 0.2$ compared to a value of 156, which would be expected if τ_2/τ_c were independent of depth. This result provides direct empirical evidence that r_λ^{-1} is depth dependent.

To estimate the depth dependence of r_λ^{-1} we plot in Figure V-16 values of $\log \tau_2/\tau_c$ vs. $\log \tau_c$ from the data in Table V-13. The values of τ_2/τ_c at each $\Delta\lambda_2$ are multiplied by $(\Delta\lambda_2/1135)^2$ in order to place them on a common scale. Near $\tau_c = 0.1$ the plotted points are represented reasonably well by

$$\tau_{1135}/\tau_c = 7 \tau_c^{0.85}. \tag{V-74}$$

Equations (V-68) and (V-74) give

$$r_{1135}^{-1} = 13 \tau_c^{0.85}. \tag{V-75}$$

This result is plotted in Figure V-16 (broken curve) together with the actual values of r_{1135}^{-1} (solid curve). The agreement between $\log \tau_c = 0.7$ and -1.7 is very good.

Near $\tau_c = 1$ the slope of the $\log \tau_2/\tau_c$ vs. $\log \tau_2$ plot is negative. In this range we approximate the rather sparse data with

$$\tau_{1135}/\tau_c = 1.8 \tau_c^{-0.25}, \tag{V-76}$$

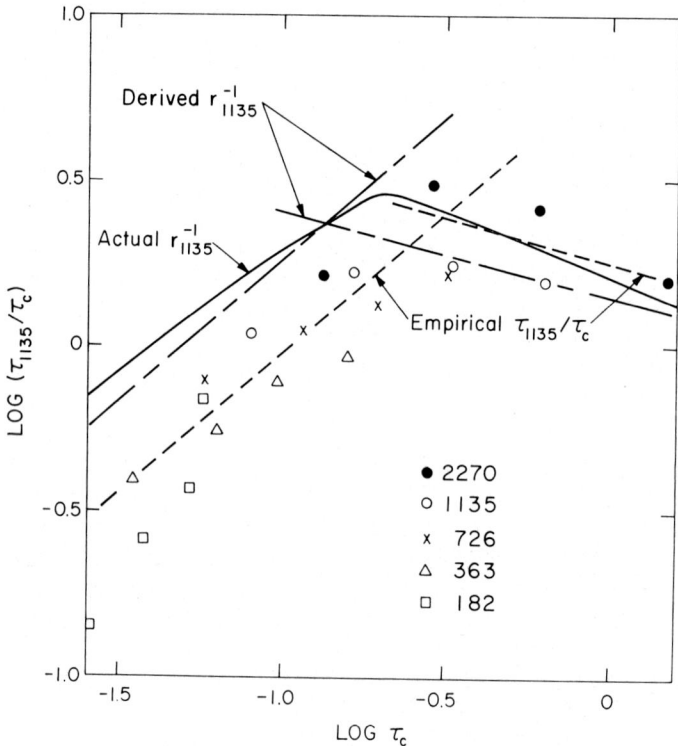

Fig. V-16. Illustration of the derivation of r_{1135}^{-1}.

which results in

$$r^{-1}_{1135} = 1.4\,\tau_c^{-0.25}. \tag{V-77}$$

These results are plotted in Figure V-16 also. Again the agreement with the actual values is reasonably good.

We could, of course, improve the results around $\tau_c = 1$ by using more of the line data in the far wings. However, we seek here only to illustrate the methodology and to demonstrate that it works reasonably well.

Having shown that r_λ can be recovered quite accurately from the data, we next ask what use we can make of r_λ. The quantities that go into r_λ are the damping parameter, a, the Doppler width, $\Delta\lambda_D$, the continuous opacity, κ_c, and the line opacity, κ_0. Both the Doppler width and κ_c can be determined from other data, so we may regard these quantities as being known. This leaves a and κ_0 as the remaining unknowns.

The damping parameter, a, is subject to theoretical calculation, of course. However, theoretical values of the damping parameters have not proven notably accurate in the past, and it is desirable to derive them from the data.

It is of interest to note from the model atmosphere in Table V-4 and from the line parameters in Tables V-4 and V-5 that a and r_0 each vary with depth much more rapidly than r_λ. Between $\tau_c = 1$ and 0.03 a and r_0 each decrease by over an order of magnitude. It is only because of the fact that r_λ is proportional to r_0/a that r_λ shows the gradual variations evidenced in Figure V-16. This same effect was noted for the solar Na D lines by Curtis and Jefferies (1965, 1967) and by Jefferies (1968).

So long as our analysis of the wing data is restricted to regions of the wing where σ is independent of wavelength, i.e., where $r_\lambda \propto \Delta\lambda^2$, there is little hope of separating a and κ_0. The line profile is completely determined, in this case, by the product of a and κ_0 and not by the two factors independently.

In the event that either a or κ_0 are known independently of the observational data, the remaining member of the pair can be determined from r_λ. If, for example, a is known κ_0 may be derived and from κ_0 the density of the absorbing atoms may be derived. Conversely, if the abundance of the element is known from other data κ_0 may be computed and the damping parameter a may be derived. Since r_λ is proportional to the product $a\kappa_0$, the error in one quantity is inversely proportional to the error in the other.

To obtain a and κ_0 separately it is necessary to extend the analysis into regions of the line wings where σ is no longer unity, i.e., where the Doppler core begins to overlap the line wings. Even here, however, we shall encounter difficulties that make the shoulder analysis difficult to use. In practice, as we shall see, the determination of r_λ in the line wings may still represent the best method for determining κ_0 even though we may be forced to use theoretical values for the damping constant. The same may be true of the damping constant, i.e., that we derive it from r_λ using theoretical values (or empirical values from other data) of κ_0. The value of κ_0 could be determined, for example, from the Doppler cores of weaker lines originating on the same ground state.

Interest in determining κ_0 stems from the desire to obtain accurate relative abundances for the various elements. If κ_0 is known and if the f-values are known, it is a simple matter to obtain the population of the lower state of the transition. This population must then be related to the total density of atoms of that specie, and this, in general, requires a knowledge of the excitation temperatures connecting the various levels. For resonance lines, of course, the lower level population gives directly the ground state population for that specie and it is not necessary to know excitation temperature. For this reason the correct analysis of the resonance lines is of primary importance for abundance determinations.

14. Analysis of Line Shoulders

As noted in the preceding section, to obtain numerical values for a and κ_0 separately we must consider regions of the profile near the transition from the Doppler core to the wing. It is only by virtue of having σ different from unity by some appreciable fraction that we are able to obtain a in addition to r_λ.

In the edges of the Doppler core we may approximate σ by

$$\sigma \approx \left(\frac{e^{-y_1^2} + a/\sqrt{\pi}y_1^2}{e^{-y_2^2} + a/\sqrt{\pi}y_2^2}\right) \frac{y_1^2}{y_2^2}. \tag{V-78}$$

In practice we will choose y_2 large enough that $e^{-y_2^2} \ll a/\sqrt{\pi}y_2^2$. Thus, we have for this case

$$a \approx \frac{\sqrt{\pi}y_1^2 e^{-y_1^2}}{(\sigma - 1)}. \tag{V-79}$$

Note that $a \propto (\sigma - 1)^{-1}$ and that we cannot expect to evaluate σ to better than about 5% accuracy. This equation demonstrates the need for σ to be appreciably different from unity before a will be accurately determined. Note also that $e^{-y_1^2}$ will usually be a small number, of the order of 10^{-3} to 10^{-4}. A value of 10^{-3} occurs at $y_1 \approx 2.6$ and a value of 10^{-4} occurs at $y_1 \approx 3.0$. Thus, to determine a to an accuracy of a factor of 2, y_1 must be known to an accuracy of about 10%. We cannot expect, therefore, to obtain precise values of a by this method. Unless y_1, hence $\Delta\lambda_D$, is very accurately known it is perhaps better to obtain a from the value of r_λ in the line wings together with an estimate of κ_0, even though the κ_0 estimate may be somewhat crude. Let us assume, however, that y_1 is accurately known in order to investigate other problems that may arise.

We show in Figure V-17a a plot of the theoretical values of σ versus y for two different values of a. The very rapid rise in σ near $y=3$ marks the transition to the Doppler core and can be identified with the shoulders on the line profiles in Figures V-10a and V-10b. When σ differs from unity we must expect that it will vary with depth. The larger σ is the more rapid the depth variation is likely to be. To facilitate the evaluation of σ it is desirable to restrict the depth variations, i.e., to work with values

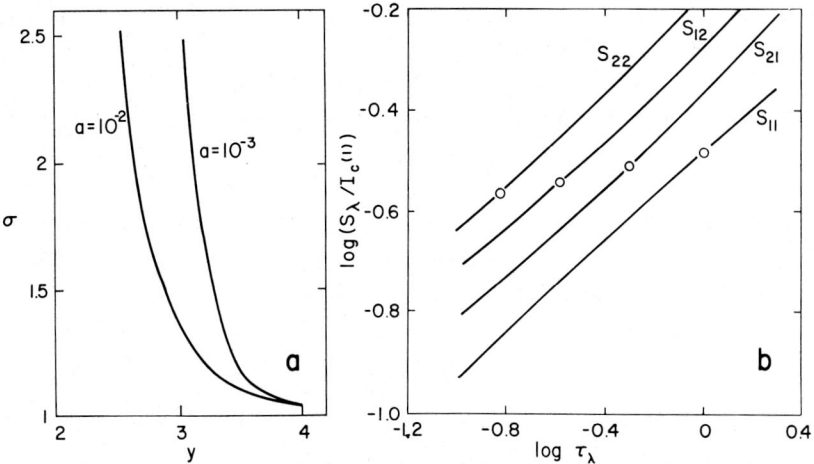

Fig. V-17. (a) – A plot of σ vs. y near the edge of the Doppler core and (b) – an illustration of the Curtis-Jefferies solution.

of σ that are not too large. This suggests that we choose $\Delta\lambda_1$ to be near the shoulders in the line profile rather than within the Doppler core, and we want this to be true even at small μ where the Doppler core is wider. By visual inspection of Figure V-10a we therefore choose $\Delta\lambda_1 = 102$ (adjacent data points are at $\Delta\lambda = 90.8$ and 114). We choose $\Delta\lambda_2 = 182$, which is large enough to make $e^{-y_2^2} \ll e^{-y_1^2}$ but small enough to ensure that $S_{\lambda 1}(\tau_{\lambda 1})$ overlaps $S_{\lambda 2}(\tau_{\lambda 2})$ sufficiently.

Since we expect that S_λ may be frequency dependent, it seems logical to attempt the Curtis-Jefferies method for evaluating σ. However, this method immediately encounters a difficulty in the particular case under consideration. The plot of $S_\lambda(\tau_\lambda)$ at $\Delta\lambda = 182$ in Figure V-14 shows that $d \log S_\lambda/d \log \tau_\lambda$ is of the order of 0.15 only. Throughout the entire range $0.125 \leqslant \tau_\lambda \leqslant 1$ in which S_λ is reliable S_λ changes by only a factor of 1.4. The Curtis-Jefferies evaluation of σ depends upon differences of the form $S_{11} - S_{21}$ and $S_{22} - S_{12}$, where the first index denotes the line and the second denotes the wavelength. Because of the small slope $d \log S_\lambda/d \log \tau_\lambda$, differences of this type will be some small percentage of S_{11}, and, consequently, will be highly inaccurate. In order for the Curtis-Jefferies method to give reliable results, the slope $d \log S_\lambda/d \log \tau_\lambda$ must be reasonably large, say, 0.4 or greater.

The particular case we are considering is further aggravated by the fact that r_λ is very small, as shown by the results in Table V-13 for $\Delta\lambda_2 = 182$. This has the consequence that S_{11} must be close to S_{21}, which requires that both S_{11} and S_{21} be accurately known. Since S_{11} and S_{21}, in fact, are not known to high accuracy, the Curtis-Jefferies method will not converge properly.

It is readily apparent from a visual inspection of the profiles for case B (Figure V-10b) that the slope $d \log S_\lambda/d \log \tau_\lambda$ is much larger in the line wings for this case than it is at corresponding values of $\Delta\lambda$ in case A (Figure V-10a). As an illustration of the Curtis Jefferies method, therefore, we apply it to case B. While it is true that the line source function differs somewhat for the two lines in case B, we are working far enough into the wings that $S_L \approx B$ for both lines.

Curves of $S_\lambda(\tau_\lambda)$ for $\Delta\lambda = 102$ and 182 are shown in Figure V-17b. Solutions of the Curtis-Jefferies equations for $r_\lambda =$ constant and for $\tau_{11} = 1$ give $\sigma \approx 1.5$ and $r_{11} \approx 0.05$. The values of τ_{ij} at which the solutions converge are marked in the figure. Although the slope $d \log S_\lambda / d \log \tau_\lambda$ is now of the order of 0.5, the accuracy in determining σ and r_{11} is poor. The differences $\log S_{11} - \log S_{21}$ and $\log S_{12} - \log S_{22}$ are approximately 0.02, which is too small to place much confidence in. The results are not unreasonable, however. At $\tau_{11} = 1$, $\log S_\lambda / I_c(1)$ is approximately -0.5. We note from Figure V-15 that $\log B/I_c(1)$ is -0.5 near $\log \tau_c = -1.1$. Thus $\log \tau_c / \tau_{11} \approx -1.1$, which is reasonably consistent with $\log r_{11} = -1.3$. Also, we have deliberately picked $\Delta\lambda_1$ such that it falls near the dividing point between the wing and the line core. This alone is enough to insure that σ will be near 2 since we expect the core term and the wing term to be contributing approximately equal amounts to $\phi_{\lambda 1}$ and since we have picked $\Delta\lambda_2$ so that only the wing term is contributing to $\phi_{\lambda 2}$.

Now that we have a specific value for σ at a specific wavelength, we can determine a by using Equation (V-79). However, $e^{-y^{12}}$ is an extremely rapidly varying function of $\Delta\lambda_D$. We have determined that τ_c is near the range 10^{-1} to 5×10^{-2}, and the Doppler widths in Figure V-13 extend only up to $\tau_c \approx 10^{-4}$, so they are of no help. We could obtain an approximate Doppler width from the relation $\Delta\lambda_D \approx \frac{1}{3} \Delta\lambda_s$, where $\Delta\lambda_s$ is the location of the line shoulder for the stronger line in case B and at $\mu = 1$. However, this is clearly not good enough since it gives only an average Doppler width from $\tau_0 = 1$ to $\tau_0 = 10^4$.

There appears, in fact, to be no way of empirically determining $\Delta\lambda_D$ accurately from these profiles at the desired depth, unless we were to synthesize the line profile accurately with a model atmosphere calculation. On the other hand, there are other spectral lines whose cores are formed in regions including $\tau_c = 10^{-1}$ and presumably we can determine $\Delta\lambda_D$ accurately from these lines. Let us assume that this is the case and that we have determined $\Delta\lambda_D$ to be 36.5 mA, which it is in the model atmosphere at $\tau_c = 10^{-1}$. Equation (V-79) then gives $a = 1.1 \times 10^{-2}$. The true value is $a = 7 \times 10^{-2}$. Thus, we have underestimated a by over a factor of 6.

The difficulties with the above determination of a are rather obvious. Neither σ nor τ_c are well determined. Since τ_c is not well determined $\Delta\lambda_D$ is not well determined even though we have used a known $\Delta\lambda_D$ at a known τ_c. Equation (V-79) is particularly sensitive to σ when σ is near unity. We could improve the situation in this regard by reducing $\Delta\lambda_1$ to, say, 90.8, which would increase σ. However, this would make r_{11} smaller and the uncertainty in r_{11} would be even larger. We cannot determine r_{11} by the Curtis-Jefferies method if it becomes small enough that the difference of $1 + r_{11}$ from unity is insignificant. From a practical point of view this means that we cannot determine r_{11} if it is less than about 0.05.

In retrospect, we have seen that although both σ and r were determined by using the data near the line shoulder, they were not determined accurately enough to give a reliable value for a. It may well be that in other situations the method would work better. Our experience suggests that it will work better if r_λ is larger. This suggests that we should work with lines whose shoulders are nearer the continuum.

The alternative to determining the damping parameter a from the line shoulders is to use the values of r_λ further in the wings of the line together with independent evaluations of κ_o, κ_c and $\Delta\lambda_D$. Both κ_o and κ_c enter r_λ linearly and $\Delta\lambda_D$ enters as the square. Thus, it is not necessary to know these quantities to high accuracy. Errors of 20 to 30%, and, in some cases, 100% ,are easily tolerated. It seems therefore that the best estimates of the damping parameters will come from the far wings of very strong lines formed by elements whose abundances are known.

15. Comments on Weak and Moderately Strong (Photospheric) Lines

It has long been argued by many astrophysicists that departures from LTE are of importance only in stellar chromospheres and coronas, in gaseous nebulae and in the interstellar medium and that within the photospheres of stars LTE is a valid approximation. As a consequence, analyses of weak and moderately strong lines have been dominated by the imposition of LTE. Many individual techniques of analysis have been employed with varying degrees of success. We shall comment on only a few of these, mainly those which have direct bearing on the mechanism of line formation.

Although it may seem to the uninitiated that analysis of weak and moderately strong Fraunhofer lines would have long since reached the point of diminishing returns, such is not the case. Much fruitful information has indeed been extracted from spectrum analyses based upon LTE. The line spectrum carries information on the chemical abundances of elements, on fluid motions, both random and systematic, on magnetic field strengths, and on temperature, pressure and gravity. Analyses of the spectra of stars have yielded useful information on each of these quantities and have provided much of the basic information upon which our knowledge of the universe rests. However, it must not be forgotten that nearly all of this information has been derived by relatively crude analytical techniques geared for mass production rather than accuracy.

In nearly all of the quantities mentioned above the uncertainties in the derived values are of the order of 100%, or larger. This point is vividly illustrated by the current state of confusion over the behavior of $\Delta\lambda_D$ in the solar photosphere. Presumably $\Delta\lambda_D$ is fixed by the motion of the atoms and should depend only upon the atomic weight of the element, the wavelength of the line, the height in the atmosphere and possibly upon direction. The myriads of solar lines together with their center-limb variations should provide far more information than is needed to fully describe the height and directional properties of $\Delta\lambda_D$. However, at the time of this writing we still do not know whether $\Delta\lambda_D$ increases or decreases with height or whether $\Delta\lambda_D$ is isotropic or anisotropic. Most authors agree that $\Delta\lambda_D$ is either isotropic and increasing with height or anisotropic and decreasing with height, but this is hardly consoling.

Why is it that after some four decades of intensive study we cannot answer such a simple question? The answer lies neither in the fundamental complexity of the problem nor in the lack of data. It lies rather in our failure to recognize that our techniques and assumptions have been crude and because they are crude they often lead to incorrect and conflicting conclusions.

Similar evidence of difficulties with analytical techniques and assumptions is found in the derivations of chemical abundances. Superficial analyses based upon equivalent widths of lines and curve-of-growth techniques may well give abundances that are approximately correct. When more refined data and techniques are used, however, one often finds that the abundance varies from center to limb and with the excitation potential of the line, indicating, in both cases, that the asumptions and techniques are faulty and that we have not yet derived an accurate abundance.

Solar spectroscopy is now sufficiently refined and computing techniques are sufficiently advanced to permit a careful analysis of the solar spectrum using accurate line profiles with their center-limb variations and more realistic theories of line formation. Even though we are still hampered by inadequate knowledge of the granular structure of the solar atmosphere and of the influence of such structure on the line profiles, we clearly are able to greatly improve upon our analysis of the line spectrum and, in so doing, to improve upon our knowledge of the solar structure. That such has not already been done is attributable to the relative newness of computing techniques of sufficient sophistication.

So far as mechanisms of line formation are concerned the fundamental distinction between weak lines and strong lines is simply in the relative values of r_0. Lines showing little or no saturation effects, i.e., those on the linear portion of the curve of growth, have values of $r_0 > 1$. Saturated lines on the shoulder of the curve of growth have values of r_0 generally between 1 and 10^{-4}. Such lines are formed below the solar temperature minimum and, strictly speaking, are photospheric in origin. More generally, however, the lines typically classed as photospheric have values of r_0 in excess of 10^{-2}.

For lines which have $r_0 > 10^{-2}$, we may safely assume that $\delta > \varepsilon$ (except in the case of forbidden lines, which are relatively rare). We recall that δ is defined by

$$\delta = \frac{2r_0}{\sqrt{\pi}} \int_0^\infty \frac{\phi_y}{\phi_y + r_0} \, dy$$

and that for $r_0 \gg 1$ $\delta = 1$. Also, for $a < 10^{-2}$ and $r_0 = 10^{-2}$ $\delta < 2r_0$. Thus 'photospheric' lines with $r_0 \geq 10^{-2}$ have values of δ between 1 and 0.02. The computations of S_L in Chapter III give $S_L = J_c$ when δ is near unity. For small δ, $S_L \approx \delta^{1/2} B$ at small τ_0 and S_L converges to B near $\tau_c = 1$. There is every reason to expect, therefore, that there are significant departures from LTE in photospheric lines, even the weak ones.

16. A Test for LTE Using Equivalent Widths

Altrock (1968) has developed a sensitive method for testing for departures from LTE using equivalent widths of weak and medium strong lines. The method gives abundances and relative levels of line formation and can be used for blended multiplets as well as individual lines. We shall present here only a brief outline of the method applied to weak lines.

Altrock's (1968) method is based upon the following assumptions:

$$B = a'' + b''\tau_c + c''\tau_c^2, \qquad (V\text{-}80)$$

$$S_L = a + b''\tau_c + c''\tau_c^2, \qquad (V\text{-}81)$$

$$r_v = r'_v \tau_c^\beta, \quad \beta < 1, \qquad (V\text{-}82)$$

and, in addition, that ϕ_v is Doppler in shape and that Δv_D is constant with depth and direction. Equations (V-80) and (V-81) allow S_L to differ from B at small values of τ_c but not at large values of τ_c. The ratio a/a'' can be derived from observational data and provides a sensitive test of the validity of the LTE assumption.

Equation (V-82) integrates to give

$$\tau_v = \int_0^{\tau_c} \frac{1 + r_v}{r_v} \, d\tau_c = \tau_c + \frac{\tau_c^{1-\beta}}{(1-\beta) r_v^1}. \qquad (V\text{-}83)$$

The equivalent width, $W(\mu)$, of a line is defined by

$$W(\mu) I_c(0, \mu) = \int_0^\infty (I_c(0, \mu) - I_v(0, \mu)) \, dv. \qquad (V\text{-}84)$$

Altrock (1968) finds after combining Equations (V-80), (V-81), (V-82), (V-83) and (V-1) with Equation (V-84) that

$$a/a'' = 1 - \frac{a'}{b' + \tfrac{1}{2}c'} \left[\frac{b''}{a''} + \frac{c''}{a''}(2 - \beta) \right], \qquad (V\text{-}85)$$

where the constants a', b', and c' are obtained by fitting the observed quantity $W(\mu) I_c(0, \mu)/I_c(0, 1)$ to an equation of the form

$$W(\mu) \frac{I_c(0, \mu)}{I_c(0, 1)} = a'\mu^{-\beta} + b'\mu^{1-\beta} + c'\mu^{2-\beta}. \qquad (V\text{-}86)$$

Appropriate values of β may be obtained from computations using a model atmosphere and the LTE assumption or by minimizing the residuals in fitting Equation (V-86) to the data. Numerical results obtained by Altrock (1968) suggest that the derived values of a/a'' are quite insensitive to the actual value of β.

Altrock (1968) found from an analysis of weak O I lines that a/a'' was near unity for all of the lines considered. However, because the wavelengths of the lines were such (≈ 6000 Å) that J_c is close to B at all depths in the atmosphere, Altrock was unable to distinguish between the results $S_L = B$ and $S_L = J_c$. For the stronger O I triplet near

$\lambda 7772$ (Altrock, 1968) and for weak lines of Si I and C I near $\tau 10\,700$ (Altrock, 1969), however, he found a/a'' to be appreciably less than unity, consistent with the assumption that S_L is less than or equal to J_c. The method has also been used by Canfield (1969) for weak lines of Ce II. He finds values of a/a'' varying from approximately 1.4 to 1.7 between $\lambda 4600$ and $\lambda 4000$ with a tendency for a/a'' to increase at shorter wavelengths. At these wavelengths J_c is greater than B and Canfield's results are consistent with the assumption that S_L lies between J_c and B as expected.

The Ce II lines studied by Canfield possess the interesting property of being in emission on the solar disk near the limb (Canfield, 1969). Many of the lines of other rare Earth elements apparently exhibit this behavior as inferred from eclipse data (Menzel, 1931) and a few lines of rare Earth's are observed in emission in the wings of the H and K lines of Ca II across the entire solar disk (f., Jensen and Orrall, 1963). For the Ce II lines the position of reversal from absorption to emission moves progressively inward from the limb as the wavelength decreases. Near $\lambda 4000$ the reversal occurs at about 2.5" inside the limb as opposed to about 1" near $\lambda 4600$.

There is no known mechanism for producing the observed emission reversals on the disk if $S_L = B$. The reversal clearly demands that $S_L > B$ at values of $\tau_c < 0.1$, which is consistent with the conclusion, based upon the inferred values of a/a'', that S_L lies between J_c and B. Both the position of the reversal and its wavelength dependence are in quantitative agreement with this conclusion (Canfield, 1969).

It is of particular interest that the emission reversals on the solar disk occur in rare-earth elements since this demonstrates quite clearly that a multiplicity of energy levels is not, by itself, a sufficient condition for LTE. The reason that the emission reversals appear to be most often associated with rare earth elements may be fortuitous. Such reversals should occur only in relatively weak lines and only in the violet end of the spectrum since it is only for these conditions that we expect $S_L > B$ near $\tau_0 = 1$. Also, the result $S_L = J_c$ may reasonably be expected to occur more frequently in lines of low excitation potential than in lines of high excitation potential, which are strongly interlocked to the photoionization continuum. Thus, it is perhaps the fact that the rare-earth elements are of low abundance and have low lying transitions in the violet end of the spectrum that accounts for the association of the reversal with these elements. In this connection, we note that Menzel's (1931) original classification of lines that extend onto the solar disk in emission includes several lines of the iron group elements and of a few other elements as well.

Canfield (1971) has shown that weak emission lines observed in the wings of the H and K lines of Ca II may result from interlocking effects. These weak emission lines are formed at small values of τ_c due to the added wing opacity of the Ca II lines. An emission line will occur provided $\int J_\nu \Phi_\nu \, d\nu$ is greater than B either in the line in question or in a strongly interlocked line. For weak lines within the wings of the Ca II lines we expect $\int J_\nu \Phi_\nu \, d\nu$ to be close to B. However, interlocked lines lying in the violet (but not within H and K) will experience $\int J_\nu \Phi_\nu \, d\nu > B$ at small τ_c provided J_ν is near J_c (i.e. the line is weak). Thus, weak emission lines in the wings of H and K are expected for atoms with strongly interlocked, weak lines, which is typical of rare earths.

17. Comparisons of Empirically Derived Line Source Functions to the Continuum Source Function

Present day knowledge of the solar photospheric model is thought to be sufficiently good to allow meaningful comparisons between source functions derived from line data and the source function derived from continuum data. Although such comparisons, in the past, have led to conflicting claims, on the one hand that $S_L = B$ and on the other that $S_L \neq B$, the $B(\tau)$ model is approaching the stage of being sufficiently well established to permit a conclusive comparison.

A meaningful comparison of $S_L(\tau)$ to $B(\tau)$ depends upon the correct evaluation of several factors. From the continuum data it is possible, in principle, to determine $B_\nu(\tau_c)$ through the use of limb-darkening data at different frequencies. To be sure, the last word has not been said as to exactly how such data should be analyzed nor as to the actual form of $B_\nu(\tau_c)$. Nevertheless, the differences between the various derivations of $B_\nu(\tau_c)$ appear to be considerably smaller than some of the differences found between $S_L(\tau_c)$ and $B_\nu(\tau_c)$. We shall assume, therefore, that $B_\nu(\tau_c)$ is known.

To arrive at a comparison of $S_L(\tau_c)$ with $B_\nu(\tau_c)$ it is first necessary to derive $S_\nu(\tau_\nu)$ from line data and then to convert τ_c to τ_ν and S_ν to S_L. If, indeed, all energy level populations are given by their LTE values and if $\Delta\lambda_D$ is known as a function of depth such conversions are relatively straightforward. On the other hand, if there are significant departures from LTE or if $\Delta\lambda_D$ is not accurately known both conversions are difficult and are subject to error. Thus, the discussion of comparisons between $S_L(\tau_c)$ and $B_\nu(\tau_c)$ generally reduces to the simple point of whether $S_L = B$ or $S_L \neq B$.

There is general agreement that in the cores of the very strong Fraunhofer lines S_L is less than B by a significant amount. The current discussion thus centers around the extent to which S_L and B differ in the weaker lines formed in the photospheric layers.

Pecker and his co-workers have carried out an extensive analysis of several elements for the purpose of comparing $S_L(\tau_c)$ to $B(\tau_c)$. The early analyses of Ti I by Pecker (1959) and of Cr I by Pecker and Vogel (1960) were based upon the rather crude data in the Utrecht atlas for the center of the solar disk. Corrections to the Atlas data are necessary in order to obtain the true central intensities of lines and such corrections are not always reliable. Furthermore, there were disagreements, at that time, about $B_\nu(\tau_c)$, about relative f-values and about the values of $\Delta\lambda_D$. The latter difficulty still persists as strongly as ever but the former two are less controversial.

Pecker (1959) and Pecker and Vogel (1960) converted τ_ν to τ_c by assuming LTE for both the derivation of the abundance and for the relative populations of energy levels. They found, respectively, that T_{ex}, obtained by equating S_L to $B(T_{ex})$ exceeded the continuum temperature, T, for the Ti I lines and fell below T for the Cr I lines for values of τ_c between about 10^{-2} and unity. Differences between T_{ex} and T amounted to several hundred degrees in some cases.

The analysis of Ti I data was extended to include center-limb variations in central line intensities by Lefevre and Pecker (1961). They confirmed the earlier conclusion

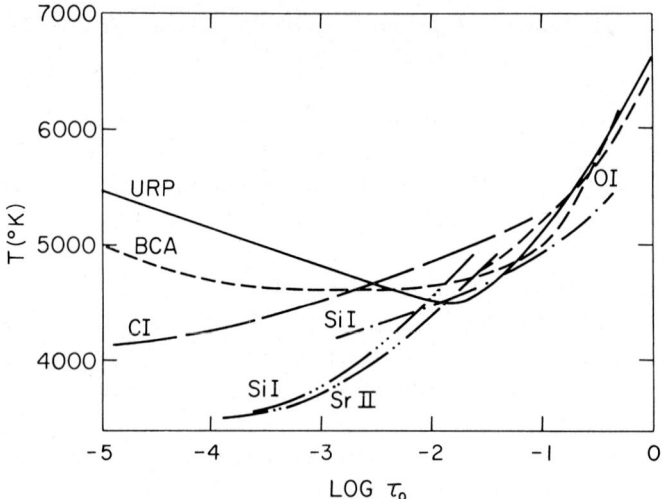

Fig. V-18. Comparison of empirically derived source functions to the Utrecht Reference Photosphere (URP) and the Bilderberg continuum atmosphere (BCA) (de Jager and Neven (1968); courtesy *Solar Physics*).

that T_{ex} exceeded T_e and found, in addition, that plots of T_{ex} vs. τ_c differed significantly from line to line to produce the so-called 'fishbone effect'.

A still further refinement of the Ti I analysis to compute the conversion of τ_v to τ_c self-consistently with the derived values of T_{ex} was made by Kandel (1960) for one multiplet. Solutions for $T_{ex}(\tau_c)$ were carried out iteratively, with and without microturbulence included in $\Delta\lambda_D$. Kandel found that the differences between T_{ex} and T were even larger than found by Pecker, and, in addition, that the self-consistent abundance of Ti is larger by a factor four (with microturbulence) to seven (without microturbulence) than the abundance obtained assuming LTE.

De Jager and Neven (1967, 1968) have analyzed infrared lines of C I, Si I, O I, and Sr II by a method similar to those used by Lefevre and Pecker and by Kandel. De Jager and Neven use center-limb data on central intensities, iterate once on the abundance to correct for differences between T_{ex} and T and convert τ_v to τ_c assuming LTE. They find differences in $T_{ex}(\tau_c)$ for different elements and for different multiplets of a given element and conclude that $S_L \neq B$ for most of the lines. Their results are generally consistent with those of Pecker and co-workers with the exception that the differences between T_{ex} and T are not as large as those found by Pecker and co-workers.

The data used by de Jager and Neven were obtained with high spatial resolution and care was taken to eliminate scattered light. Thus, their data are better than the data that were available to Pecker and his coworkers. Some of the results found by de Jager and Neven are summarized in Figure V-18. The curves labeled URP and BCA represent the Utrecht Reference Photosphere and the Bilderberg continuum atmosphere (Gingerich and de Jager, 1968).

Without attempting to evaluate the many interesting implications and controversies stimulated by these results, we turn to a consideration of the work by Holweger (1967), which is much more thorough, in some respects, than the work cited in the preceding. Holweger's analysis differs from those cited earlier in that he does not attempt to compare a line model giving $S_L(\tau_c)$ to a continuum model giving $B(\tau_c)$. He chooses instead to find a single model which best represents both $S_L(\tau_c)$ and $B(\tau_c)$. He assumes LTE, consistently, in the determinations of abundances, the conversion of τ_v to τ_c and in setting $S_L \equiv B$.

The data used by Holweger include center-limb observations of continuum intensities for $\lambda\lambda 4000$ to 16000 and $\lambda\lambda 0.1$ to 1 mm and center-limb profiles for some 900 Fraunhofer lines. In the deeper layers of the photosphere, $\tau_c \geqslant 0.1$, the temperature is determined primarily from the continuum data. In higher layers the temperature is determined from the line data by equating the observed central intensity of the line at disk position μ to $B(\tau)$ at $\tau_v = \mu$. The microturbulent velocity is deduced from the line profiles and is used in relating τ_v to τ_c.

Somewhat to the surprise and consternation of those who have criticized the LTE analysis of Fraunhofer lines, Holweger does find, in fact, that a single model gives an acceptable representation of both the continuum and the lines, with the notable exception of the resonance lines. The fact that Holweger is able to fit both the continuum and some 900 line profiles with a single $T(\tau_c)$ model is impressive indeed. Taken at first impression such a result seems somewhat unexpected if departures from LTE are important. One would expect, in the absence of LTE, that a moderately wide range of excitation temperatures would exist at any given depth. On the other hand, we have seen repeatedly that conclusions based upon consistency alone are often incorrect. It behooves us, therefore, to look more closely at the Holweger model.

Figure V-19 exhibits a plot of Holweger's $T(\tau_c)$ model together with the Harvard-Smithsonian (Gingerich, private communication) reference model and, for comparison, the model in Table IV-1. The Harvard-Smithsonian model is taken as representative of recent models derived from continuum data alone. It does not differ appreciably from Elste's (1968) model for $\tau_c > 10^{-3}$, which is based upon similar data. At values of τ_c near 10^{-2} to 10^{-3} Holweger's model gives temperatures approximately 200° higher than the Harvard-Smithsonian model. If one considers the Holweger model as a model of $S_L(\tau_c)$ and the Harvard-Smithsonian model as a more nearly correct $B(\tau_c)$ model, the disparity between $S_L(\tau_c)$ and $B(\tau_c)$ for the two models is very similar to the disparity found by Pecker and co-workers for Ti I and by de Jager and Neven for C I. We must, in all fairness, admit that the $T(\tau_c)$ model is relatively uncertain at optical depths less than about 10^{-1} and it is difficult to argue in support of apparent differences between T_{ex} and T of only 200°. Nevertheless, we should note that differences between $T(\tau_c)$ and $T_{ex}(\tau_c)$ of the type indicated in Figure V-19 have been found by other workers as noted above.

At optical depths less than about 10^{-4} the disparity between Holweger's model and current models of the low chromosphere is very large. Eclipse data showing strong Balmer continuum emission in the low chromosphere where $\tau_c \approx 10^{-5}$, for example, are incompatible with Holweger's suggestion that $T = 4200°$ at these heights (cf., Tho-

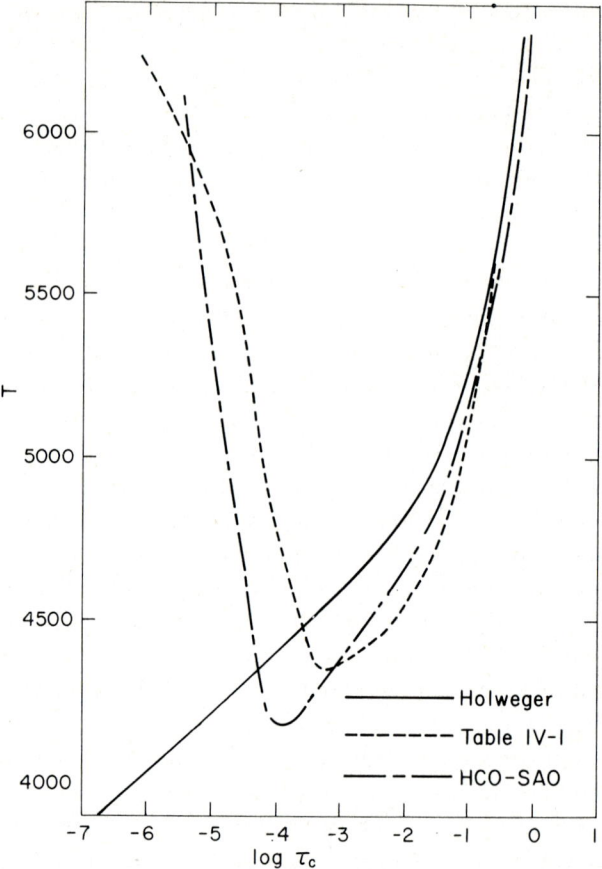

Fig. V-19. Comparison of Holweger's source function model to the temperature model used in this chapter and the Harvard-Smithsonian Reference Atmosphere.

mas and Athay, 1961). Furthermore at a wavelength of 1 mm, which is expected to originate near $\tau_c \leqslant 10^{-5}$, the radiation temperature of the Sun is of the order of 5500° or higher. This is compatible with the Balmer continuum data but not with Holweger's model. Also, the XUV continuum data for the Sun clearly indicate a temperature minimum near $\tau_c = 10^{-4}$ with a minimum value of T near 4300 to 4400° (Parkinson and Reeves, 1969). Nowhere in the continuum are temperatures as low as 4000° observed yet many strong Fraunhofer lines show radiation temperatures near 4000°, as indicated by Holweger's model for $\tau_c < 10^{-5}$.

It seems imminently clear that Holweger's model for $\tau_c \leqslant 10^{-5}$ is in sharp disagreement with continuum data at mm and XUV wavelengths and with Balmer continuum data observed at eclipse. At these heights the model is clearly a model for $T_{ex}(\tau_c)$ rather than $T(\tau_c)$ and it is clear that T_{ex} differs from T by a large amount. Thus, the consistency argument, which is the only supporting argument for the Holweger model, fails once again. The problem, then, is to explain why a single curve of $T_{ex}(\tau_c)$ apparently explains so well so many lines.

If we accept that the Holweger model does not give $T(\tau_c)$ for $\tau_c \leqslant 10^{-5}$, there is no good reason to suppose that it gives $T(\tau_c)$ for $\tau_c > 10^{-5}$. Although, as we indicated earlier, our first impression is that T_{ex} should not be uniquely defined at a given depth, there are circumstances under which such a result is expected. We recall from Figure III-1b, for example, that for $r_0 > \varepsilon$ the source function for $\tau_0 > 10^{-1}$ is essentially independent of the strength of the line (as measured by r_0) at a given value of τ_c. If, in fact, we pick from Figure III-1b the values of the source functions at $\tau_0 = 1$ and $\tau_0 = 0.3$, the resulting points do define a single curve. Since this mimics, qualitatively, the result found by Holweger, we shall pursue this point further.

The analogy between the results in Figure III-1b and Holweger's results is misleading in one sense. The uniqueness of $S_L(\tau_c)$ in Figure III-1b results because we have used the same $B(\tau_c)$ for each of the computations and it is only the ratio $B(\tau_c)/S_L(\tau_c)$ that is unique. In reality of course B is a function of wavelength and thus $B_\lambda(\tau_c)$ has a different slope at each wavelength. As the slope of $B_\lambda(\tau_c)$ changes there will be changes in the slope of $S_L(\tau_c)$. However, there are other considerations which tend to offset this effect.

$S_L(\tau_c)$ does not respond very much to changes in $B_\lambda(\tau_c)$ at values of $\tau_0 \leqslant 1$ and Holweger has used only the values of S_L at $\tau_0 = 1$ and $\tau_0 = 0.3$. Also, of the lines that Holweger analyzes approximately 25% lie between $\lambda 4000$ and $\lambda 5000$, 45% between $\lambda 5000$ and $\lambda 6000$ and 30% between $\lambda 6000$ and $\lambda 7000$. Thus, the distribution of lines is strongly and rather symmetrically peaked near $\lambda 5500$. The spread in the wavelengths of the lines would introduce in Figure III-1b an apparent 'scatter' in the values of S_L at $\tau_0 = 1$, but the concentration of lines between $\lambda 5000$ and $\lambda 6000$ would tend to concentrate the points near the center of the distribution.

It is clear from the results in Chapter III that any attempt to explain the difference between $T_{ex}(\tau_c)$ and $T(\tau)$ indicated in Figure V-19 with a two-level atom will fail. In all such cases we expect $T_{ex} \leqslant T$ rather than the other way around. In several of the illustrations in Chapter IV using more complex atoms we found cases where $T_{ex} > T$ by a significant amount. An increase in T from 4500° to 4700° at $\lambda 5000$ represents only a 30% increase in B, which is not difficult to produce in specific situations. The variety of results in Chapter IV and the large number of cases in which $S_L < B$ at $\tau_0 \leqslant 1$ clearly indicates, however, that there is no ready explanation of Holweger's result. The problem is a challenging one and demands further attention.

Finally we note that Holweger's results are at considerable disparity with results found by other workers. The curves in Figure V-18, for example show that for some of the infrared lines T_{ex} reaches values as low as 4000° at $\tau_c \approx 10^{-3}$ where Holweger's results give 4600°. A satisfactory explanation of these differences has not been given.

18. Analysis and Restoration of Line Profile Data

Since astrophysics has now advanced to the stage where careful, accurate analysis of line profiles is both possible and necessary, it is mandatory that the data should be correspondingly accurate. It is not sufficient simply to photograph the spectrum, microphotometer the line profiles and apply a crude correction for instrumental broadening. The raw data obtained from photographic film or from photoelectric scanners inevitably contains noise and distortion which may seriously degrade the information content unless the data are properly handled. Powerful, convenient techniques for restoration of noisy, distorted data are available and should be utilized whenever it is practical to do so (cf., Brault and White, 1971).

References

Altrock, R. C.: 1968, *Solar Phys.* **5**, 260.
Altrock, R. C.: 1969, *Solar Phys.* **7**, 3.
Athay, R. G.: 1961, *Astrophys. J.* **134**, 365.
Athay, R. G.: 1965, *Astrophys. J.* **142**, 755.
Athay, R. G.: 1970a, *Solar Phys.* **12**, 175.
Athay, R. G.: 1970b, *Solar Phys.* **11**, 347.
Athay, R. G. and Canfield, R. C.: 1969, *Astrophys. J.* **156**, 695.
Athay, R. G. and Canfield, R. C.: 1970, *Spectrum Formation in Stars with Steady State Extended Atmospheres*, NBS Special Publication 332, p. 65.
Athay, R. G. and Skumanich, A.: 1968a, *Solar Phys.* **4**, 176.
Athay, R. G. and Skumanich, A.: 1968b, *Solar Phys.* **3**, 181.
Athay, R. G. and Skumanich, A.: 1968c, *Astrophys. J.* **152**, 141.
Athay, R. G. and Skumanich, A.: 1968d, *Astron. J.* **73**, S2.
Athay, R. G. and Thomas, R. N.: 1958, *Astrophys. J.* **127**, 96.
Athay, R. G., Brault, J. W., Lites, B., and White, O. R.: 1971, *Solar Phys.*, in press.
Auer, L. H. and Mihalas, D.: 1969a, *Astrophys. J.* **156**, 157.
Auer, L. H. and Mihalas, D.: 1969b, *Astrophys. J.* **156**, 681.
Auer, L. H. and Mihalas, D.: 1970, *Astrophys. J.* **160**, 233.
Avrett, E. H.: 1965, *Proceedings 2nd Harvard-Smithsonian Conference on Stellar Atmospheres*, Smithsonian Astrophys. Obs. Space Report 174.
Banos, G.: 1968, *Ann. Astrophys.* **31**, 501.
Beebe, H.: 1971, *Solar Phys.* **17**, 304.
Beebe, H., and Johnson, H. R.: 1968, *Resonance Lines in Astrophysics*, National Center for Atmospheric Research, Boulder.
Brault, J. W. and White, O. R.: 1971, in press.
Canfield, R. C.: 1969, *Astrophys. J.* **157**, 425.
Canfield, R. C.: 1971, *Astron. Astrophys.* **10**, 64.
Chamaraux, P.: 1967, *Ann. Astrophys.* **30**, 67.
Corliss, C. H. and Tech, J. L.: 1968, *National Bureau of Standards Monograph* **108**, Washington, D.C.
Cuny, Y.: 1967, *Ann. Astrophys.* **30**, 143.
Cuny, Y.: 1968, *Resonance Lines in Astrophysics*, National Center for Atmospheric Research, Boulder.
Curtis, G. W. and Jefferies, J. T.: 1965, *Proceedings of 2nd Harvard-Smithsonian Conference on Stellar Atmospheres*, Smithsonian Astrophys. Obs. Space Report 174.
Curtis, G. W. and Jefferies, J. T.: 1967, *Astrophys. J.* **150**, 1061.
De Jager, C.: 1952, *Rech. Astron. de L.Obs. d'Utrecht* Vol. XIII, Part I.
De Jager, C. and Neven, L. 1967, *Solar Phys.* **1**, 27.
De Jager, C. and Neven, L.: 1968, *Solar Phys.* **3**, 159.
Dumont, S.: 1967a, *Ann. Astrophys.* **30**, 421.
Dumont, S.: 1967b, *Ann. Astrophys.* **30**, 861.
Dunn, A. R. and Olson, E. C.: 1971, *Solar Phys.* **16**, 272.
Elste, G.: 1968, *Solar Phys.* **3**, 106.
Finn, G. D., Muggleston, D., and Young, R. L.: 1967, *Monthly Notices Roy. Astron. Soc.* **137**, 445.
Gingerich, O. and de Jager, C.: 1968, *Solar Phys.* **3**, 5.

Goldberg, L.: 1958, *Astrophys. J.* **127**, 308.
Hearn, A. G.: 1967a, *Monthly Notices Roy. Astron. Soc.* **135**, 305.
Hearn, A. G.: 1967b, *Monthly Notices Roy. Astron. Soc.* **136**, 417.
Holweger, H.: 1967, *Z. Astrophys.* **65**, 365.
Huang, S. S. and Struve, O.: 1960, in J. L. Greenstein (ed.), *Stellar Atmospheres*, Univ. of Chicago, Press, Chicago, Illinois.
Hummer, D. G. and Rybicki, G.: 1968a, *Astrophys. J. Letters* **153**, L107.
Hummer, D. G. and Rybicki, G.: 1968b, *Resonance Lines in Astrophysics*, National Center for Atmospheric Research, Boulder, Colorado.
Jefferies, J. T.: 1968, *Spectral Line Formation*, Blaisdell, Waltham, Mass.
Jefferies, J. T. and Thomas, R. N.: 1960, *Astrophys. J.* **131**, 695.
Jensen, E. and Orrall, F. Q.: 1963, *Publ. Astron. Soc. Pacific* **75**, 162.
Johnson, H. R.: 1964, *Ann. Astrophys.* **27**, 695.
Kalkofen, W.: 1970, *Spectrum Formation in Stars with Steady Extended Atmospheres*, NBS Special Pub. 332, p. 120.
Kandel, R.: 1960, *Ann. Astrophys.* **23**, 995.
Lefevre, J. and Pecker, J.-C.: 1961, *Ann. Astrophys.* **24**, 238.
Linsky, J. L.: 1968, Smithsonian Astrophys. Obs. Spec. Report No. 274.
Linsky, J. L. and Avrett, E. H.: 1970, *Publ. Astron. Soc. Pacific* **82**, 169.
Menzel, D. H.: 1931, *Publ. Lick Obs.* **17**, 1.
Mihalas, D. and Auer, L. H.: 1970a, *Astrophys. J.* **160**, 1161.
Mihalas, D. and Auer, L. H.: 1970b, *Astrophys. J.* **161**, 1129.
Miyamoto, S.: 1953a, *Z. Astrophys.* **31**, 282.
Miyamoto, S.: 1953b, *Publ. Astron. Soc. Japan*, **5**, 142.
Miyamoto, S.: 1954a, *Publ. Astron. Soc. Japan*, **6**, 140.
Miyamoto, S.: 1954b, *Publ. Astron. Soc. Japan*, **6**, 196.
Morton, D. L. and Widing, K. G.: 1961, *Astrophys. J.* **133**, 596.
Parkinson, W. H. and Reeves, E. M.: 1969, *Solar Phys.* **10**, 342.
Pecker, J.-C.: 1959, *Ann. Astrophys.* **22**, 499.
Pecker, J.-C. and Vogel, L.: 1960, *Ann. Astrophys.* **23**, 594.
Roddier, F.: 1965, *Ann. Astrophys.* **28**, 478.
Schoolman, S.: 1969, Thesis Univ. of Colorado, Boulder.
Thomas, R. N. and Athay, R. G.: 1961, *Physics of the Solar Chromosphere*, Interscience, New York.
Unno, W.: 1959a, *Astrophys. J.* **129**, 375.
Unno, W.: 1959b, *Astrophys. J.* **129**, 388.
Waddell, J. H. III.: 1962, *Astrophys. J.* **136**, 231.
Waddell, J. H. III.: 1963, *Astrophys. J.* **137**, 1210.
White, O. R.: 1963, *Astrophys. J.* **137**, 1217.

CHAPTER VI

TOTAL INTENSITIES OF LINES

1. Introduction

In many situations in astrophysics we are limited to data giving only the total energy emitted or absorbed by a spectral line with little or no information concerning the shape of the line. This is particularly true for stellar spectra and for present data for the solar XUV spectrum. It is useful in these situations to seek methods for relating the total energy, or equivalent width, of a spectral line to the thermodynamic and chemical properties of the atmosphere. Such analyses cannot be expected to yield the same quality or quantity of information as the line profiles, but, correspondingly, analyses of equivalent widths can be reduced to much simpler form than the analyses of profiles. Thus, in working with equivalent widths we sacrifice information but we gain simplicity. The added simplicity permits the handling of much more data, which helps to offset the loss of information in individual lines.

Extensive discussions of the equivalent widths of absorption lines and their interpretation are available in most text books on stellar atmospheres, astrophysics or spectrum analysis and will not be repeated here. Instead, we shall restrict our discussion to the relation of the preceding chapters to analyses of this type. For a discussion of the normal interpretation of equivalent widths of absorption lines using the standard curve-of-growth techniques the reader is referred to Aller (1960).

2. Curve-of-Growth for Absorption Lines

The concept of a curve-of-growth normally implies the use of many spectral lines from different levels of excitation and, in some cases, ionization. It is implicit in the use of such curves that within a given atomic constituent the relative populations of energy levels can be described by a single temperature parameter, T_{ex}, which, in fact, almost universally is set equal to the kinetic temperature, T.

In the deeper layers of the photosphere we have seen that T_{ex} is close to T. Formally, we find that $S_L = J_c$ is a better solution for faint lines than is the solution $S_L = B_c$ However, near $\tau_c = 1$ J_c and B differ by only a small amount. For this reason, the approximation $T_{ex} = T$ may be reasonably accurate on the linear portion of the curve-of-growth, i.e., for faint lines formed near $\tau_c = 1$.

On the other hand, for strong lines formed higher in the photosphere T_{ex} is clearly not equal to T. In this situation the concept of a curve-of-growth loses its normal meaning, as noted by Jefferies (1968). Instead of a single curve-of-growth there is now a continuous family of such curves and each spectral multiplet may lie on a different curve. The discovery by Holweger (1967) that a very large number of solar Fraunhofer lines are well represented by a single curve of $T_{ex}(\tau_c)$ lends support to the idea that the family of curves-of-growth, at least in the solar case, are relatively closely packed around a single curve and that the concept of a single curve-of-growth for a large number of solar lines may have more meaning than we would otherwise suspect. It should be noted, however, that Holweger's result specifically excludes the resonance lines, which implies that T_{ex} for these lines is substantially different from T_{ex} for the subordinate lines. Thus, even though it may turn out that many solar lines can be treated with a single curve-of-growth and that excitation temperatures for different multiplets may agree it is still not permissible to use these excitation temperatures to derive ground state populations from the populations of excited levels. Since this is a commonly used procedure in abundance determinations, we must regard abundances derived by this method as almost certainly in error. An error of 10% in T_{ex} at $T_{ex} \approx 5000°$ amounts to an abundance error of a factor of 2 if the abundance is derived from excited levels lying 3 eV above the ground state.

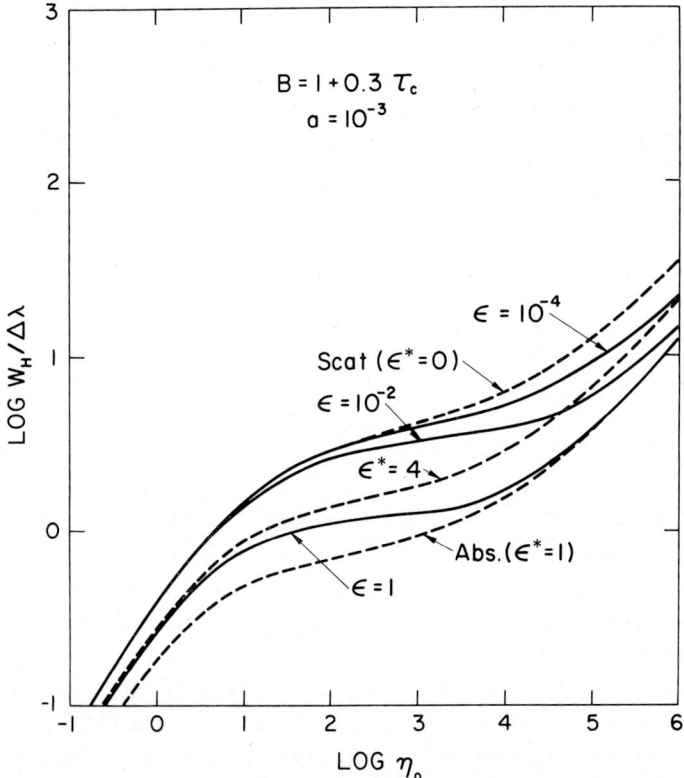

Fig. VI-1. Curves-of-growth for $\beta = 0.3$ and $a = 10^{-3}$. The dashed curves were computed by Wrubel (1949, 1950) (Athay and Skumanich, 1968, courtesy *Astrophysical Journal*, copyright 1968 by the University of Chicago. All rights reserved).

An error of 10% in the ionization temperature at $T_{ex} \approx 5000$ gives a factor of 5 error in ionization for an ionization potential of 7 eV. Errors in T_{ex} of this order are easily possible.

As we have noted, lines on the linear portion of the curve-of-growth may be reasonably well represented by the LTE approximation. That this is not strictly true is strongly attested by the emission reversals in rare-earth lines near the limb (Menzel, 1931; Canfield, 1969) and by such weak lines as λ 10830 of He I, which is chromospheric in origin. Also, we note that the argument for LTE on the linear portion of the curve-of-growth is based on the concept that the lines are formed near $\tau_c = 1$. Near the limb, of course, this is not true. The weak lines are formed near $\tau_c = \mu$ and at values of τ_c near 10^{-1} departures from LTE for even weak lines may be substantial.

We have noted in earlier chapters that the wings of strong lines also are likely to be in LTE. Thus, lines whose equivalent widths are dominated by the wings may again be explainable on an LTE basis. The lines whose equivalent widths will be most affected by departures from LTE are those that are formed reasonably high in the atmosphere but which do not have well developed wings. These are the lines whose Doppler cores are saturating, i.e., the lines on the shoulder of the curve-of-growth.

This effect is illustrated in Figures VI-1 and VI-2 (Athay and Skumanich, 1968). The ordinate in these figures is the equivalent width measured in the net flux from the star divided by the Doppler width $\Delta\lambda_D$ and the abscissa is defined by $\eta_0 = r_0^{-1} =$ constant. The dashed lines shown in the figures are curves-of-growth computed by Wrubel (1949, 1950) for coherent scattering with ε^* defined by

$$S_L = (1 - \varepsilon^*) J_\nu + \varepsilon^* B. \qquad \text{(VI-1)}$$

Thus, ε^* is not strictly equivalent to ε, which is defined by

$$S_L = \frac{\int J_\nu \Phi_\nu \, d\nu + \varepsilon B}{1 + \varepsilon}. \qquad \text{(VI-2)}$$

With the exception that J_ν in Equation (VI-1) is replaced by $\int J_\nu \Phi_\nu \, d\nu$ in Equation (VI-2), $\varepsilon^* = \varepsilon/(1+\varepsilon)$.

Note in Figure VI-1 that the solid curves merge with the pure absorption curve, $\varepsilon^* = 1$, for large values of $\log \eta_0$ even for small ε. This results, as indicated earlier, from the fact that the line wings tend to be formed in LTE. The reason for the failure of Wrubel's curves to approach the pure absorption curve at large $\log \eta_0$ is not readily apparent and evidently reflects an error in his computations.

On the linear portion of the curve-of-growth in Figure VI-1, $\log \eta_0 \leqslant 1$, we note that the curves which include some scattering do not coincide with the pure absorption curve. This contradicts our earlier statement that this portion of the curve-of-growth could perhaps be treated as in LTE. However, note that the slope $dB/d\tau_c$ in Figure VI-1 is only 0.3, which is an exaggerated case. For this small value of $dB/d\tau_c$, J_c is substantially less than B even for $\tau_c \approx 0.3$ to 1. Such low values of $dB/d\tau_c$ occur in

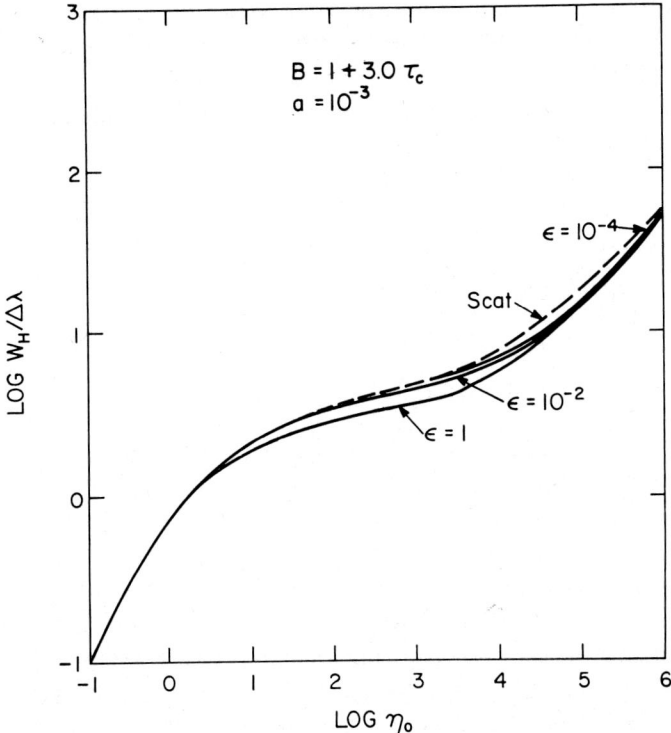

Fig. VI-2. Curves-of-growth for $\beta = 3$ and $a = 10^{-3}$ (Athay and Skumanich, 1968, courtesy *Astrophysical Journal*, copyright 1968 by the University of Chicago. All rights reserved).

stars only in the far infrared. Throughout the visual and violet spectral regions in stars $dB/d\tau_c$ is of order unity or greater. In Figure VI-2, where $dB/d\tau_c = 3$, the linear portions of the curve-of-growth do, in fact, coincide for different values of ε.

As expected, the largest differences in the curves-of-growth with different values of ε occur in the shoulder region. For the cases considered here, the effect of small ε is to raise the shoulder, i.e., to increase the equivalent width.

One of the frequent uses of the curve-of-growth is in the derivation of a so-called 'microturbulence'. This is accomplished by first deriving T_{ex} by constructing an empirical curve-of-growth. Using this value of T_{ex} one then derives a theoretical curve-of-growth, which, in general, will be shifted vertically from the empirical curve. The vertical shift gives the difference between $\log \Delta\lambda_D$(theor.) and $\log \Delta\lambda_D$(emp.), which is assumed to be given by

$$\log \frac{\Delta\lambda_D(\text{emp.})}{\Delta\lambda_D(\text{theor.})} = \frac{(2kT_{ex}/m)^{1/2}}{(2kT_{ex}/m + \zeta^2)^{1/2}}. \tag{VI-3}$$

This allows a derivation of the 'microturbulence', ζ.

The increase in $W_H/\Delta\lambda_D$ resulting from $\varepsilon < 1$ is closely analogous to the effect that is normally interpreted as 'microturbulence', i.e., it elevates the shoulder of the curve-of-growth. Thus, the normal curve-of-growth analyses based upon the assumptions that $T_{ex}=T$ will mistakenly interpret the effects of $\varepsilon \leqslant 1$ as a 'microturbulence', which may well account for the fact that 'microturbulence' velocities derived from the curve-of-growth sometimes are larger than those inferred from line widths and sometimes differ from element to element (cf., Aller, 1960).

The reader should be cautioned that the curves-of-growth shown in Figures VI-1 and VI-2 are computed for the idealized case of $\varepsilon = $ const and for a strictly two-level atom. The change in η_0 is accomplished by changing the abundance of the element, which is very different from changing, say, the excitation potential.

3. Emission Line Fluxes

The appearance of emission lines in a stellar spectrum is not generally associated with photospheric phenomena. Weak lines may appear in emission in the photosphere under certain circumstances such as those noted in Chapter V in connection with solar rare-earth lines. Also, lines may develop photospheric emission wings in the violet end of the spectrum where J_ν exceeds B_ν. Such cases are not basically different from the formation of absorption lines, however, and will not be discussed further.

More commonly the occurrence of emission lines is associated with either external layers of stars (chromospheres and coronas) or with gaseous nebulae. It is this class of phenomena which we shall discuss in the following. Specifically, we shall restrict the discussion to cases where the emission and absorption of photons by continuum processes may be ignored. Also, for algebraic simplicity we ignore stimulated emissions.

In the absence of continuum processes the transfer equation may be written as

$$\mu(dI_\nu/d\tau_0) = \phi_\nu (I_\nu - S_L). \tag{VI-4}$$

We operate on this equation with $\int d\omega/4\pi$ to obtain

$$dH_\nu/d\tau_0 = \phi_\nu (J_\nu - S_L). \tag{VI-5}$$

We next integrate over frequency to obtain

$$\frac{dH}{d\tau_0} = M^{-1} \int J_\nu \Phi_\nu \, d\nu - M^{-1} S_L. \tag{VI-6}$$

To a sufficiently good approximation $M^{-1} = \sqrt{\pi} \, \Delta \nu_D$.

Equation (VI-6) may be rewritten in the form

$$\frac{dH}{d\tau_0} = -M^{-1} S_L \left(1 - \frac{\int J_\nu \Phi_\nu \, d\nu}{S_L} \right), \tag{VI-7}$$

which, from Equation (II-16) is

$$\frac{dH}{d\tau_0} = -M^{-1}\varrho S_L.\qquad\text{(VI-8)}$$

Thus, the total flux escaping an atmospheric layer of thickness τ_1 in a given line is

$$H_0 = \int_0^{\tau_1} M^{-1}\varrho S_L\, d\tau_0.\qquad\text{(VI-9)}$$

We recall from Equation (II-88) that

$$\varrho S_L = \varepsilon^* B - \varepsilon^\dagger S_L.$$

Hence, we have

$$H_0 = \int_0^{\tau_1} M^{-1}\varepsilon^* B\left(1 - \frac{\varepsilon^\dagger S_L}{\varepsilon^* B}\right) d\tau_0,\qquad\text{(VI-10)}$$

or, ignoring stimulated emissions, we have

$$H_0 = \int_0^{\tau_1} M^{-1}\varepsilon^* B\left(1 - \frac{\varepsilon^\dagger b_j}{\varepsilon^* b_i}\right) d\tau_0,\qquad\text{(VI-11)}$$

where, as before, j and i refer, respectively, to the upper and lower levels.

Equation (VI-11) is valid line-by-line, but the quantities $\varepsilon^\dagger$ and ε^* depend upon other transitions in the atom. An alternate form for Equation (VI-11) may be obtained from Equation (VI-9) together with Equations (II-20) and (II-21). This yields

$$H_0 = \int_0^{\tau_1} M^{-1}\varepsilon B\left[1 - \frac{b_j}{b_i} + \frac{1}{c_{ij}W_i}\sum_m\left(C_{im}W_i - \frac{b_m}{b_i}R_{mi}W_m\right)\right] d\tau_0,\qquad\text{(VI-12)}$$

and the integrand now involves the other transitions explicitly.

It is sometimes convenient to sum Equation (VI-9), or its equivalent, over all the transitions in a particular spectral series. This gives

$$\sum_j H_{0,ji} = \int_0^{\tau_{1,ji}} \sum_j M_{ji}^{-1}\varrho_{ji} S_{ji}\, d\tau_0.\qquad\text{(VI-13)}$$

For a given energy level to be in a steady state it is necessary that the net rate of radiative transitions out of the level equal the net rate of collisional transitions into the level. Hence, we may write the equilibrium equation in the form

$$\sum_j \varrho_{ji} S_{ji} = \sum_j \varepsilon_{ji} B_{ji}(1 - b_j/b_i),\qquad\text{(VI-14)}$$

and to the approximation that M_{ji}^{-1} is constant Equation (VI-13) may be written

$$\sum_j H_{0, ji} = \langle M_{ji}^{-1} \rangle \int_0^{\tau_{1, ji}} \sum_j \varepsilon_{ji} B_{ji} \left(1 - \frac{b_j}{b_i}\right) d\tau_0. \tag{VI-15}$$

For the atom, as a whole, conservation of energy requires that the net energy emitted in all spectral lines (and continua) must be derived from the imbalance in collisional excitation energy. Hence, we must have that

$$\sum_i \sum_j H_{0, ji} = \sum_i \sum_j \int_0^{\tau_{1, ji}} M_{ji}^{-1} \varepsilon_{ji} B_{ji} \left(1 - \frac{b_j}{b_i}\right) d\tau_0. \tag{VI-16}$$

Note that Equation (VI-12) is valid line-by-line whereas Equation (VI-16) is valid only for the sum over all transitions. Note, also, that Equation (VI-15) is only an approximation (requiring M_{ji}^{-1} = const) and is valid only for the sum over a given spectral series with a common lower level.

As τ_1 increases, the flux H_0 in a given line increases following an emission curve-of-growth somewhat akin to absorption curves-of-growth. There is, however, one difference that is immediately obvious. Recalling the definition of the degradation length and the definition of ε^* from Chapter II, we rewrite Equation (VI-11) in the approximate form

$$H_0 = \int_0^{\tau_{deg}} M^{-1} \varepsilon^* B \, d\tau_0 + \int_{\tau_{deg}}^{\tau_1} M^{-1} \varepsilon^* B \left(1 - \frac{\varepsilon^\dagger b_j}{\varepsilon^* b_i}\right) d\tau_0. \tag{VI-17}$$

In other words, we set the cut-off term

$$\left(1 - \frac{\varepsilon^\dagger b_j}{\varepsilon^* b_i}\right)$$

equal to unity over the first degradation length in the atmosphere, which follows simply from the definition of τ_{deg}. Hence, we see that the linear portion of the emission curve-of-growth extends to $\tau_0 = \tau_{deg}$, and we note that τ_{deg} may exceed unity by a large factor. By contrast, the absorption curve-of-growth is linear only to $\tau_0 \approx 1$. We note further that τ_{deg} may vary markedly from line to line, even in a given atom, and that the family of emission curves-of-growth, as a result, is inherently more varied than the family of absorption curves-of-growth.

The emission line analogue of the horizontal and wing branches of the absorption curve-of-growth are contained in the second term in Equation (VI-17). Although the cut-off factor in this term is small, by definition, the smallness of this factor may be partially compensated for by the increasing size of τ_0. In the limit of very large τ_0 most of the flux, H_0, will in fact be escaping from deep in the atmosphere in the line wings.

In general, it is not possible to evaluate the cut-off factor in the second term in Equation (VI-17) without solving the radiative transfer and statistical equilibrium equations in detail for $\varepsilon^\dagger$, ε^*, b_j and b_i. As we have seen in Chapter III, the detailed manner in which S_L approaches B in the deep layers of an atmosphere depends very much upon wavelength and upon ε^*, $\varepsilon^\dagger$, r_0 and a, and upon the depth variations of these latter quantities as well. We shall restrict our discussion in the following, therefore, to the first term in Equation (VI-17), i.e., to the linear segment of the curve-of-growth where $\tau_1 \leqslant \tau_{\text{deg}}$. Also, we choose M, ε, a and B to be constant with depth. This latter simplification is done to facilitate comparisons between the different cases considered but need not be done in practical applications, of course. With these approximations in mind, we write

$$H_0 = M^{-1}\varepsilon^* B \tau_1, \quad \tau_1 \leqslant \tau_{\text{deg}}. \tag{VI-18}$$

We note now that Equation (VI-18) simply equates H_0 to the total energy of all of the line photons created within the layer of thickness τ_1.

It is implicit in Equations (VI-17) and (VI-18) that $\varepsilon^* B$ represents all of the photon sources and that all of the photon sinks are contained in $\varepsilon^\dagger$. In other words, ε^* and $\varepsilon^\dagger$ must be obtained by solving the full set of equilibrium equations rather than a sub-set of these equations.

4. The Two Level Atom

In the simple case of a two-level atom $\varepsilon^* = \varepsilon$ and $\tau_{\text{deg}} = \tau_{\text{th}}$. Thus,

$$H_0' = M^{-1}\varepsilon B \tau_1, \quad \tau_1 \leqslant \tau_{\text{th}}. \tag{VI-19}$$

Since $\tau_1 \leqslant \tau_{\text{th}}$, we refer to this case as the effectively thin case. Equation (VI-19), with $\tau_1 < \tau_{\text{th}}$ and without stimulated emissions, is the exact equivalent of

$$H_0' = hv C_{12} n_1 h_1, \tag{VI-20}$$

where h_1 is the geometrical thickness of the layer, as well as the exact equivalent of

$$H_0' = hv A_{21} n_2 h_1. \tag{VI-21}$$

Equation (VI-20) is the familiar coronal approximation commonly used in discussions of the solar corona. It has the interesting implication that two spectral lines with approximately equal values of C_{12} and with equal values of n_1 and v will have approximately the same flux regardless of the value of A_{21}. Optically forbidden transitions in complex atoms and ions do in fact have values of C_{12} comparable to those for permitted transitions and one finds in the spectrum of the solar corona that permitted and forbidden lines are of comparable intensity. The upper level populations, n_2, for the forbidden lines build up to the point where $A_{21} n_2$ for these transitions is comparable to $A_{21} n_2$ for permitted transition in spite of the fact that A_{21} is orders of magnitude larger for permitted transitions than for forbidden transitions.

The most notable difference between forbidden and permitted lines is that the thermalization length for the forbidden lines is comparatively short whereas for the permitted lines it is comparatively long. Electron densities in the solar corona are low enough ($n_e \lesssim 3 \times 10^8$ cm^{-3}) that $\tau_1 < \tau_{th}$ even for the forbidden lines.

At the higher densities found in the upper solar chromosphere ($n_e \approx 3 \times 10^{10}$) the collision rates increase to where $\varepsilon \gg 1$ for the forbidden transitions. However the reverse inequaly remains true for the permitted transitions. As ε becomes larger the factor $(1 - b_2/b_1)$ in Equation (VI-14) and the quantity ϱ_{21} in Equation (VI-13) both approach zero and Equation (VI-20) no longer gives the correct flux in the line. The flux will, in fact, be much smaller than given by Equation (VI-20). For the permitted line, however, where ε is still small compared to unity we find $(1 - b_2/b_1) \approx 1$ and Equation (VI-20) gives a good estimate of the flux. Thus, at higher densities than those found in the corona the strengths of the permitted lines grow relative to the forbidden lines.

We saw in Chapters II and III that τ_{th} is given by ε^{-1} when $\varepsilon > a$ and by $a\varepsilon^{-2}$ when $\varepsilon < a < 1$. In the former case H'_0 increases to a limit of

$$H'_0 = M^{-1} B \qquad \text{(VI-22)}$$

and in the latter case it increases to a limit of

$$H'_0 = M^{-1} \frac{a}{\varepsilon} B \qquad \text{(VI-23)}$$

for effectively thin atmospheres. Note that in Equations (VI-19), (VI-22), and (VI-23) H'_0 is directly proportional to B even though H'_0 itself is much below the LTE values for an atmosphere of the same optical thickness.

In the case of Equation (VI-19) the proportionality of H'_0 to B is misleading. It is true only if τ_1 is held constant. For an atmosphere of fixed geometrical thickness, the two-level approximation gives $H'_0 \propto n_1 C_{12}$ (Equation (VI-20)) and if the temperature is in a range where n_1 is varying because of excitation or ionization the product $n_1 C_{12}$ may be nearly independent of temperature (cf., Athay, 1965, 1969).

Note that Equation (VI-22) equates H'_0 to the energy under a Gaussian profile of central intensity B. It follows that when $\varepsilon > a$ the linear portion of the curve-of-growth ends at the same flux level, independently of ε, and that this flux level is the same as would occur in LTE. What is changed when LTE is not valid, therefore, is the value of τ_0 required to reach a given flux level. The increase in τ_0 is inversely proportional to ε. For the case $a > \varepsilon$ (Equation (VI-23)), the limiting flux on the linear portion of the emission curve-of-growth increases beyond the LTE level by a factor a/ε.

In Equation (VI-19) both ε^* and τ_1 are proportional to n_e. Thus, $H'_0 \propto n_e^2$. However, in Equation (VI-22) H'_0 is independent of n_e and in Equation (VI-24) H'_0 is proportional to $a n_e^{-1}$. If, in this latter case, a is dominated by radiation damping, $H'_0 \propto n_e^{-1}$, if a is dominated by collision damping H'_0 is independent of n_e; and if a dominated by Stark damping $H'_0 \propto n_e$. These latter proportionalities occur, of course, only for a range of values of τ_1 around τ_{th}. The point is, however, that H'_0 may exhibit a markedly different response to changes in n_e, ranging from a proportionality to n_e^2

to n_e^{-1} depending upon the opacity of the layer and the line broadening mechanism.

Still other dependences upon n_e are possible for lines for which the two-level approximation is not valid. In this case the equivalent of Equation (VI-22) is

$$H'_0 = M^{-1} \frac{\varepsilon^* B}{\varepsilon^\dagger}, \qquad \text{(VI-24)}$$

and one may find, as in the case of the solar Lyman-α and Lyman-β lines (Athay, 1969) that $\varepsilon^* \propto n_e$ whereas $\varepsilon^\dagger$ is dominated by radiative quantities and is independent of n_e. Thus, H'_0 is again proportional to n_e.

5. Upper Level Doublets

The two lines in an upper level doublet are characterized by having the same values of ε and τ_{th} but different values of τ_0 (cf., Chapter IV, Section 2). In the effectively thin case, therefore, the two line fluxes will be in proportion to the f-values of the lines when τ_1 is small and will merge to common values at $\tau_1 = \tau_{th}$. The common value of H'_0 will be approached at a rate that depends upon the coupling between the lines and will be decreased somewhat over the value that the stronger line would have if considered by itself. This latter effect results from the combined effects of increased thermalization length of the multiplet due to the lower opacity in one of the lines, as discussed in Chapter IV, Section 2, and the net loss of flux to the weaker line.

Throughout the effectively thin regime the combined line fluxes are given by

$$H'_0(3\text{-}1) + H'_0(2\text{-}1) = h\nu n_1 (c_{12} + c_{13}) h_1. \qquad \text{(VI-25)}$$

In general, c_{12}/c_{13} will equal f_{12}/f_{13} and will differ from unity. The limit $h_1 = h_{th}$ will increase over its single line value h_{th}^*, due to the coupling between the lines. At the limit $h_1 = h_{th}$ each line will have a flux equal to

$$H'_0 = h\nu n_1 \left(\frac{c_{12} + c_{13}}{2}\right) h_{th}. \qquad \text{(VI-26)}$$

However,

$$\left(\frac{c_{12} + c_{13}}{2}\right) h_{th}$$

will be less than either $c_{12} h_{th}^*$ or $c_{13} h_{th}^*$.

6. Lower Level Doublet and Metastable Level

We consider in this section three energy levels, 1, 2 and 3, with radiative transitions 3-1 and 3-2 and with level 2 lying intermediate to levels 1 and 3. When level 2 is close to level 1 we refer to the configuration as a lower-level doublet, and when level 2 is separated from level 1 by a substantial fraction of the energy difference 3-1 we refer to the configuration as a metastable level.

The distinguishing feature of this type of configuration is that fluxes in the two lines are automatically shared regardless of the origin of the electrons in the common excited level. If, for example, we take the metastable case where $A_{31} = A_{32}$, $c_{12} = \gamma c_{13}$, $c_{23} \ll c_{21}$ and where radiative excitations can be ignored, the equilibrium equations are then given by

$$(c_{13} + c_{12}) n_1 = c_{21} n_2 + A_{31} n_3 \qquad \text{(VI-27a)}$$

and

$$c_{21} n_2 = c_{12} n_1 + A_{32} n_3, \qquad \text{(VI-27b)}$$

which reduce to

$$(1 + \gamma) c_{13} n_1 = \gamma c_{13} n_1 + 2 A_{31} n_3$$

or

$$\tfrac{1}{2} c_{13} n_1 = A_{31} n_3. \qquad \text{(VI-28)}$$

The combined flux in the 3-1 and 3-2 lines throughout the effectively thin regime is given by

$$H_0'(3\text{-}1) + H_0'(3\text{-}2) = h \nu_{31} c_{13} n_1 h_1 \qquad \text{(VI-29)}$$

Hence, for the conditions we have assumed, the 1-3 collisional excitations feed both lines. For $A_{31} = A_{32}$ the photons divide evenly between the lines. However, if A_{32} were less than A_{31} the photons would divide in the ratio A_{32}/A_{31} so long as τ_1 remains small in both lines. Since $\tau_1(3\text{-}1)$ is larger than $\tau_1(3\text{-}2)$ ($n_2 < n_3$ and $A_{32} < A_{31}$), however, as soon as $\tau_1(3\text{-}1)$ exceeds unity by a substantial amount 3-1 photons will show a net scatter into the 3-2 line. This process will continue until the photon fluxes are again equally divided.

In the limit $h_1 = h_{\text{th}}$ we will again have

$$h_{\text{th}}^* < h_{\text{th}} \leqslant 2 h_{\text{th}}^*$$

(cf., Figure IV-6).

For the doublet configuration it is more likely that $c_{13} = c_{23}$ and that $n_1/n_2 = \omega_1/\omega_2$. Hence we will have $c_{13} n_1 / c_{23} n_2 = \omega_1/\omega_2 = A_{31}/A_{32}$. The total flux in the lines will be

$$H_0'(3\text{-}1) + H_0'(3\text{-}2) = h \nu n_1 c_{13} (1 + \tilde{\omega}_2/\tilde{\omega}_1) h_1, \qquad \text{(VI-30)}$$

which will initially divide in the ratio $H_0(3\text{-}1)/H_0(3\text{-}2) = \tilde{\omega}_1/\tilde{\omega}_2$ and will saturate to equal values given by half the right hand side of Equation (VI-30). Again, however,

$$h_{\text{th}}^* < h_{\text{th}} < (1 + \tilde{\omega}_2/\tilde{\omega}_1) h_{\text{th}}^*.$$

7. Three Line Loop

The three-line closed loop of transitions represented by 3-1, 3-2 and 2-1 presents a new possibility for the redistribution of photons among the lines. In the metastable configuration the return of electrons from level 3 to level 1 via level 2 is relatively inefficient compared to the direct 1-3 route. In the three-line closed loop configuration, however, the two alternate routes from 3-1 are of comparable efficiency. This, coupled with the fact that τ_{32} is generally much less than τ_{31}, provides an easy escape route for 3-1 photons and suppresses the flux in the 3-1 line when the opacity becomes large.

To illustrate this effect, consider the case where $A_{31}=A_{32}=A_{21}$ and $c_{12}=c_{13}=c_{23}$ and where radiative excitations can be ignored. The equilibrium equations can be written as

$$(A_{31} + A_{32}) n_3 = c_{13}n_1 + c_{23}n_2$$

and

$$A_{21}n_2 = c_{12}n_1 + A_{32}n_3.$$

These equations reduce to

$$A_{21}n_2 = \tfrac{3}{2}c_{12}n_1,$$

$$A_{31}n_3 = \tfrac{1}{2}c_{12}n_1,$$

and

$$A_{32}n_3 = \tfrac{1}{2}c_{12}n_1$$

if we set $A_{21}n_2 \gg c_{23}n_2$. The total observed flux in the three lines is given by

$$H_0'(2\text{-}1) + H_0'(3\text{-}1) + H_0'(3\text{-}2) = h\nu_{12}c_{12}n_1 h + h\nu_{13}c_{13}n_1 h + \\ + h\nu_{23}c_{23}n_2 h, \qquad (\text{VI-31})$$

where $c_{23}n_2 \ll c_{12}n_1$. So long as τ_1 in all three lines is low the photons will divide among the three lines in the ratio 2-1:3-2:3-1 = 3:1:1. As τ_1 increases photons will continue to accumulate in the 2-1 transition because once an electron is in the level $n=2$ it can only return to the level $n=1$ ($c_{23} \ll A_{21}$). On the other hand photons in the 3-1 line will branch into the 3-2 line at every other scattering and because the 3-2 opacity is much less than the 3-1 opacity ($n_2 \ll n_1$) the 3-2 photon will escape much more easily than a 3-1 photon. Thus, the photon flux will continue to build up in both the 3-2 and 2-1 lines but not in the 3-1 line. If the thermalization lengths are large, this process may continue until both $H_0'(3\text{-}2)$ and $H_0'(2\text{-}1)$ are very much larger than $H_0'(3\text{-}1)$. In this case the line fluxes will be given to close approximation by

$$H_0'(2\text{-}1) = h\nu_{12}n_1(c_{12} + c_{13}) h,$$

and

$$H_0'(3\text{-}2) = h\nu_{32}n_1 c_{13} h,$$

with $H_0'(3\text{-}1) \ll H_0'(3\text{-}2)$. This effect is clearly illustrated in the solar Lyman lines of hydrogen where the Lyman-α flux exceeds the Lyman-β flux by a factor of about 10^2 even though the line cores are clearly saturated due to large opacity.

An interesting application of the three-line loop configurations has been made by Gabriel and Jordan (1969) in the solar corona for cases where the gas is optically thin, $\tau_0 < 1$. In helium like ions of multiply ionized heavy elements in the solar corona the transitions $2^1P\text{--}1^1S$, $2^3P\text{--}1^1S$ and $2^3S\text{--}1^1S$ are each observed and the individual lines have comparable intensities. Since the transition 2^3P-2^3S is permitted, the system $1^1S-2^3S-2^3P$ represents a closed loop of transitions. The lines from the transitions 2^3P-1^1S and 2^3S-1^1S are referred to, respectively, as the intercombination and forbidden lines.

To simplify the notation we label the three levels as follows: 1^1S-1, 2^3S-2 and 2^3P-3. If we treat these three levels as an independent system (following Gabriel and Jordan), which ignores interactions with 2^1S and 2^1P as well as higher levels, and if we ignore radiative excitations in the intercombination and forbidden lines but not in the 2-3 permitted line and ignore all collisional de-excitations, the equilibrium equations become

$$(A_{21} + A_{23} + c_{23})n_2 = A_{32}n_3 + c_{12}n_1 \tag{VI-32}$$

and

$$(A_{32} + A_{31})n_3 = (A_{23} + c_{23})n_2 + c_{13}n_1. \tag{VI-33}$$

Let

$$\gamma = \frac{c_{12} + c_{13}}{c_{13}}, \tag{VI-34}$$

$$\alpha = \frac{A_{31} + A_{32}}{A_{31}}, \tag{VI-35}$$

and

$$R = \frac{A_{21}n_2}{A_{31}n_3}. \tag{VI-36}$$

The latter quantity is equal to the intensity ratio of the forbidden line to the intercombination line, since for the type of configuration assumed $v_{21} \approx v_{23}$ and since the gas is assumed optically thin.

Equations (VI-32), (VI-33), and (VI-36) give

$$R = \frac{A_{21}}{A_{31}} \frac{(c_{12} + c_{13})A_{32} + A_{31}c_{12}}{(A_{23} + c_{23})(c_{12} + c_{13}) + A_{21}c_{13}}.$$

Dividing top and bottom by c_{13} and writing $A_{32}/A_{31} = \alpha - 1$, we obtain

$$R = \frac{A_{21}}{A_{21} + \gamma(A_{23} + c_{23})} [\gamma(\alpha - 1) + \gamma - 1]$$

or

$$R = \frac{A_{21}}{A_{21} + \gamma(A_{23} + c_{23})} (\gamma\alpha - 1). \tag{VI-37}$$

Each of the quantities in Equation (VI-37) is independent of electron density except c_{23}, which is directly proportional to n_e. Thus, equation (VI-37) is useful, under certain conditions, for determining n_e.

The conditions under which Equation (VI-37) is of interest are when $c_{23} > A_{23}$ and $c_{23} > A_{21}$. Since A_{23} is proportional to the photon density, A_{23} will be relatively small when the 2-3 transition is in the far ultraviolet. Also, c_{23} will be large when the temperature and electron density are high. For O VII in the solar corona, for example, Gabriel and Jordan (1969) estimate $c_{23} = 1.2 \times 10^{-8} n_e$ s^{-1} cm^3, $A_{23} = 0.37$ s^{-1}, $A_{21} = 32$ s^{-1}, $\gamma = 1.35$ and $\alpha = 3.4$. At normal coronal densities ($n_e \leqslant 3 \times 10^8$ cm^{-3}), therefore, we expect $c_{23} \gg A_{23}$ but $c_{23} < A_{21}$. In this case R is nearly independent of n_e. In coronal active regions, however, n_e may be as high as 10^{10} cm^{-3} and c_{23} is then large compared to A_{21} and R is inversely proportional to n_e.

For C V, Gabriel and Jordan (1969) give $c_{23} = 2.8 \times 10^{-8} n_e$ s^{-1} cm^3, $A_{23} = 260$ s^{-1}, $A_{21} = 2.5$ s^{-1}, $\gamma = 1.35$ and $\alpha = 9$. Thus, c_{23} should not exceed A_{23} until $n_e \geqslant 10^{10}$ and for the normal sun R should be about 0.08. The observed value of R, however, is approximately 2, which could only occur if either A_{21}, A_{23} or c_{23} is seriously in error.

Estimates of n_e based upon Equation (VI-37) are often greater than estimates of n_e obtained in other ways, which suggests either that c_{23} is underestimated or that the larger of the quantities A_{21} and A_{23} are overestimated. An alternative possibility is that the 2^3P and 2^3S levels are interacting with a fourth energy level in such a way that the value of R is increased over the value predicted by Equation (VI-37).

The addition of a fourth energy level to the preceding system of three energy levels leads to the equilibrium equation

$$R' = \frac{A_{21} n_2}{A_{31} n_3} = \frac{A_{21}}{A_{31}} \times \frac{R_{32} R_{11} R_{44} + R_{34} R_{42} R_{11} + R_{31} R_{12} R_{44} + R_{31} R_{14} R_{12} + R_{34} R_{41} R_{12}}{R_{23} R_{11} + R_{44} + R_{24} R_{43} R_{11} + R_{21} R_{13} R_{44} + R_{21} R_{14} R_{43} + R_{24} R_{41} R_{13}}.$$

Divide numerator and denominator by $c_{13} A_{31} R_{44}$ and define

$$\gamma' = \frac{c_{12} + c_{13} + c_{14}}{c_{13}} = \frac{R_{11}}{c_{13}}, \tag{VI-38}$$

$$\delta = c_{12}/c_{13}, \tag{VI-39}$$

$$\beta = c_{14}/c_{13}, \tag{VI-40}$$

$$r_{4j} = R_{4j}/R_{44}, \tag{VI-41}$$

and

$$\omega = R_{34}/A_{31}, \tag{VI-42}$$

to obtain

$$R' = \frac{A_{21} [\alpha \gamma' - 1 - \beta(1 - r_{42}) + \omega \gamma' r_{42} + \omega \delta r_{41}]}{A_{21} (1 + \beta r_{43}) + (R_{23} + R_{24} r_{43}) \gamma' + R_{24} r_{41}}. \tag{VI-43}$$

Since each of the quantities in R' is positive and since $r_{42} < 1$, the only terms in Equation (VI-43) that can increase R' over the value given by Equation (VI-37) are those containing ω. The quantities δ, γ', r_{42}, and r_{41} are each expected to be of order unity or smaller. It follows from Equation (VI-43) that if ω is of order unity or smaller the addition of a fourth energy level cannot increase R' by more than a factor of order unity.

The reason Equation (VI-37) gives a low value for R for Cv is that level 2 is depopulated preferentially to level 3 rather than to level 1. To offset this effect, a fourth level would have to either depopulate level 3 to level 1 via level 4 or depopulate level 3 back to level 2 via level 4. The relevant combinations of rates in the two cases are $(R_{41}/R_{44})(R_{34}/A_{31}) = \omega r_{41}$ and $(R_{42}/R_{44})(R_{34}/A_{31}) = \omega r_{42}$. It is easy to pick a fourth level such that either r_{41} or r_{42} is near unity (for example 2^1P or 3^3P), but there are no choices for level 4 that give $\omega \gg 1$.

Gabriel and Jordan (1969) give $A_{31} = 9 \times 10^6$ s^{-1} for Cv and $A_{31} = 1.8 \times 10^8$ s^{-1} for Ovii. Unless the transition rate R_{34} is large compared to A_{31}, ω will be of order unity or smaller. For helium-like ions in the solar corona there are no known transitions from 2^3P that are as frequent as those to 1^1S other than those to 2^3S. Thus, there is no fourth level for which ω is large, unless, as noted earlier, the published values of A_{31} are much too large. The only known explanation for the failure of Equation (VI-37) for Cv and the tendency for values of n_e given by Equation (VI-37) to be too high for other helium-like ions in the corona seems to lie in errors in the transition rates A_{31}, R_{23}, and A_{21}.

Lines in the Li and Be isoelectronic sequences can be used in a similar way.

8. Added Comments

The preceding examples somewhat oversimplify the problem of interpreting emission line fluxes in multiplets, in metastable configurations and in closed loops. In addition, we have said nothing at all about more complex configurations, mainly because it is difficult to reduce the more complex cases to a form that is suitable for discussion.

Perhaps equally seriously with the omissions just named is our neglect of wavelength effects. In metastable and closed loop configurations large effects may be present when one of the transitions lies in the far ultraviolet, as was illustrated, for example, in Figure IV-6. One effect of a far ultraviolet transition is to drive S_L above B. This enhances the effect of the cut-off factors in Equations (VI-10), (VI-11), and (VI-12) and may even drive them to negative values. In this latter case there would be a net loss of flux in the deeper layers in some lines. It is difficult to make predictions about relative line fluxes in such cases without detailed computations. This could readily be done for selected atomic configurations and selected model atmospheres. However, there is a point at which such sample calculations become of little practical value and where the effort should be directed to a more realistic interpretation of data.

References

Aller, L. H.: 1960, in J. L. Greenstein (ed.), *Stellar Atmospheres*, Univ. of Chicago, Chicago, Ill.
Athay, R. G.: 1965, *Astrophys. J.* **142**, 755.
Athay, R. G.: 1969, *Astrophys. J.* **157**, 281.
Athay, R. G. and Skumanich, A.: 1968, *Astrophys. J.* **152**, 211.

Canfield, R. G.: 1969, *Astrophys. J.* **157**, 425.
Gabriel, A. H. and Jordan, C.: 1969, *Monthly Notices Roy. Astron. Soc.* **145**, 241.
Holweger, H.: 1967, *Z. Astrophysik* **65**, 365.
Jefferies, J. T.: 1968, *Spectral Line Formation*, Ginn and Blaisdell, Waltham, Mass.
Menzel, D. H.: 1931, *Publ. Lick Obs.* **17**, 1.
Wrubel, M. H.: 1949, *Astrophys. J.* **109**, 66.
Wrubel, M. H.: 1950, *Astrophys. J.* **111**, 157.

CHAPTER VII

THE LINE BLANKETING EFFECT

1. Definition of Terms

The presence of lines in the spectrum of a star influences both the energy balance in the external layers of the star and the distribution of energy in the continuum spectrum radiated by the star. The latter effect may be divided into two parts: (a) a real redistribution of continuum spectral energy induced by the change in the energy balance in the star, and (b) an apparent redistribution of spectral energy caused by the 'obscurration' of the continuum by many overlapping lines. We shall not concern ourselves with this second effect except for the brief comments in the following two paragraphs.

In stars with sufficiently many spectral lines (or molecular bands) that the lines (or bands) blend together for appreciable distances in the spectrum, the 'continuum' spectrum of the star may be difficult or impossible to identify observationally in the regions of blending. The lines (or bands) are then said to 'block' the continuum. If by either extrapolating the 'continuum' from adjacent wavelengths or adopting the continuum from a theoretical model we can establish the apparent level that the continuum would have if the lines were not present, we can then determine quantitatively the fraction of the continuum energy removed by line blocking. Such a procedure, of course, is entirely without meaning in the strict physical sense since the 'continuum' we are establishing is purely fictitious.

The total energy radiated by a star is governed by energy generation in the deep interior and is independent of line blocking effects. Thus, energy blocked in one part of the spectrum must appear elsewhere in the spectrum, which implies that the entire spectrum should be treated as an integral quantity. In this sense the division of the spectrum into separate 'continuum' and 'line' spectra is a rather artificial procedure. It is useful for some purposes, nevertheless, to make such artificial distinctions. In stars with large numbers of spectral lines it is still impractical to treat the entire spectrum integrally and we must, for purely practical reasons, continue to divide the spectrum into continuum and line components.

The addition of line absorption to an otherwise continuous source of opacity in a stellar atmosphere increases the 'mean' opacity of the atmosphere. This added opacity may be thought of in the Milne-Eddington case as an increase in the Rosseland opacity of the atmosphere. One then argues that, since radiative equilibrium requires (cf., Milne, 1930) the temperature gradient with regard to the Rosseland optical depth to be constant sufficiently deep in the atmosphere (i.e. for a given effective temperature) the effect of the added opacity is to increase the temperature gradient with regard to the continuum optical depth by the fractional increase in the Rosseland opacity. Later, we shall consider the effects more quantitatively. Here, we wish only to note that in this simple, somewhat superficial, picture the increased temperature gradient due to the lines results in an increase in the temperature in the deeper layers of the photosphere and to a decrease in the temperature in the higher layers. The former effect is referred to as 'back warming' and the latter is referred to as 'surface cooling'. One may think of the back warming effect as resulting from the partial 'blocking' of the radiation flow in the deeper layers due to the added opacity of the lines and of the surface cooling as a 'pumping' of the energy flow in the higher layers due to the added emissivity of the lines. The specific nature of these effects will be clarified in the following discussion.

The term 'blanketing effect' has been widely used in the literature to describe 'blocking', 'backwarming' and 'surface cooling' both individually and collectively, i.e., it has been used inconsistently. We shall use the term 'blanketing effect' in its physical sense to denote the combined effects of backwarming and surface cooling.

2. Historical Summary

Prior to the availability of large computers the blanketing effect was discussed primarily in terms of LTE or pure scattering models of line formation. The few attempts that were made to consider more realistic cases were severely restricted and largely qualitative in nature.

Chandrasekhar (1935) considered the blanketing effect for the special case of a gray atmosphere of the Milne-Eddington (M-E) type with lines of equal r_0 ($r_0 \approx 10^{-1}$) distributed uniformly over the spectrum (the 'picket fence' model). The lines were further assumed to have rectangular profiles. Each of the assumptions: r_0 constant with depth, r_0 constant from line to line, lines uniformly distributed, gray atmosphere, and rectangular profiles restricts the nature of the blanketing effect (Athay and Skumanich, 1969) as we shall later show. Nevertheless, Chandrasekhar (1935) successfully demonstrated the basic nature of backwarming and surface cooling. In particular, he found that the backwarming effect in the photosphere is relatively independent of the mechanism of line formation in the two extremes of LTE and pure coherent scattering but that the surface cooling effect, which is strong in LTE, essentially disappears in the pure scattering case. Subsequent work by Münch (1946) and Pecker (1951) confirmed and refined Chandrasekhar's results.

Thomas (1965) showed by an appeal to the general form of the terms in the flux-divergence equation that lines formed via a mixture of pure absorption (LTE) and non-coherent scattering would produce a smaller thermal effect (backwarming and surface cooling) than the pure absorption case. He argued, in fact, that the surface cooling effect would be reduced by a factor of ε and that because $\varepsilon \lesssim 10^{-2}$ the surface cooling effect could be largely ignored. These arguments were expanded by Cayrel (1966) and by Frisch (1966). The latter concluded that the only significant thermal effect in the sun is a cooling for $\tau_c < 0.01$ that arises mainly from H and K of Ca II and the corresponding resonance lines of Mg II. These conclusions conflict with those of Chandrasekhar and are incorrect as shown by the more detailed studies by Athay and Skumanich (1969) and Athay (1970).

Auer and Mihalas (1969a, b, c) and Mihalas and Auer (1970) have included the blanketing effect in model atmosphere calculations from O and early B stars without imposing the restriction of LTE. Their results confirm the conclusion that the lines strongly modify the temperature structure in the external layer.

A detailed treatment of the line blanketing effect requires an explicit evaluation of the line source functions, S_L, the energy in the line, J_v, and the ratio of continuum to line opacity, r_0, each as a function of depth (Athay and Skumanich, 1969). It is necessary, therefore, that the blanketing effect be discussed in parallel with a theory of line formation.

3. Mathematical Derivation of Blanketing Terms

In order to derive an explicit algebraic formalism for the blanketing effect, we restrict the following discussion to a single spectral line whose source function is (Equation (II-31)):

$$S_L = \frac{\int J_v \Phi_v \, dv + \varepsilon^* B}{1 + \varepsilon^\dagger}. \tag{VII-1}$$

We write the equation of radiative transfer in the flux-divergence form (Equation (II-43)):

$$\frac{dH_v}{d\tau_c} = \frac{(\phi_v + r_0)}{r_0} J_v - \frac{\phi_v S_L}{r_0} - B_v. \tag{VII-2}$$

Integration of Equation (VII-2) over frequency gives

$$\frac{dH}{d\tau_0} = \frac{M^{-1}}{r_0} \int J_v \Phi_v \, dv + \int J_v \, dv - \frac{M^{-1}}{r_0} S_L - \int B_v \, dv. \tag{VII-3}$$

To the right hand side of Equation (VII-3) we add and subtract the quantity $\int J_c \, dv$, where J_c is the fictitious continuum mean intensity obtained by bridging the observed continuum smoothly across the spectral lines. After adding and subtracting this term and combining appropriate terms, we obtain

$$\frac{dH}{d\tau_c} = \frac{M^{-1}}{r_0} \left[\int J_v \Phi_v \, dv - S_L \right] - \int (J_c - J_v) \, dv + \int (J_c - B_v) \, dv. \tag{VII-4}$$

The quantity $(\int J_v \Phi_v \, dv - S_L)$ is given by Equation (VII-1) as

$$\int J_v \Phi_v \, dv - S_L = \varepsilon^\dagger S_L - \varepsilon^* B \tag{VII-5}$$

or from the definition of ϱ (Equation (II-16)) as

$$\int J_\nu \Phi_\nu \, d\nu - S_L = -\varrho S_L. \qquad \text{(VII-6)}$$

Also, we note that in the absence of spectral lines Equation (VII-2) becomes

$$dH_\nu^c/d\tau_c = J_c - B_\nu \qquad \text{(VII-7)}$$

so that

$$\int \frac{dH_\nu^c}{d\tau_c} \, d\nu = \frac{dH_c}{d\tau_c} = \int (J_c - B_\nu) \, d\nu. \qquad \text{(VII-8)}$$

We next invoke the condition of radiative equilibrium, $dH/d\tau_c = 0$, and make use of Equations (VII-5) and (VII-8) to obtain

$$\frac{dH_c}{d\tau_c} = \frac{M^{-1}}{r_0} \varrho S_L + \int (J_c - J_\nu) \, d\nu. \qquad \text{(VII-9)}$$

Following Athay and Skumanich (1969), we designate the terms on the right hand side of Equation (VII-9) as*

$$t = \frac{M^{-1}}{r_0} \varrho S_L, \qquad \text{(VII-10)}$$

$$c = \int (J_c - J_\nu) \, d\nu, \qquad \text{(VII-11)}$$

and

$$e = t + c \qquad \text{(VII-12)}$$

so that

$$dH_c/d\tau_c = t + c = e. \qquad \text{(VII-13)}$$

Equation (VII-13) states, simply, that a spectral line induces a net flow of energy between the line and the continuum. The flow may be in either sense and may change markedly with depth. Thus the 'continuum' flux is no longer constant with depth, which it would be if there were no lines in the spectrum and if radiative equilibrium was present. It is this divergence of the 'continuum' flux that accounts for the change in the temperature structure of the atmosphere with respect to τ_c.

It is immediately obvious from the derivation of Equation (VII-13) that if more than one line exists in the spectrum we may simply sum the terms t and c over all the lines. The interlocking effects among the different lines of a given atom are properly accounted for in ϱ and S_L. Furthermore, we note that since we have used the general multilevel form for S_L Equations (VII-9) and (VII-13) are valid independently of the atomic model.

In the work by Athay and Skumanich (1969), Thomas (1965), Cayrel (1966), and Frisch (1966) the discussion was restricted to a two-level atom. For this case Equations

---

* Note that the t defined here is the negative of that defined by Athay and Skumanich (1969).

(VII-5) and (VII-6) reduce to

$$\varrho S_L = \varepsilon (B - S_L) \tag{VII-14}$$

and the term represented by t becomes

$$t = M^{-1} \frac{\varepsilon}{r_0} (B - S_L). \tag{VII-15}$$

4. The c Term

Although the c and t terms in Equation (VII-13) are interrelated, they may also operate somewhat independently of each other. In addition, they represent somewhat different physical effects, and it is instructive to consider the two terms individually.

The equivalent width, W, of a spectral line is defined in flux units as

$$W_{H,0} = \frac{1}{H_0^c} \int (H_0^c - H_v) \, dv, \tag{VII-16}$$

where the notation H_0^c means the flux measured in the continuum adjacent to the line as it emerges from the star. H_0^c is assumed constant over the width of the line. Let us now define an equivalent width in terms of the local mean intensity. Thus, we define

$$W_J = \frac{1}{J_c} \int (J_c - J_v) \, dv, \tag{VII-17}$$

where J_c and J_v are the local values of J_c near the line and J_v in the line at optical depth τ_c. With this definition of W_J Equation (VII-11) gives

$$c = W_J J_c \tag{VII-18}$$

Because c is proportional to the equivalent width W_J it will have the same numerical value for two lines of the same W_J irrespective of the mechanism of line formation. When we impose a particular depth dependence upon r_0 (as in the Milne-Eddington and Schuster-Schwarzschild atmospheres) and assume a given value for W_J outside the star we, in fact, are imposing a severe restriction upon the depth variation of W_J. The mechanism of line formation is then of secondary importance in determining the local value of W_J at a given τ_c. As we shall see later on the quantity c is primarily responsible for the backwarming in the photosphere and it is for this reason that Chandrasekhar (1935) and others find that the backwarming is independent of the line forming mechanism when the depth variation of r_0 is specified. In reality, of course, r_0 does depend upon the mechanism of line formation and thus c depends upon the mechanism of line formation.

Let us now consider the meaning of the c term more explicitly. If W_J is positive, i.e., if the line is in absorption in J_v, then c is positive. This means that $dH_c/d\tau_c$ is positive, or, in other words, that the continuum flux is experiencing a 'sink' as it flows outward. This is a local cooling effect and will contribute to the surface cooling at small τ_c.

A local heating, or backwarming, effect requires that c be negative, i.e., that the line be in emission in J_ν. Whether or not a line that is in absorption at small τ_c reverses to an emission line at larger τ_c depends critically upon wavelength, upon the line strength, and upon the depth variation of r_0 (Athay and Skumanich, 1969). Lines in the violet end of the spectrum where J_c lies above B tend to stay in absorption until they eventually disappear whereas those in the red end of the spectrum where J_c lies below B tend to reverse to emission lines before merging with the continuum. A rapid increase in r_0 in the deeper layers, as in the S-S atmosphere, enhances the tendency of the lines to go into emission and may result in emission lines in the deeper layers even in the violet end of the spectrum (Athay and Skumanich, 1969).

The value of c at small τ_c is readily estimated from Equation (VII-18). We may assume that W_J is approximately equal to W_H at small τ_c since, in this case, J and H are closely related. In fact, for B of the form

$$B = (1 + \beta\tau_c) B(0) \tag{VII-19}$$

the transfer equation gives (Kourganoff, 1963)

$$J_0^c = \tfrac{1}{4}(2 + \beta) B(0) \tag{VII-20}$$

and

$$H_0^c = \tfrac{1}{6}(1.5 + \beta) B(0). \tag{VII-21}$$

Thus, if we set $W_J = W_H$ we see that

$$\begin{aligned} c_0 &= W_J J_0^c \approx 2 W_H H_0^c, \quad \beta = 0 \\ &\approx 1.5 W_H H_0^c, \quad \beta \gg 2. \end{aligned} \tag{VII-22}$$

Values of $c/W_H H_0^c$ computed by Athay and Skumanich (1969) are shown in Figure VII-1. For the representation of the Schuster-Schwarzschild atmosphere in Figure VII-1, r_0 is set equal to

$$r_0 = r_{0,1}(1 + m\tau_c)^5. \tag{VII-23}$$

Assumed values of m are given in the figure. Note the wide range in behavior of c at τ_c near 10^{-1} to 10^{-2}.

To understand the nature of the c term and its effect upon the atmosphere it is helpful to consider the rates of continuum emission and absorption at the wavelength of the line. The continuum emissivity is $\kappa_c B$ and is independent of the line strength. The absorptivity, on the other hand, is $\kappa_c J_\nu$ and depends directly upon the line strength. If, in the absence of the line, $\kappa_c B = \eta \kappa_c J_c$ an absorption line in J_ν will make $\kappa_c B > \eta \kappa_c J_\nu$ and an emission line will make $\kappa_c B < \eta \kappa_c J_\nu$. Hence an absorption line results in an excess of continuum emissions over continuum absorp-

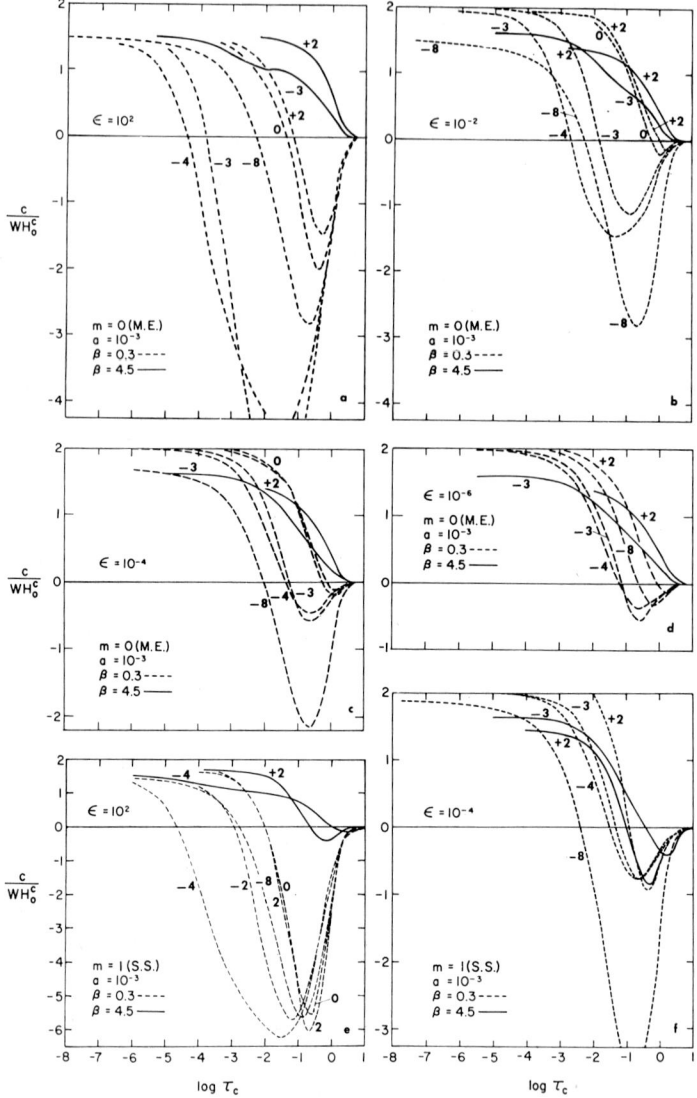

Fig. VII-1. Variation of the c term with depth (Athay and Skumanich, 1969, courtesy *Astrophysical Journal*, copyright 1969 by the University of Chicago. All rights reserved).

tions at the line and an emission line results in a deficiency of continuum emissions relative to continuum absorptions at the line. The continuum processes act, in other words, in such a way as to attempt to restore the continuum distribution of energy. The effect comes simply from the fact that the spectrum is perturbed from a normal continuum distribution by the presence of the line. It has nothing to do with the mechanism of line formation except as this affects the local value of W_J. The same is not at all true for the t term, however, which we now consider.

5. The t Term

Although the t term in Equation (VII-13) is defined for the general case it is simpler to discuss its meaning in terms of the two-level atom. For this case, Equation (VII-15) gives

$$t = M^{-1} \frac{\varepsilon}{r_0} (B - S_L).$$

Since

$$S_L = \frac{\int J_\nu \Phi_\nu \, d\nu + \varepsilon B}{1 + \varepsilon},$$

we may rewrite the expression for t in the form

$$t = M^{-1} \frac{\varepsilon}{r_0} \frac{1}{1 + \varepsilon} \left(B - \int J_\nu \Phi_\nu \, d\nu \right), \tag{VII-24}$$

which we may rewrite without loss of generality as

$$t = M^{-1} \frac{\varepsilon}{r_0} \frac{1}{1 + \varepsilon} \int \Phi_\nu (B - J_\nu) \, d\nu. \tag{VII-25}$$

The purpose in writing t in this latter form is to demonstrate that t does not vanish for large ε. In fact t increases with increasing ε and reaches a limiting value for $\varepsilon \gtrsim 10$. Strictly speaking, the only condition under which the assumption of LTE is justified is the condition of large ε and we see that t tends to be the largest in this case (Athay and Skumanich, 1969).

It is of interest to estimate surface values of t as compared to the preceding estimates of the surface values of c, viz., $c/WH_0^c \approx 1.5$ to 2. To a sufficiently good approximation for strong lines and for $\varepsilon \ll 1$ we may set $J_\nu \ll B$ in Equation (VII-25) to obtain

$$t(0) = M^{-1} \frac{\varepsilon}{r_0} \frac{1}{1 + \varepsilon} B(0) \tag{VII-26}$$

and

$$\frac{t(0)}{WH_0^c} = M^{-1} \frac{\varepsilon}{r_0} \frac{1}{1 + \varepsilon} \frac{B(0)}{WH_0^c} \tag{VII-27}$$

These approximate forms for $t(0)$ and $t(0)/WH_0^c$ are valid only in the strict two-level case and when the lines in question are in absorption. The more general form for S_L for a multilevel atom gives in place of Equation (VII-25)

$$t = \frac{M^{-1}}{r_0} \frac{\varepsilon^\dagger}{1 + \varepsilon^\dagger} \int_0^\infty \Phi_\nu \left(\frac{\varepsilon^*}{\varepsilon^\dagger} B - J_\nu \right) d\nu. \tag{VII-25a}$$

Since the ratio $\varepsilon^*/\varepsilon^\dagger$ may be small compared to unity $(\varepsilon^*/\varepsilon^\dagger) B$ may be less than

J_ν even when the line is in absorption. Thus, t may be negative for some strongly interlocked lines and will be poorly represented by Equation (VII-26) for these cases.

We recall from Chapter II, Equation (II-68) that the equivalent width W is related to δ through the approximate relation

$$W \approx \frac{M^{-1}\delta}{r_0}, \tag{VII-28}$$

where

$$\delta \approx \pi^{1/4}(r_0 a)^{1/2}, \quad \text{wing},$$
$$\approx (1 + \ln r_0^{-1}) r_0, \quad \text{shoulder}, \tag{VII-29}$$

and

$$\approx 1, \quad \text{linear},$$

on the regions of the curve-of-growth indicated. Thus, on the three regions of the curve-of-growth we have

$$W \approx M^{-1} \pi^{1/4} \frac{a^{1/2}}{r_0}, \quad \text{wing},$$
$$\approx M^{-1}(1 + \ln r_0^{-1}), \quad \text{shoulder},$$

and

$$\approx \frac{M^{-1}}{r_0}, \quad \text{linear}, \tag{VII-30}$$

The approximation $J_\nu \ll B$ is not valid, of course, on the linear portion of the curve-of-growth. This is particularly true in the violet end of the spectrum where J_c exceeds B and where J_ν will also exceed B in weak lines. In the red end of the spectrum where $J_c < B$ the approximation is better. Also, when $\varepsilon \gg 1$ we cannot assume $J_\nu \ll B$. We shall continue to apply the approximation to the lines on the linear portion of the curve-of-growth and to $\varepsilon \gg 1$ but the reader should keep in mind that it fails badly for the weak lines in the violet. Although the specific assumption of $J_\nu \ll B$ is not valid when $\varepsilon \gg 1$, we shall show later that it does not lead to large errors.

An approximate value of H_0^c is given by $\frac{1}{4}B(T_{\text{eff}})$ where T_{eff} is the effective temperature of the star. This result follows from Equation (VII-21) where $B(T_{\text{eff}}) \approx B(\tau_c = 1.5)$. With this value for H_0^c and the above values of W we obtain for the three segments of the curve-of-growth (Athay, 1970)

$$\frac{t(0)}{WH_0^c} \approx \frac{4\pi^{1/4}}{1+\varepsilon} \frac{\varepsilon}{(ar_0)^{1/2}} \frac{B(0)}{B(T_{\text{eff}})}, \quad \text{wing},$$
$$\approx \frac{4}{(1+\varepsilon)} \frac{1}{(1+\ln r_0^{-1})} \frac{\varepsilon}{r_0} \frac{B(0)}{B(T_{\text{eff}})}, \quad \text{shoulder}, \tag{VII-31}$$

and

$$\approx \frac{4}{1+\varepsilon} \varepsilon \frac{B(0)}{B(T_{\text{eff}})}, \quad \text{linear}.$$

Several points concerning Equations (VII-31) are worthy of note. Firstly, we note that $t(0)/WH_0^c$ is proportional to the boundary temperature through the quantity

$B(0)$. Since $B(0)$ is a strong function of temperature, the evaluation of $t(0)/WH_0^c$ will be very sensitive to the model assumed. Note that all of the estimated values of $t(0)/WH_0^c$ are positive and, as in the case of $c(0)$, positive values result in surface cooling. If $B(0)$ is overestimated in computing $t(0)$ the values of $t(0)$ will be too large and will produce excessive surface cooling. Similarly, if $B(0)$ is too small $t(0)$ will be too small and will produce too little surface cooling. It follows that an iterative scheme for successively estimating $B(0)$ and $t(0)$ will converge to a stable solution.

Secondly, we note that $t(0)$ may, in certain cases, exceed $c(0)$ by a large amount and in other cases lie below $c(0)$. The quantity $B(0)/B(T_{\text{eff}})$ is generally less than unity. If, in a very approximate way, we set $4\pi^{1/4}(B_0/B(T_{\text{eff}}))$ and $4B(0)/B(T_{\text{eff}})$ each equal to unity, we obtain

$$\frac{t(0)}{WH_0^c} \approx \frac{1}{1+\varepsilon}\left(\frac{\varepsilon}{ar_0}\right)^{1/2}, \quad \text{wing},$$

and

$$\approx \frac{1}{1+\varepsilon}\frac{\varepsilon}{(1+\ln r_0^{-1})r_0}, \quad \text{shoulder}, \quad \text{(VII-32)}$$

$$\approx \frac{\varepsilon}{1+\varepsilon}, \quad \text{linear}.$$

For $\varepsilon \ll 1$ this gives $t(0)/WH_c^0 \approx \varepsilon$ for the lines on the linear portion of the curve-of-growth and we therefore have $t(0) \ll c(0)$. However, $t(0)$ increases markedly for lines on the upper shoulder and wing portions of the curve-of-growth and may easily exceed $c(0)$ for such lines. With $\varepsilon = 10^{-4}$, $a = 10^{-3}$ and $r_0 = 10^{-5}$, for example, we obtain

$$\frac{t(0)}{WH_0^c} \approx 10^2.$$

These values are not unreasonable for such lines as the Na D lines and Mg b lines in the solar spectrum. The large values of $t(0)$ for the stronger lines means that the surface cooling effects will be largely with these lines and will result from the t term. This is simply another way of saying that the cooling in the outer layers is caused by the effect of added emissivity in the strong lines, which are formed in these higher layers.

Since the cooling effects of the lines are linear the relatively large number of lines on the upper shoulder of the curve-of-growth may combine to predominate over the relatively less numerous lines with strongly developed wings even though the latter are individually more important.

In the deeper layers of the atmosphere the ratio c/t is given by

$$\frac{c}{t} \approx \frac{r_0}{\varepsilon}\frac{\int (J_c - J_v)\,dv}{\int (B - J_v)\phi_v\,dv}, \quad \text{(VII-33)}$$

where we have taken $\varepsilon \ll 1$ and we have combined M^{-1} with Φ_ν to obtain ϕ_ν. For weak lines J_ν approaches J_c before it approaches B so the integral in the numerator is generally smaller in magnitude than that in the denominator. However, this is more than offset by the fact that r_0/ε is a large number. Thus, we obtain $c/t \gg 1$ for these lines. For strong lines the integral in the denominator becomes very small while the integral in the numerator remains relatively large and even though r_0/ε is small or of order unity we still find $c/t \gg 1$. These conclusions are fully verified by numerical computations.

In the violet end of the spectrum there is a tendency for S_L to exceed B in the layers just above $\tau_c = 1$ (Figure III-7). This leads to negative values of t and adds to the backwarming effect. In the red end of the spectrum the opposite tends to be true. However strong gradients in r_0 (Figure III-7b) may modify these results markedly.

Since B and J_c are generally not equal it is often the case in the deeper layers that J_ν is greater than B but less than J_c, or vice versa. Thus, t and c are often of opposite sign in the deeper layer. It is most often the case, however, that t and c are both positive in the higher layers, but, for the stronger lines, $t \gg c$. In a very real sense, therefore, one may think of c as the backwarming term and of t as the surface cooling term.

Again, however, the reader is cautioned that for some strongly interlocked lines t is negative even in the layers where τ_0 is small and these lines will produce surface heating.

Thirdly, we note that for the stronger lines $t(0)$ is proportional to ε/r_0 or $(\varepsilon/r_0)^{1/2}$ rather than to ε by itself. Departures from the LTE condition that result from $\varepsilon \ll 1$ most often lead to the result that r_0 is also much smaller than it is in LTE which is the case when the b_j coefficient is large compared to unity. Thus, the quantity ε/r_0 is not smaller by a factor of ε than it is in LTE, as has been argued by some authors. In fact, in some cases, ε/r_0 is almost independent of ε and the surface cooling effect for small ε is nearly as large as for large ε.

In the previous section we illustrated the effect of c in terms of an imbalance between continuum emissions and absorptions that were induced by the line. In a like manner the quantity t expresses the imbalance in the radiative rates in the line transition, as follows directly from the definition of ϱ in Chapter II. For the special case of a two-level atom this imbalance in radiative rates is exactly equal to the imbalance in collision rates. Thus, we may write t in the form

$$t = \frac{M^{-1}}{r_0} \varepsilon B \left(1 - \frac{S_L}{B}\right)$$
$$= \frac{M^{-1}}{r_0} \varepsilon B \left(1 - \frac{b_2}{b_1}\right). \qquad \text{(VII-34)}$$

Also, for εB we may write (Chapter II)

$$\varepsilon B = \frac{2h\nu^3}{c^2} \frac{\tilde{\omega}_1}{\tilde{\omega}_2} \frac{C_{12}}{A_{21}}, \qquad \text{(VII-35)}$$

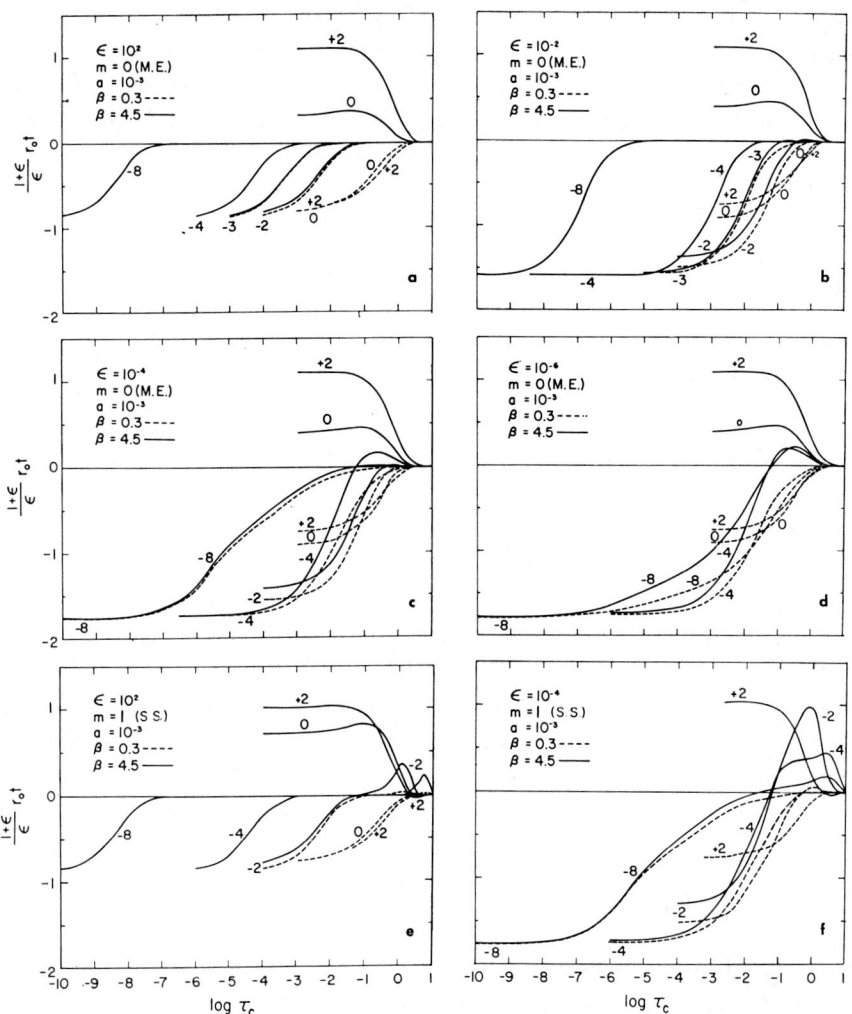

Fig. VII-2. Variation of the t term with depth. The quantity t plotted here follows the convection used by Athay and Skumanich and is the negative of the t used in the text. (Athay and Skumanich, 1969, courtesy *Astrophysical Journal*, copyright 1969 by the University of Chicago. All rights reserved).

and we see t is proportional to $C_{12}(1-b_2/b_1)$, which is just the net rate of collisional transitions in the line. The multilevel case is complicated by interlocking effects but t still contains a term of this same type in addition to the interlocking terms.

Numerical computations of t for the two level atom, for B linear in τ_c and for prescribed values of r_0 and ε are shown in Figure VII-2 (Athay and Skumanich, 1969). Note that the units on the ordinate are

$$t\frac{1+\varepsilon}{\varepsilon}\frac{r_0\pi^{1/2}}{M^{-1}B(0)}.$$

According to Equation (VII-26), we should expect, therefore, that the surface values are equal to $\pi^{1/2}$ for the strong lines. This is apparently true for all cases where $\varepsilon < 1$. For large ε the approximation $J_v \ll B$, or equivalently $S_L \ll B$, is not valid as previously noted. Note, however, that the surface values of t in Figure VII-2 for the cases $\varepsilon \gg 1$ are of the order of $\frac{1}{2}\pi^{1/2}$. Thus, the results obtained by assuming $J_v \ll B$ are not grossly in error.

6. Integrated Quantities

For some purposes it is useful to consider the integrated quantities

$$E(\tau_c) = \int_0^{\tau_c} e \, d\tau_c, \tag{VII-36}$$

$$C = \int_0^\infty c \, d\tau_c, \tag{VII-37}$$

and

$$T = \int_0^\infty t \, d\tau_c. \tag{VII-38}$$

The latter two quantities can be evaluated, generally, only by explicitly evaluating c and t as functions of τ_c.

Pecker (1951) has worked with a quantity $w(\tau_c)/W$, where $w(\tau_c)$ is the local equivalent width in flux units at τ_c and W is the equivalent width at $\tau_c = 0$. The quantity $w(\tau_c)/W$ is related to $E(\tau_c)$ through the equation

$$E(\tau_c) = w(\tau_c) H^c(\tau_c) - W H_0^c, \tag{VII-39}$$

which follows from the fact that

$$e = dH_c/d\tau_c$$

and that

$$dH_c = d[w(\tau_c) H^c(\tau_c)] \tag{VII-40}$$

in order to conserve flux. Equation (VII-39) can be rewritten as

$$\frac{w(\tau_c)}{W} = \frac{H_0^c}{H^c(\tau_c)} \left\{ 1 + \frac{E(\tau_c)}{W H_0^c} \right\}, \tag{VII-41}$$

the left hand side of which corresponds to the quantity computed by Pecker (1951).

The behavior of $w(\tau_c)$ depends explicitly on the depth variation of r_0 (as well as other parameters). Consider, for example, the value of $E(\tau_c)$ at $\tau_c = \infty$. For convenience, we denote this simply by E. Equation (VII-39) gives

$$E = W_\infty H_\infty^c - W H_0^c. \tag{VII-42}$$

Consider the case where B is linear in τ_c and where ϕ_ν and r_0 are constant with τ_c. We then have

$$B = B_0(1 + \beta\tau_c) \qquad \text{(VII-43)}$$

and

$$B = B_0\left\{1 + \beta\left(1 - \frac{\phi_\nu}{\phi_\nu + r_0}\right)\tau_\nu\right\}. \qquad \text{(VII-44)}$$

The flux at $\tau = \infty$ is given by (Kourganoff, 1963)

$$H_\infty^c = (\beta/3)B_0 \qquad \text{(VII-45)}$$

and

$$H_\infty^\nu = \frac{\beta}{3}\left(1 - \frac{\phi_\nu}{\phi_\nu + r_0}\right)B_0. \qquad \text{(VII-46)}$$

Thus, for $W_\infty H_\infty^c$ we have

$$W_\infty H_\infty^c = \int (H_\infty^c - H_\infty^\nu)\,d\nu$$

$$= B_0\frac{\beta}{3}\int \frac{\phi_\nu}{\phi_\nu + r_0}\,d\nu$$

$$= M^{-1}B_0\frac{\beta}{3}\frac{\delta}{r_0}. \qquad \text{(VII-47)}$$

We recall from Equation (VII-28) that

$$M^{-1}\frac{\delta}{r_0} = bW, \qquad \text{(VII-48)}$$

where b is a number of order unity when β is large. When β is of order unity or smaller b increases to a number appreciably larger than unity and in the special case of LTE and $\beta = 0$, b increases to infinity. This latter result follows from the disappearance of W when $S_L = B$ and $B = $ constant. With this understanding of b, we write

$$W_\infty H_\infty^c = B_0\frac{\beta}{3}bW. \qquad \text{(VII-49)}$$

From Equation (VII-21), we note that

$$B_0\frac{\beta}{3} = 2H_0^c - \frac{B_0}{2}. \qquad \text{(VII-50)}$$

Hence we find

$$W_\infty H_\infty^c = \left(2H_0^c - \frac{B_0}{2}\right)bW, \qquad \text{(VII-51)}$$

and Equation (VII-42) then gives

$$E = (2b - 1)WH_0^c - \frac{bWB_0}{2}. \qquad \text{(VII-52)}$$

Dividing by $W H_0^c$ and again using Equation (VII-21), we find

$$\frac{E}{WH_0^c} = \frac{(2b-1)\beta - 1.5}{\beta + 1.5}. \tag{VII-53}$$

According to Equation (VII-53) E/WH_0^c will be -1 at $\beta = 0$ and $\varepsilon \ll 1$ since b will remain finite under these conditions. As β increases E/WH_0^c will increase through zero and reach a value of approximately $+1$ for large β. This behavior is illustrated in Figure VII-3a from the work of Athay and Skumanich (1969). Note from this figure that for LTE the computed values of E/WH_0^c remain at $+1$ for all values of β.

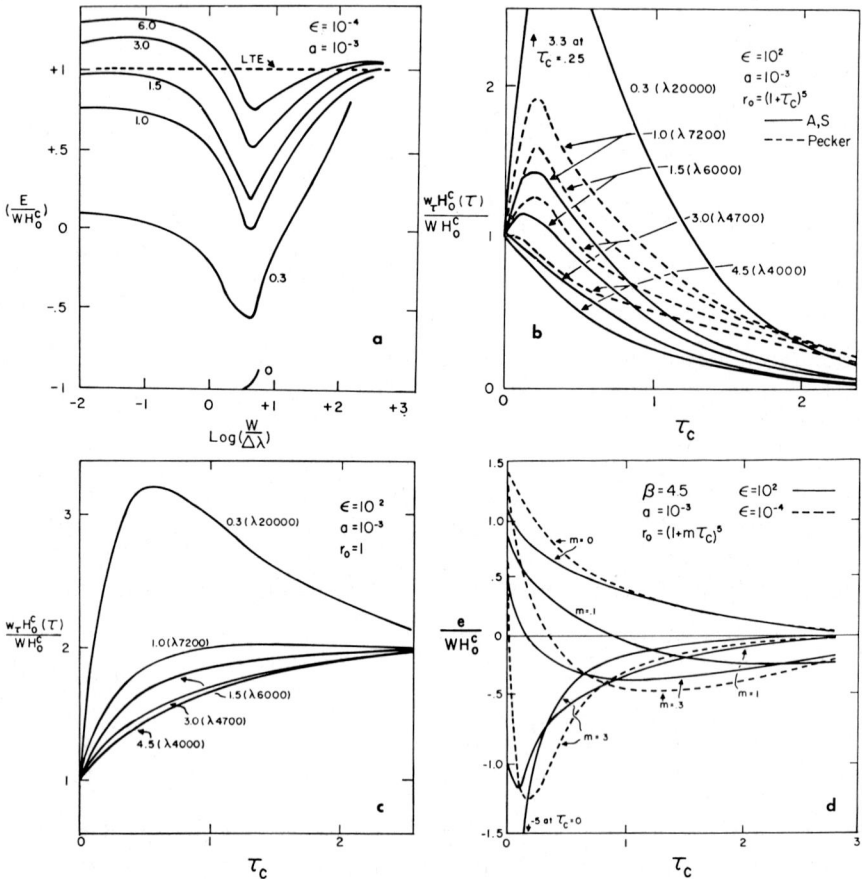

Fig. VII-3. a – The integrated blanketing term for $a = 10^{-3}$, $\varepsilon = 10^{-4}$ and $\beta = 0 - 6.0$. b – Comparison of the fractional equivalent width at τ_c obtained by Athay and Skumanich (1969) to the values obtained by Pecker for a Schuster-Schwarzschild model. c – The fractional equivalent width at τ_c for a Milne-Eddington atmosphere. d – The depth dependence of the blanketing function for a Schuster-Schwarzschild atmosphere as a function of thickness of the line forming layer (Athay and Skumanich, 1969, courtesy *Astrophysical Journal*, copyright 1969 by the University of Chicago. All rights reserved).

TABLE VII-1

Values of e/WH_0^c for $a = 10^{-3}$, $\varepsilon = 10^2$ (LTE), M-E Model

$r_0 \backslash \tau_c$	2.5	1.5	0.8	0.4	0.2	10^{-1}	10^{-2}	10^{-4}	10^{-6}	10^{-8}	0	$W/\Delta\lambda_D$	C/WH_0^c	T/WH_0^c	E/WH_0^c
							$\beta = 0.3$								
10^{+2}	−0.28	−0.44	0.01	1.58	3.5	5.6	9	10	—	—	10	0.0030	−2.9	3.9	1.00
1	−0.27	−0.67	−0.64	1.20	4.4	7.8	14	16	—	—	16	0.181	−3.3	4.3	1.00
10^{-2}	−0.16	−0.52	−1.45	−2.41	−2.3	0.77	102	382	—	—	382	0.74	−3.7	4.7	0.92
10^{-4}	−0.20	−0.58	−1.10	−1.41	−1.2	−0.47	26	5660	19800	—	19800	1.43	−3.7	4.8	1.15
10^{-8}	−0.26	−0.60	−0.88	−0.19	2.1	5.9	39	338	3770	666000	2330000	121	−3.7	4.8	1.16
							$\beta = 1.0$								
10^{+2}	0.022	0.081	0.34	0.80	1.34	1.84	2.8	3.0	—	—	3.0	0.0070	0.50	0.51	1.01
1	−0.007	0.011	0.18	0.79	1.67	2.62	4.4	4.9	—	—	4.9	0.43	0.134	0.87	1.00
10^{-2}	−0.010	−0.042	−0.13	−0.21	−0.03	0.97	32	118	—	—	118	1.71	−0.37	1.35	0.98
10^{-4}	−0.010	−0.019	−0.02	0.04	0.18	0.51	8.4	1730	6030	—	6030	3.4	−0.24	1.26	1.03
10^{-8}	−0.002	0.009	0.08	0.39	1.22	2.21	11.9	102	1130	200000	700000	290	−0.02	1.07	1.06
							$\beta = 3.0$								
10^{+2}	0.10	0.23	0.44	0.58	0.68	0.76	0.95	1.01	—	—	1.01	0.0117	1.50	−0.49	1.01
1	0.07	0.21	0.41	0.68	0.89	1.08	1.51	1.65	—	—	1.65	0.72	1.13	−0.128	1.00
10^{-2}	0.04	0.10	0.27	0.45	0.65	1.03	11.2	40	—	—	40	2.8	0.63	0.37	1.00
10^{-4}	0.05	0.14	0.29	0.46	0.59	0.74	3.3	580	2020	—	2020	5.6	0.77	0.22	0.99
10^{-8}	0.07	0.18	0.36	0.56	0.84	1.12	4.1	34	377	66600	233000	480	1.02	0.05	1.03
							$\beta = 4.5$								
10^{+2}	0.12	0.26	0.45	0.54	0.57	0.59	0.64	0.68	—	—	0.70	0.0132	1.67	−0.66	1.01
1	0.08	0.24	0.45	0.65	0.79	0.84	1.02	1.10	—	—	1.10	0.80	1.29	−0.29	1.00
10^{-2}	0.04	0.13	0.34	0.56	0.76	1.04	7.7	27	—	—	27	3.2	0.80	0.20	1.00
10^{-4}	0.06	0.17	0.35	0.53	0.66	0.84	2.5	386	1350	—	1350	6.3	0.94	0.04	0.98
10^{-8}	0.08	0.21	0.40	0.59	0.77	0.94	2.8	23	176	44500	156000	540	1.20	−0.174	1.02

TABLE VII-2

Values of e/WH_0^c for $a = 10^{-3}$, $\varepsilon = 10^{-4}$ M-E Model

$r_0 \backslash \tau_c$	2.5	1.5	0.8	0.4	0.2	10^{-1}	10^{-2}	10^{-4}	10^{-6}	10^{-8}	0	$W/\Delta\lambda_D$	C/WH_0^c	T/WH_0^c	E/WH_0^c
							$\beta = 0.3$								
10^{+2}	−0.10	−0.14	−0.01	0.37	0.82	1.20	1.80	1.94	—	—	1.94	0.0054	0.10	0.00	0.10
1	−0.09	−0.19	−0.15	0.21	0.70	1.13	1.82	1.97	—	—	1.97	0.39	−0.07	0.00	−0.07
10^{-2}	−0.04	−0.15	−0.36	−0.50	−0.30	0.15	1.58	2.00	—	—	2.00	2.8	−0.49	0.00	−0.49
10^{-4}	−0.05	−0.14	−0.27	−0.37	−0.36	−0.17	1.48	2.97	3.0	—	3.00	5.5	−0.55	0.01	−0.45
10^{-8}	−0.21	−0.48	−0.70	−0.12	1.70	4.8	29	141	312	389	393	149	−2.8	3.6	0.75
							$\beta = 1.0$								
10^{+2}	0.029	0.10	0.32	0.62	0.92	1.15	1.53	1.61	—	—	1.61	0.0079	0.79	0.00	0.79
1	0.004	0.04	0.22	0.53	0.87	1.14	1.58	1.68	—	—	1.68	0.54	0.61	0.00	0.61
10^{-2}	−0.005	−0.02	−0.05	−0.03	0.16	0.52	1.52	1.81	—	—	1.81	3.1	0.075	0.00	0.07
10^{-4}	−0.002	−0.005	−0.001	0.03	0.10	0.26	1.40	2.42	2.5	—	2.5	6.2	0.05	0.04	0.09
10^{-8}	−0.002	0.010	0.08	0.38	1.01	2.19	10.2	50	109	136	138	309	0.012	0.92	0.93
							$\beta = 3.0$								
10^{+2}	0.10	0.24	0.51	0.76	0.97	1.13	1.37	1.43	—	—	1.43	0.0107	1.19	−0.00	1.19
1	0.07	0.18	0.45	0.74	0.98	1.15	1.43	1.49	—	—	1.49	0.70	1.04	−0.00	1.05
10^{-2}	0.03	0.08	0.23	0.40	0.61	0.87	1.47	1.64	—	—	1.64	3.6	0.59	0.00	0.59
10^{-4}	0.04	0.11	0.24	0.38	0.50	0.64	1.34	1.94	1.96	—	1.96	7.0	0.58	−0.00	0.58
10^{-8}	0.07	0.18	0.35	0.56	0.79	1.11	3.77	17.7	39	48	49	492	1.01	−0.01	0.99
							$\beta = 4.5$								
10^{+2}	0.12	0.27	0.55	0.79	0.98	1.12	1.34	1.39	—	—	1.39	0.0116	1.27	−0.00	1.27
1	0.08	0.22	0.50	0.78	1.00	1.16	1.40	1.45	—	—	1.45	0.75	1.14	−0.00	1.14
10^{-2}	0.04	0.11	0.29	0.51	0.73	0.96	1.46	1.60	—	—	1.60	3.7	0.72	0.00	0.72
10^{-4}	0.05	0.14	0.30	0.48	0.61	0.74	1.32	1.81	1.82	—	1.82	7.4	0.73	−0.01	0.71
10^{-8}	0.07	0.19	0.40	0.58	0.74	0.94	2.69	12.1	27	33	33	550	1.14	−0.17	0.98

TABLE VII-3
Values of e/WH_0^c for $a = 10^{-3}$, $\varepsilon = 10^2$ (LTE), S-S Model

$r_0\backslash\tau_c$	2.5	1.5	0.8	0.4	0.2	10^{-1}	10^{-2}	10^{-4}	10^{-6}	10^{-8}	0	$W/\Delta\lambda_D$	C/WH_0^c	T/WH_0^c	E/WH_0^c
								$\beta = 0.3$							
10^{+2}	−0.28	−1.04	−2.6	−2.8	1.6	10.0	31	36	—	—	36	00.0074	−5.6	4.7	−0.93
1	−0.28	−1.04	−2.6	−2.8	1.6	10.1	34	40	—	—	40	0.066	−5.4	4.5	−0.94
10^{-2}	−0.38	−1.16	−2.5	−3.8	−3.8	−1.0	115	440	—	—	440	0.65	−4.8	3.9	−0.92
10^{-4}	−0.48	−0.88	−2.0	−3.2	−3.6	−2.2	33	26000	—	—	26000	1.08	−5.2	4.2	−0.93
10^{-8}	−0.33	−1.02	−2.7	−3.8	−1.8	4.7	70	640	7400	1260000	4300000	66	−5.4	4.7	−0.73
								$\beta = 1.0$							
10^{+2}	−0.11	−0.41	−0.99	−1.21	−0.26	1.63	6.6	7.8	—	—	7.8	0.00175	−1.72	0.73	−0.99
1	−0.11	−0.33	−0.99	−1.28	−0.39	1.64	7.8	9.6	—	—	9.6	0.157	−1.67	0.70	−0.99
10^{-2}	−0.19	−0.70	−1.10	−1.51	−1.24	−0.11	39	146	—	—	146	1.43	−1.79	0.76	−1.04
10^{-4}	−0.36	−0.40	−0.60	−0.76	−0.76	−0.21	10	2300	8000	—	8000	2.6	−1.76	0.63	−1.14
10^{-8}	−0.18	−0.49	−1.07	−1.45	−0.93	0.86	20	192	2200	380000	1290000	160	−1.68	0.72	−0.96
								$\beta = 3.0$							
10^{+2}	−0.065	−0.22	−0.51	−0.75	−0.85	−0.84	−0.45	−0.23	—	—	−0.23	0.0029	−0.59	−0.41	−1.00
1	−0.068	−0.23	−0.53	−0.79	−0.89	−0.75	0.22	0.64	—	—	0.64	0.26	−0.58	−0.43	−0.99
10^{-2}	−0.192	−0.55	−0.88	−0.71	−0.37	−0.16	1.3	49	—	—	49	2.4	−0.80	−0.34	−1.13
10^{-4}	−0.330	−0.26	−0.18	−0.04	0.08	0.39	3.1	780	2700	—	2700	4.2	−0.76	−0.44	−1.20
10^{-8}	−0.144	−0.33	−0.62	−0.79	−0.69	−0.23	5.6	64	770	127000	430000	270	−0.62	−0.40	−1.03
								$\beta = 4.5$							
10^{+2}	−0.051	−0.20	−0.44	−0.68	−0.92	−1.24	−1.6	−1.55	—	—	−1.55	0.0032	−0.42	−0.58	−0.99
1	−0.056	−0.18	−0.45	−0.71	−0.97	−1.15	−1.0	−0.84	—	—	−0.84	0.29	−0.40	−0.60	−1.0
10^{-2}	−0.164	−0.53	−0.81	−0.58	−0.16	0.20	8.6	32	—	—	32	2.7	−0.63	−0.52	−1.15
10^{-4}	−0.324	−0.24	−0.11	−0.10	−0.22	0.45	4.6	449	1790	—	1790	4.8	−0.59	−0.62	−1.21
10^{-8}	−0.138	−0.31	−0.54	−0.68	−0.64	−0.42	3.2	43	520	85000	290000	300	−0.45	−0.59	−1.04

TABLE VII-4

Values of e/WH_0^c for $a = 10^{-3}$, $\varepsilon = 10^{-4}$, S-S Model

$r_0 \backslash \tau_c$	2.5	1.5	0.8	0.4	0.2	10^{-1}	10^{-2}	10^{-4}	10^{-6}	10^{-8}	0	$W/\Delta\lambda_D$	C/WH_0^c	T/WH_0^c	E/WH_0^c
							$\beta = 0.3$								
10^{+2}	−0.08	−0.28	−0.70	−0.92	−0.54	0.19	2.0	2.5	—	—	2.5	0.00192	−0.96	0.00	−0.96
1	−0.08	−0.30	−0.65	−0.92	−0.56	0.18	1.91	2.4	—	—	2.4	0.174	−0.97	0.00	−0.97
10^{-2}	−0.07	−0.28	−0.58	−0.79	−0.68	−0.20	1.50	2.0	—	—	2.0	2.5	−0.97	0.00	−0.97
10^{-4}	−0.11	−0.18	−0.43	−0.68	−0.74	−0.54	1.36	3.1	3.2	—	3.2	4.9	−1.04	0.04	−1.00
10^{-8}	−0.23	−0.72	−1.9	−2.6	−1.2	3.3	45	230	506	626	636	93	−3.7	2.9	−0.82
							$\beta = 1.0$								
10^{+2}	−0.08	−0.28	−0.67	−0.87	−0.60	0.08	1.58	1.98	—	—	1.98	0.0022	−0.98	0.00	−0.98
1	−0.08	−0.29	−0.64	−0.87	−0.54	0.10	1.56	1.92	—	—	1.92	0.20	−0.99	0.00	−0.99
10^{-2}	−0.12	−0.37	−0.59	−0.59	−0.36	0.11	1.45	1.86	—	—	1.86	2.7	−1.03	0.00	−1.03
10^{-4}	−0.16	−0.20	−0.28	−0.34	−0.32	−0.11	1.29	2.6	2.6	—	2.6	5.3	−1.07	0.02	−1.07
10^{-8}	−0.17	−0.44	−0.95	−1.28	−0.80	0.78	16.2	87	193	240	242	178	−1.55	0.58	−0.97
							$\beta = 3.0$								
10^{+2}	−0.09	−0.29	−0.66	−0.84	−0.59	−0.015	1.21	1.55	—	—	1.55	0.0026	−0.99	−0.00	−0.99
1	−0.09	−0.30	−0.64	−0.83	−0.52	0.035	1.25	1.55	—	—	1.55	0.23	−1.00	−0.00	−1.00
10^{-2}	−0.16	−0.46	−0.60	−0.40	−0.03	0.39	1.39	1.68	—	—	1.68	3.0	−1.09	−0.00	−1.09
10^{-4}	−0.23	−0.21	−0.16	−0.04	0.11	0.33	1.21	2.1	2.1	—	2.1	5.6	−1.13	−0.00	−1.13
10^{-8}	−0.14	−0.33	−0.60	−0.75	−0.64	−0.19	5.1	31	70	87	88	276	−0.69	−0.34	−1.03
							$\beta = 4.5$								
10^{+2}	−0.07	−0.30	−0.67	−0.85	−0.59	−0.045	1.14	1.44	—	—	1.44	0.0026	−1.00	−0.00	−0.99
1	−0.09	−0.30	−0.63	−0.82	−0.51	0.018	1.17	1.45	—	—	1.45	0.24	−1.01	−0.00	−1.01
10^{-2}	−0.17	−0.49	−0.61	−0.34	0.06	0.47	1.38	1.62	—	—	1.62	3.1	−1.10	−0.00	−1.10
10^{-4}	−0.25	−0.22	−0.13	0.12	0.22	0.46	1.19	1.91	1.93	—	1.93	5.7	−1.14	−0.01	−1.14
10^{-8}	−0.14	−0.31	−0.54	−0.65	−0.61	−0.37	3.0	20	48	60	60	308	−0.54	−0.50	−1.04

The preceding discussion of E/WH_0^c is valid only for the strict Milne-Eddington condition r_0 =constant. If r_0 varies with depth, other results will be found. In the case of the Schuster-Schwarzschild atmosphere where r_0 increases to infinity, it is clear from Equation (VII-47) that W_∞ goes to zero. There is no longer any line opacity and no mechanism for retaining the line in the spectrum. In this case, we must have

$$E = - WH_0^c$$

and

$$E/WH_0^c = - 1. \tag{VII-54}$$

It follows that the depth variation of e must be radically different for the Milne-Eddington and Schuster-Schwarzschild cases, as illustrated earlier for the individual terms t and c.

Figure VII-3d illustrates the variation of e/WH_0^c with τ_c for the special case of $\beta = 4.5$ and $r_0 = (1 + m \, \tau_c)^5$ for different values of m. The Milne-Eddington case corresponds to $m = 0$ and gives positive values of e/WH_0^c at all depths. For $m \neq 0$, e/WH_0^c becomes negative at some depths in the atmosphere. The strongest negative region moves to progressively smaller values of τ_c as m is increased. These results are qualitatively the same regardless of the value of ε, i.e., for LTE and for strong departures form LTE.

The quantity $w(\tau_c) \, H^c(\tau_c)/WH_0^c$ given by Equation (VII-41) is plotted in Figures VII-3b and VII-3c for the Schuster-Schwarzschild (b) ($m=1$) Milne-Eddington (c) cases. Similar computations made by Pecker (1951) for the case of LTE and for r_0 'essentially constant' are shown in Figure VII-3b for comparison. The two sets of results in Figure VII-3b are not directly comparable because the explicit depth dependence of r_0 is not specified in Pecker's work.

Tables VII-1, VII-2, VII-3, and VII-4 contain values of e/WH_0^c as functions of τ_c and β and values of C/WH_0^c, T/WH_0^c and E/WH_0^c for the Schuster-Schwarzschild ($m=1$) and Milne-Eddington cases, and for $\varepsilon = 10^2$ (LTE) and $\varepsilon = 10^{-4}$ (Athay and Skumanich, 1969). Note that in the LTE cases for small β, C and T are of opposite sign and of large amplitude and that for $\varepsilon = 10^{-4}$ both C and T change markedly. Note also the shift in relative values of C and T as β, ε, r_0 and m are changed.

The damping parameter a also plays an important role in fixing the detailed behavior of $e(\tau_c)$ (Athay and Skumanich, 1969), for lines such that r_0 lies between unity and a, i.e., for lines that lie on the shoulder of the curve-of-growth. On the linear portion of the curve-of-growth the value of a makes little difference to either S_L or J. Hence, t and c are relatively unaffected by the value of a. For lines with strong wings each of the quantities c, t and W are essentially proportional to $a^{1/2}$ so that e/WH_0^c, E/WH_0^c, C/WH_0^c and T/WH_0^c are again relatively independent of a. Detailed differences occur, but these are of secondary importance.

The situation is quite different for lines on the shoulder of the curve-of-growth since the values of a and r_0 are of the same order. Under these conditions a has a different influence on each of the quantities t, c and W, and, as a result, e/WH_0^c and E/WH_0^c are quite dependent upon the exact value of a. The strong dips in the curves of

E/WH_c^0 in Figures VII-3a at $\log(W/\Delta\lambda_D) \approx 0.5$ largely disappear at $a = 10^{-1}$, for example, and E/WH_0^c becomes much more nearly equal to $+1$ for all $\beta \geq 1$.

As a general rule, all lines with well developed wings ($r_0 \ll a$) behave similarly with regard to e/WH_0^c.

7. Influence on Temperature Structure

To illustrate the influence of the blanketing terms on the temperature structure, we adopt the approximate treatment of Athay and Skumanich (1969). This will be followed by an illustration of the more detailed numerical effects resulting from model atmosphere computations.

The basic equation we wish to solve is Equation (VII-13) with t and c summed over all lines. We consider the Milne-Eddington case. The boundary conditions are

$$\left. \begin{array}{l} S \to B \\ B \to 3H(T_e)_{\tau_R} \end{array} \right\} \quad \tau_c \gg 1, \tag{VII-55}$$

where τ_R is the Rosseland opacity, and

$$J_\nu(0) = \text{const.} \, H_\nu(0), \quad \tau_c = 0. \tag{VII-56}$$

We require further that

$$\int H_\nu^b(0) \, d\nu = \int H_\nu^n(0) \, d\nu \tag{VII-57}$$

and

$$\int J_\nu^b(0) \, d\nu = \int J_\nu^n(0) \, d\nu. \tag{VII-58}$$

Superscripts b and n refer to the blanketed and non-blanketed cases, respectively. The first of these last two conditions preserves the total energy radiated by the stars and the second preserves the effective temperature of the stars. Conservation of flux (radiative equilibrium) requires

$$\int \{B_\nu^n(0) - J_\nu^n(0)\} \, d\nu = 0 \tag{VII-59}$$

and

$$\int \{B_\nu^b(0) - J_\nu^b(0)\} \, d\nu = \sum_i t_i(0). \tag{VII-60}$$

The last three equations combine to give

$$B^b(0) - B^n(0) = \sum_i t_i(0), \tag{VII-61}$$

again illustrating that the surface cooling is caused by the t term.

In pure scattering $t_i = 0$ and Equation (VII-61) predicts that no net surface cooling will occur. This agrees with Chandrasekhar's (1935) results and is based upon the same approximations. The more accurate solutions by Münch (1946) show, however, that even for pure scattering there is a small net cooling at the surface.

We seek a solution of Equation (VII-13) together with the boundary conditions that is of the form

$$B^b(\tau_c) = B_{C.I.}(\tau_c) + B_{P.I.}(\tau_c), \quad \text{(VII-62)}$$

where $B_{C.I.}(\tau_c)$ is the complementary solution and $B_{P.I.}(\tau_c)$ is the particular solution For the particular solution we write (Kourganoff, 1963)

$$B_{P.I.}(\tau_c) = -\sum_i e_i(\tau_c). \quad \text{(VII-63)}$$

At the surface we write Equations (VII-61), (VII-62), and (VII-63) as

$$B^b(0) = B^n(0) + \sum_i t_i(0),$$

$$B^b(0) = B_{C.I.}(0) + B_{P.I.}(0),$$

and

$$B_{P.I.}(0) = -\sum_i e_i(0),$$

which give

$$B_{C.I.}(0) = B^n(0) + \sum_i c_i(0). \quad \text{(VII-64)}$$

We can now write the homogeneous solution as

$$B_{C.I.}(\tau_c) = B^n(0) + \sum_i c_i(0) + \left\{3H(\tau_{\text{eff}})\frac{\kappa_R}{\kappa_c}\right\}_{\tau_c} \quad \text{(VII-65)}$$

and the total solution as

$$B^b(\tau_c) = B_{C.I.}(\tau_c) - \sum_i e_i(\tau_c). \quad \text{(VII-66)}$$

A schematic construction of the iterated solution is illustrated in Figure VII-4. The roles of the c and t terms are made clear by such an illustration.

Equation (VII-61) provides a convenient means for estimating the cooling effect of the lines at the surface. Suppose, for example, that in a particular star whose net flux $H(0)$ the lines block an apparent fraction ω of the continuum energy and that the value of $\sum_i t_i(0)/(W H_0^c)_i$ is $\alpha(0)$. It then follows from Equation (VII-61) that

$$B^b(0) - B^n(0) = \alpha(0)\omega H(0) \quad \text{(VII-67)}$$

or, since $H(0) \approx \tfrac{1}{2}B^b(0)$, and $B(0) = \sigma T^4(0)$, it follows that

$$\frac{T_b^4(0) - T_n^4(0)}{T_b^4(0)} = \tfrac{1}{2}\alpha(0)\omega.$$

Finally, we obtain

$$T_b(0) = T_n(0)\left(\frac{1}{1 + \alpha(0)\omega/2}\right)^{1/4}. \quad \text{(VII-68)}$$

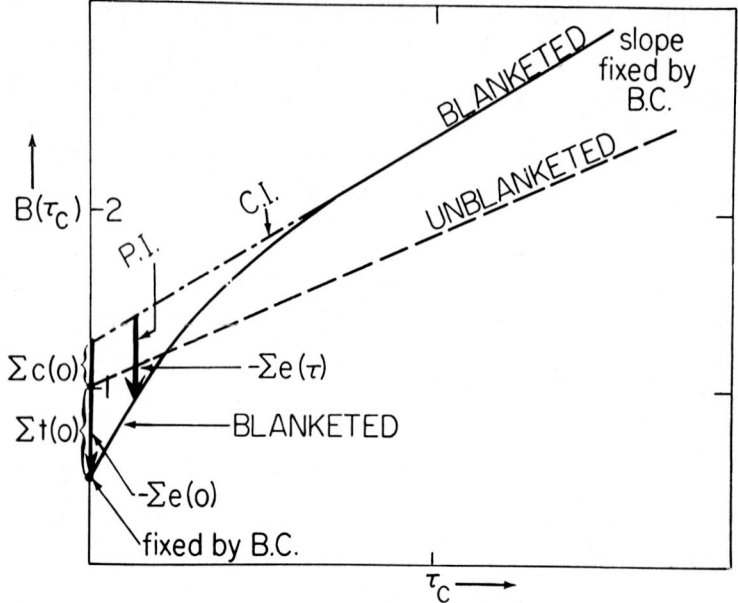

Fig. VII-4. A schematic illustration of the role of the c and t terms in changing the radiative equilibrium temperature distribution (Athay and Skumanich, 1969, courtesy *Astrophysical Journal*, copyright 1969 by the University of Chicago. All rights reserved).

For $\omega = 0.1$, this equation gives $T_b(0) = 0.95 \, T_n(0)$ at $\alpha(0) = 5$, $T_b(0) = 0.90 \, T_n(0)$ at $\alpha(0) = 10$ and $T_b(0) = 0.84 \, T_n(0)$ at $\alpha(0) = 20$. The corresponding decreases in the values of $T(0)$ for $T_n(0) = 5000°$ are 250°, 500°, and 800°.

Note from the results in Tables VII-1 to VII-4 that a single line with $r_0 = 10^{-8}$ has values of t/WH_0^c ranging from about 10^5 in LTE to about 10^2 when $\varepsilon = 10^{-4}$. Such a line, in the visual spectrum, will block a fraction ω of the continuum energy of about 10^{-3}. Hence, the product $\omega \, \alpha(0)$ for such a line amounts to about 10^2 for LTE and to about 10^{-1} for $\varepsilon = 10^{-4}$. One such line, formed in LTE, would produce remarkably strong surface cooling. Even for $\varepsilon = 10^{-4}$, ten lines with $\omega \, \alpha(0) = 10^{-1}$ (or their equivalent) would produce a 10% drop in $T(0)$.

8. Multilevel Effects

Interlocking effects in multilevel atomic models couple the values of J_ν in different spectral regions and may strongly modify the local values of J_ν in a given line. This, in turn, may modify both the escape coefficient and the line source function. Thus, the interlocking may alter both c and t. The results in Chapter IV suggest that, in some cases, the interlocking effects may be large.

The effects of interlocking on line blanketing have been investigated in a few specific cases only. Auer and Mihalas (1969a, b, c) and Mihalas and Auer (1970) have investigated the effects of the Lyman, Balmer, Paschen, and Brackett lines of hydrogen up to level $n=5$ in O and early B type stars. Although these authors have not formulated the problem explicitly in terms of t and c and have not separated the line and continuum fluxes, it is possible to infer some of the multilevel effects in hydrogen in the early type stars. The reader should be aware by now that the effects will change depending upon the energy level configuration and the temperature structure of the atmosphere itself.

When only the lowest two energy levels of hydrogen are allowed to depart from their LTE populations, Auer and Mihalas (1969a) find that in the higher layers of the atmosphere b_1 and b_2 are each less than unity but that $b_2 < b_1$. This means that the source functions for the Lyman and Balmer continua exceed the Planck function but that the Lyman-α source function is less than the Planck function. As a result $t(0)$ is negative ($\varrho < 0$) for the continua and positive for Lyman-α, i.e., the continua produce a local heating and Lyman-α produces cooling.

When higher energy levels are allowed to depart from LTE (Auer and Mihalas, 1969b, c; Mihalas and Auer, 1970) the blanketing effects of Lyman-α and of the continua are not seriously altered. However, the Balmer, Paschen, and Brackett lines interlock strongly with the continua and produce an increase in the net heating in the higher layers. The heating again results from the fact that the escape coefficients are negative (Equation (VII-10) and this can arise only when interlocking effects are strong. If each line were treated individually, without interlocking, the results in Chapter III clearly indicate that we could find in the outer layers that $S_L < B$, which requires that the escape coefficient would be positive and that a net cooling would occur. This, then, is a clear demonstration that interlocking effects can be of major importance.

As noted earlier the heating due to Balmer-α will depend upon the temperature distribution in the atmosphere of the star and upon the effective temperature of the star. In cooler stars the continua will interact with the lines in a somewhat different way, and, in the sun, the Balmer-α line produces a net cooling at all depths (Athay, 1970).

Blanketing functions computed for doublet and triplet configurations such as those represented by the Na D, Mg b and Ca II H and K lines in the solar atmosphere (Athay, 1970) show that in the deeper layers e/WH_0^c is the same for each line in the multiplet. In the surface layers, however, e/WH_0^c becomes proportional to $(\tilde{\omega} f)$ or $(\tilde{\omega} f)^{1/2}$ (note that $\tilde{\omega} f \propto r_0^{-1}$) as indicated by Equation (VII-31).

For two lines in series (3-2 and 2-1) in the solar atmosphere there is very little difference in the blanketing functions for the individual lines when the wavelengths, equivalent widths, and atomic cross-sections are the same. This suggests that the interlocking effects are minimal in this case, as might be expected from the results in Chapter IV.

There are energy level configurations for which interlocking can lead to a heating effect in one or more lines in the higher layers of the solar atmosphere. An example of such a case is shown in Figure VII-5 for two lines with a common upper level (3) and different lower levels (1 and 2). Level 1 is the ground state and level 2 is a metastable level lying 0.83 eV above the ground state. The ionization potential is 6.2 eV. The two lines are at wavelengths of $\lambda 2500$ and $\lambda 3000$. Their transition probabilities are $A_{31} = 10^8 \text{s}^{-1}$ and $A_{32} = 10^7 \text{ s}^{-1}$, and their excitation cross-sections are $Q_{12} = 3 \pi a_0^2$, $Q_{13} = 10 \pi a_0^2$ and $Q_{23} = 10 \pi a_0^2$.

Examples are given in Figure VII-5 for three different relative abundances, i.e., three equivalent widths, for Gaussian profiles and for a fourth case with a Voigt profile. When the 3-2 line has a small equivalent width it produces heating at all

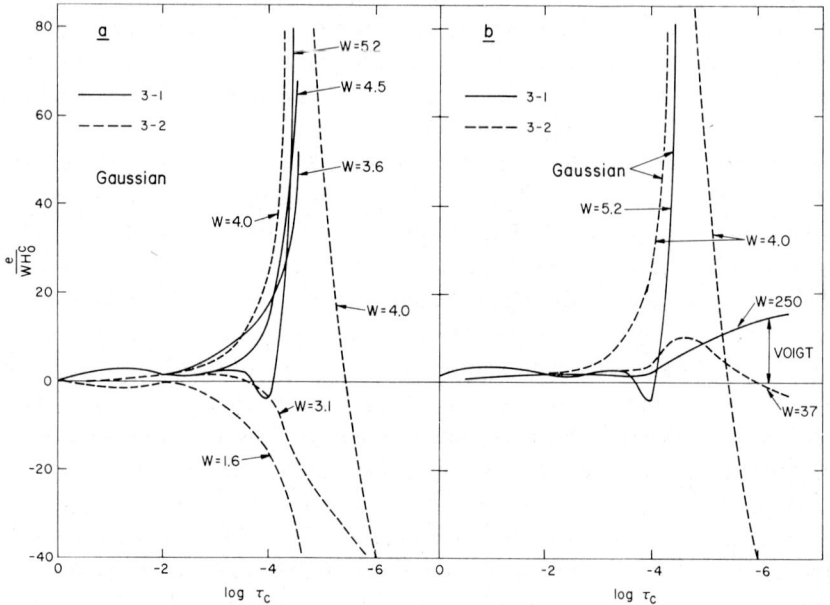

Fig. VII-5. Depth dependence of the blanketing functions for a two-line metastable configuration illustrating the occurrence of surface heating due to the 3-2 transition and cooling due to the stronger 3-1 transition. The lines are at $\lambda_{31} = 2500$ and $\lambda_{32} = 3000$.

depths, but as its equivalent width increases it produces cooling at all depths except $\tau_c > 1$ and τ_c very small. The 3-1 line, by contrast, produces cooling at all depths where $\tau_c < 1$ except for a small region around $\tau_c = 10^{-4}$ and for the case where the Doppler core is heavily saturated. In all cases, the net effect of the two lines is to produce cooling at small τ_c.

Effects similar to those illustrated in Figure VII-5 occur in more complex configurations involving metastable levels and closed loops of transitions. However, the heating effects at small τ_c appear always to be confined to one or two of the weaker lines and the net effect of all the lines from a given energy level configuration is to produce cooling at small τ_c. Thus, there appears to be no counterpart in stars of solar type to the Balmer-α heating found in stars of early spectral class.

9. Early Stellar Types

The effects of line blanketing in stars of early spectral type have been treated by Mihalas and Auer (1970) for effective temperatures from 25000° to 40000°. They consider departures from LTE in the lines and continua of hydrogen up to level $n=4$. Comparison of their results for cases in which the lines and continua are included to results where only the continua are included shows an *increase* in temperature in the external layers of zero to 10%. In the cases considered there are competing effects with some lines producing cooling (Lyman lines) and some producing heating (Balmer and Paschen lines). The free-bound continua also produce a heating, which increases the boundary temperature by 10 to 15%. Heating is pronounced at all depths where m, the mass in a cm^2 column, is less than 10^{-2} gm. In the hotter stars a net cooling occurs, mainly due to the free-bound continua, at depths where m is between 1 and 10^{-2}.

Earlier computations by Auer and Mihalas (1969c) for an effective temperature of 15000° gave a combined heating due to Balmer-α, Balmer-β and Paschen-α of about 13% of $T(0)$. Approximately half of this amount is due to Balmer-α. The free-bound continua alone produce an additional 7% increase in $T(0)$. In LTE the same lines and continua produce strong cooling effects. Thus, in these early type stars the blanketing effect of just a few strong lines is sufficient to produce marked changes in the temperature structure in the outer layers where the lines form.

These blanketing effects are based upon the assumption of radiative equilibrium. As indicated by Equations (VII-31), the surface effects depend upon $T(0)$ itself. If there is mechanical energy dissipation in early type stars, as there is in solar type stars, $T(0)$ will be increased above the radiative equilibrium value. This effect is pronounced in the sun and probably plays some role in stars of all spectral types.

10. The Solar Case

In stars of solar type it is not feasible to treat all of the lines individually. There are some 40000 known lines in the solar spectrum blocking out about 12% of the apparent continuum. No single line nor class of lines dominates the line spectrum, at least in any obvious way.

To treat line blanketing in solar type stars with large numbers of spectral lines it is necessary to find an adequate classification of lines that permits the effect of the lines to be simulated in some tractable way. Such a classification must adequately allow for the depth variations of r_0, for the equivalent widths and wavelengths of the lines and for an appropriate range of atomic parameters, including, damping widths, collision cross-sections and transition probabilities. Depth variations of r_0 will depend upon ionization potential. Also, each of the three broad regions of the curve-of-growth (the linear, shoulder and wing regions) exhibit different relationships between $t(0)$ and r_0 as shown by Equations (VII-31). Thus, a minimum classification of solar lines must include at least six cases: one for each region of the curve-of-growth for neutral metals and a similar set for ionized metals. Some lines, notably those of hydrogen, are not well represented by any of these classes and should form a separate class.

Lines on the shoulder of the curve-of-growth encompass values of r_0 ranging from about 0.3 to 10^{-4}. As r_0 decreases the line cores are formed progressively higher in the atmosphere and have a progressively stronger effect upon surface cooling. Since $t(0)/WH_0^c$ is proportional to r_0^{-1} for these lines (Equation (VII-31)), it is difficult to represent such a large range in r_0 by a single class of lines. This suggests that lines on the shoulder of the curve-of-growth should more properly be represented by two or more classes. Still further refinement in classification is perhaps needed to properly allow for the effects of wavelength and interlocking. The wavelength effects enter through the depth dependence of r_0 and through the slope of the Planck function, hence the relationship between J_ν and J_c. Interlocking effects were considered in an earlier Section 8 of this chapter.

Because of the many effects that must be taken into account in the solar case, it is not clear exactly how the problem should be treated. As a first attempt at the solar blanketing problem Athay (1970) considered the six classes of lines represented by the three regions of the curve-of-growth for a neutral metal and an ionzed metal plus some dozen or so strong lines of Na, MgI, MgII, CaII and hydrogen.

Values of α ($\alpha = e/WH_0^c$) for ten of the individual lines treated are given in Table VII-5. Note that for $\tau_5 \gtrsim 1$ most of the values are negative (back warming), as is to be expected. The negative values result primarily from the c term and reflect the fact that the lines are locally in emission at these depths. Exceptions occur in the case of Hα and the two subordinate lines of CaII and MgII. These lines have relatively large lower excitation potentials and this strongly modifies the depth variation of r_0.

For $\tau_5 \lesssim 10^{-4}$ most of the values of α lie above ten. According to Equation (VII-68) a value of $\alpha = 10$ for all of the solar lines would produce a surface cooling of about 10%. Thus, the results in Table VII-5 clearly suggest that the majority of the strong solar lines will contribute to the surface cooling effect.

The lines illustrated in Table VII-5 have been considered with sufficient care to obtain reasonably close agreement between computed and observed profiles. Care has been taken to choose the best available atomic parameters and to use a reasonably realistic model atmosphere. The model used is essentially identical to that given in Chapter IV with the exception that $T_{\min}$ is 4600° for the blanketing computations. Since $t(0)$ is proportional to $B(0)$, the values of α at small τ_5 are sensitive to the model. The exceptionally large values of α at $\tau_5 = 10^{-6}$ and 10^{-8} result from the chromospheric rise in temperature and reflect the fact that the temperature rise leads to an increased loss of energy. Note that the cooling at $\tau_5 < 10^{-6}$ is dominated almost entirely by the Balmer lines with a somewhat smaller contribution from the MgII resonance lines.

At small values of τ_5 the cooling indicated in Table VII-5 must be replaced by mechanical heating, i.e., the chromospheric temperature rise would disappear if heating were not present. The amount of heat energy required is given by $4\pi \Delta H = 4\pi \alpha WH_0^c \Delta\tau$ and amounts to about 1.8×10^6 ergs cm^{-2} s^{-1} for $\tau_5 \leqslant 10^{-4}$, most of

TABLE VII-5

Blanketing Functions for Selected Solar Lines

Line	τ_5 2.5	1.5	0.8	0.4	0.2	10^{-1}	10^{-2}	10^{-3}	10^{-4}	10^{-6}	10^{-8}	C/WH_0^c	T/WH_0^c	E/WH_0^c	$W(Å)$
Na D$_2$	−0.18	−0.32	−0.38	−0.22	+0.06	0.46	0.71	5.9	31	270	340	−1.0	+0.3	−0.7	1.3
Na 8195	−0.17	−0.28	−0.21	+0.07	+0.48	0.89	2.5	11	24	56	70	−1.3	+1.0	−0.3	0.3
Mg b$_1$	−0.26	−0.42	−0.40	0	+0.70	1.6	2.3	5.0	14	45	45	−1.0	+0.4	−0.6	3.8
Mg 3838	−0.23	−0.33	−0.29	−0.04	+0.30	1.0	1.6	3.3	7.3	56	70	−0.4	−0.2	−0.6	4.5
Mg II 2795	0	0	+0.02	+0.24	+0.65	0.94	1.2	2.7	12	250	3000	+1.1	−0.1	+1.0	55
Mg II 2790	0	+0.05	+0.15	+1.2	+1.6	1.6	1.4	1.3	1.5	90	300	+2.3	+0.3	+2.6	...
Ca II 3933	−0.25	−0.55	−0.24	+0.12	+0.90	1.2	1.5	2.5	12	55	160	−0.3	0	−0.3	35
Ca II 8542	−0.23	−0.66	−0.96	−0.85	+0.39	0.90	2.8	11	16	150	400	−1.9	+0.3	−1.6	7.8
Ca II 3179	−0.12	−0.53	+0.01	+0.49	+1.4	1.5	1.4	2.0	4.7	60	250	+0.1	+0.3	+0.4	0.6
Hα	+0.01	+2.0	+1.9	+3.0	+1.9	1.6	1.1	1.2	30	4000	1.1×10^5	+0.6	+3.8	+4.4	26(4)

which comes from the Balmer lines. Since the total solar flux is 6.18×10^{10} ergs cm^{-2} s^{-1} the heat energy that must be supplied to sustain the chromospheric temperature rise near $\tau_5 = 10^{-4}$ represents a fraction 3×10^{-5} of the solar flux.

For the remainder of the solar lines not represented in Table VII-5 it is impractical to attempt to match profiles. Nevertheless, it is necessary to make allowances for the differences in wavelengths, equivalent widths and atomic parameters. The total equivalent width of all solar lines between $\lambda 3061$ and $\lambda 8770$ is approximately 1000 Å. These divide into the six classes on the curve-of-growth plus the lines in Table VII-5 as shown in Table VII-6. There is a strong clustering of the lines in the violet end of

TABLE VII-6

Total equivalent widths (A) of lines in six classes plus Table VII-5

Class	Linear	Shoulder	Wing
Neutral Metal	340	340	70
Ionized Metal	60	60	7
Table VII-5	–	–	134

the spectrum. However, in the ultraviolet H_0^c decreases rapidly and the value of WH_0^c decreases even though W is increasing. The lines in the vicinity of $\lambda 4000$ are particularly important to the blanketing effect because the cumulative W is large H_0^c is still relatively large.

Athay (1970) computed blanketing functions for the six classes of lines on the curve-of-growth for a neutral metal with the ionization potential of Fe I and a metal ion with the ionization potential of Ti II. The model atoms included three bound energy levels with the 3-2 and 2-1 transitions permitted. Level 3 was at an excitation potential of 5.8 eV and level 2 was varied from 2.5 to 4.1 eV. This gave a resonance line and a subordinate line with wavelengths varied from $\lambda 3000$ to $\lambda 7500$, although most of the cases computed were for lines with wavelengths $\lambda 3750$ to $\lambda 5000$. The atomic parameters, including collision cross-sections, transition probabilities, photoionization cross-sections and damping parameters were varied by factors of 10^{-1} and 10 from the reference values. Additional changes were made in relative abundances in order to obtain a range of equivalent widths.

The changes in wavelengths, atomic parameters and abundances represented enough different cases to provide approximately 15 separate blanketing functions for each of the six classes of lines. These cases were then averaged to obtain average blanketing functions for each of the six classes. Finally, the six classes were combined with the lines in Table VII-5 to obtain a composite blanketing function by weighting each class in proportion to its contribution to the total equivalent width (Table VII-6).

This process was repeated for three initial model atmospheres with selected boundary temperatures of 4600°, 4300° and 4000°. The models follow that given in Chapter IV for $\tau_5 > 10^{-2}$ but omit the chromospheric temperature rise. The 4300° boundary

TABLE VII-7
Average and Composite Blanketing Functions, α

Parent	Curve-of-Growth	τ_5 2.5	1.5	0.8	0.4	0.2	0.1	10^{-2}	10^{-3}	10^{-4}	10^{-6}	10^{-8}	Weight
Fe I	Linear	−0.23	−0.42	−0.57	−0.47	−0.16	+0.34	+1.2	+1.7	1.8	2.1	2.3	...
	Shoulder	−0.19	−0.27	−0.25	−0.07	+0.17	+0.50	+1.4	+5.8	13	32	45	...
	Wing	−0.18	−0.29	−0.35	−0.23	−0.10	+0.18	+0.07	+2.2	9.5	150	290	...
Ti II	Linear	−0.43	−0.45	+0.02	+0.70	+1.1	+1.2	+1.3	+1.4	1.8	3.2	15	...
	Shoulder	−0.22	−0.17	+0.12	+0.55	+0.78	+0.96	+1.3	+7.0	81	100	230	...
	Wing	−0.22	−0.14	+0.31	+0.72	+0.82	+0.88	+1.0	+2.9	24	190	320	...
Fe I	Linear	−0.08	−0.14	−0.19	−0.16	−0.05	+0.12	+0.41	+0.58	0.61	0.71	0.78	0.34
	Shoulder	−0.06	−0.09	−0.09	−0.02	+0.06	+0.17	+0.48	+2.0	4.4	11	15	0.34
	Wing	−0.01	−0.02	−0.02	−0.02	−0.01	−0.01	0	+0.15	0.66	10	20	0.07
Ti II	Linear	−0.03	−0.03	0	+0.04	+0.07	+0.07	+0.08	+0.08	0.11	0.16	0.90	0.06
	Shoulder	−0.01	−0.01	+0.01	+0.03	+0.05	+0.06	+0.08	+0.42	4.9	6	14	0.06
	Wing	0	0	0	+0.01	+0.01	+0.01	+0.01	+0.02	0.17	1.3	2.2	0.007
Table VII-5	...	−0.01	+0.02	+0.03	+0.08	+0.10	+0.12	+0.15	+0.30	1.8	26	540	W/1000 Sum of weighted results
Composite	...	−0.20	−0.27	−0.26	−0.04	+0.23	+0.56	+1.2	+3.6	13	56	590	

temperature is reached at $\tau_5 \approx 2.5 \times 10^{-3}$ and the 4000° boundary temperature is reached at $\tau_5 \approx 5 \times 10^{-4}$.

Table VII-7 contains the individual blanketing functions for the 4600° case. The first six rows give the average functions for each class and the second six rows give the weighted contributions to the composite curve. The unweighted functions are plotted in Figure VII-6. Composite blanketing functions for the three models plus an additional model with a chromospheric temperature rise are illustrated in Figure VII-7.

It is of interest to note that the major contribution to the cooling at $\tau_5 = 10^{-4}$ comes from lines on the shoulder of the curve-of-growth and that the cooling is shared equally between neutral and ionized metals even though the neutral metal lines contribute nearly six times as much to the total equivalent width as do the ionized metal lines. It is of further interest to note that the composite blanketing curve for the 4600° case is very similar to the individual curves in Table VII-5 at optical depths near 10^{-2} to 10^{-4} where most of the surface cooling occurs.

The back warming in the deeper photospheric layers around $\tau_5 = 1$ is produced mainly by the neutral metal lines on the linear and shoulder portions of the curve-of-growth.

From the composite curves in Figure VII-7 and Equation (VII-68) it is clear that the 4600° model leads to relatively strong surface cooling and the 4000° model leads to relatively weak surface cooling. Model solar atmospheres obtained by solving

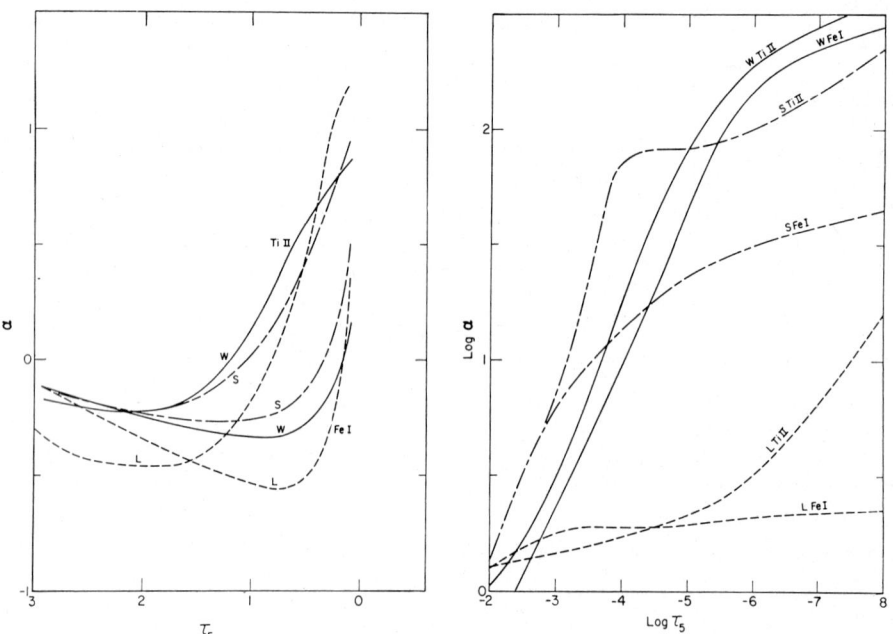

Fig. VII-6. The depth dependence of the blanketing functions α for the six classes of Fe I and Ti II lines for the 4600° model (Athay, 1970, courtesy *Astrophysical Journal*, copyright 1970 by the University of Chicago. All rights reserved).

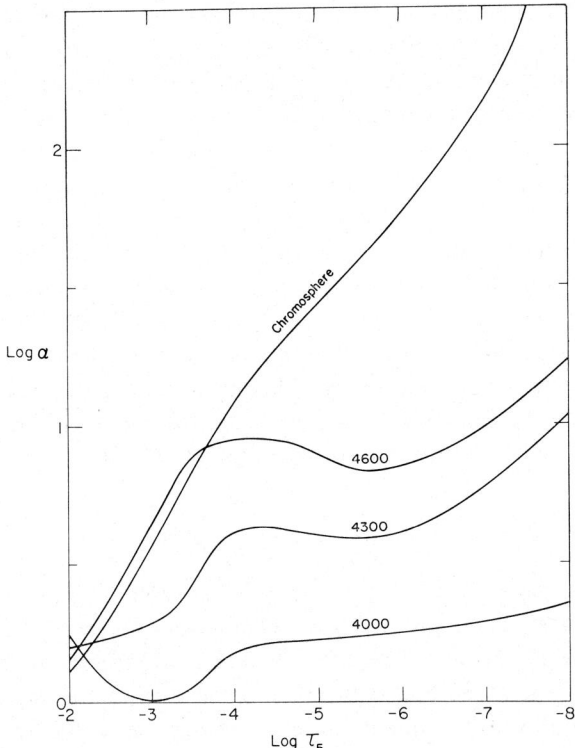

Fig. VII-7. The depth dependence of the composite blanketing function α for four model atmospheres with different boundary temperatures (Athay, 1970, courtesy *Astrophysical Journal*, copyright 1970 by the University of Chicago. All rights reserved).

Equation (VII-13) with $e=0$ give a boundary temperature of 4660°. However, when e is set equal to $\alpha\,WH_0^c$ with $\omega=0.12$ and values of α given by the three initial model atmospheres the solutions to Equation (VII-13) give (Athay, 1970):

Assumed $T(0)$	4600	4300	4000
Computed $T(10^{-4})$	4030	4420	4650

A self-consistent solution is found for $T(10^{-4})=4330$. Thus, the effect of line blanketing in the sun appears to lower the temperature at $\tau_5=10^{-4}$ by a little over 300°. At $\tau_5=10^{-3}$ the cooling amounts to about 170° and near $\tau_5=1$ the back warming amounts to about 200° (Athay, 1970).

Although the effects of line blanketing in the sun appear to be small they nevertheless are of great importance in solar physics. Much of the current concern in solar physics is with the effects of mechanical energy dissipation at small optical depths and the onset of the chromospheric temperature rise. The cooling at $\tau_5=10^{-3}$ of 170° results from an equivalent change in the continuum energy flux given by Equation (VII-13) of

$$\Delta H/H = \alpha\omega\Delta\tau.$$

For $\alpha = 3.6$, $\omega = 0.12$ and $\Delta\tau = 10^{-3}$ we find $\Delta H/H = 4.3 \times 10^{-4}$. It follows that a rise in T of 170° above the radiative equilibrium value at $\tau_5 = 10^{-3}$ would require an equivalent amount of mechanical energy dissipation. By comparison, the mechanical energy required to produce the much larger rise in T in the chromosphere at $\tau_5 \leqslant 10^{-6}$ was found in the preceding discussion to be equivalent to $\Delta H/H = 3 \times 10^{-5}$. Thus, the decrease in T of 170° at $\tau_5 = 10^{-3}$ is much more significant in terms of energy considerations than is the chromospheric rise in T.

Obviously, the model atoms considered (Athay, 1970) for the solar blanketing problem are too simple to provide a completely accurate description of the blanketing effect. More complex configurations in which the type of effects illustrated in Figure VII-5 are present are needed. Also, the importance of the lines on the shoulder of the curve-of-growth in the composite blanketing curve together with the strong dependence of α upon r_0 and the wide range in r_0 for these lines requires that these lines be further subdivided in at least two or three classes.

11. The Cayrel Mechanism

Cayrel (1966) has pointed out that in the external layers of stars where collisional effects become unimportant the radiative equilibrium temperature will rise to the color temperature of the star. The effect is exactly analogous to that in planetary nebulae where, in the absence of collisions, the nebula reaches an equilibrium temperature that conforms to the 'quality' of the radiation field rather than to the photon energy density. This effect is aside from our central theme of photon transport in spectral lines. However, because the actual response of the stellar atmosphere to the Cayrel mechanism is closely coupled with the line blanketing effects we shall discuss the Cayrel mechanism briefly at this point.

The Cayrel mechanism arises from a breakdown in LTE in the continuum. We write the continuum source function in the form

$$S_c = \frac{2h\nu^3}{c^2} \frac{1}{b_c e^{h\nu/kT} - 1}, \qquad \text{(VII-69)}$$

which we approximate by

$$S_c = B - B(1 - 1/b_c), \qquad \text{(VII-70)}$$

i.e., we ignore stimulated emissions. We then have in place of Equation (VII-13)

$$dH_c/d\tau_c = e - \int_0^\infty B(1 - 1/b_c) \, d\nu$$

$$= e - \sigma T^4 (1 - 1/b_c), \qquad \text{(VII-71)}$$

or dividing by $H_0^c = \sigma T_{\text{eff}}^4$ we obtain

$$\frac{1}{H_0^c} \frac{dH_c}{d\tau_c} = \omega\alpha - \frac{T^4}{T_{\text{eff}}^4} (1 - 1/b_c). \qquad \text{(VII-72)}$$

TABLE VII-8

Values of $\omega\alpha$ and $(T(0)^4/T_{\text{eff}})^4 (1 - 1/b_c)$ at small τ_c

	$T(0)$		
	4600	4300	4000
$\omega\alpha$	1.1	0.49	0.19
$(T(0)/T_{\text{eff}})^4 (1 - 1/b_c)$	0.20	0.15	0.11

Comparison of Equation (VII-22) to Equation (VII-13) shows that the departures from LTE in the continuum produce an effect equivalent to that of line blanketing. When $b_c = 1$ the effect disappears and when b_c becomes large the effect becomes relatively large. We do not expect b_c to become less than unity (note that b_c is averaged over the entire continuum) since this would imply that the continuum processes radiated more efficiently than a black-body. Since most of the radiant energy is in the continuum, this is not possible.

Table (VII-8) contains estimates of the last term on the right hand side of Equation (VII-72) for the sun together with values of $\omega\alpha$ obtained from Figure VII-7 with $\omega = 0.12$. For these estimates we have set $b_c = 2$ and $T_{\text{eff}} = 5785°$. Estimates of b_c for the sun (Gebbie and Thomas, 1970) generally give $b_c < 2$ so the continuum term in Table VII-8 is overestimated.

We note from the results in Table VII-8 that the line blanketing term is considerably larger than the continuum term and that the ratio of the two terms increases as $T(0)$ increases. Thus, any elevation of $T(0)$ by the Cayrel mechanism is strongly resisted by the cooling effect of the lines. Since the continuum term in Equation (VII-72) is proportional to $T^4(0)$, the cooling effect of the lines actually suppresses the Cayrel mechanism in addition to overriding it.

In the case of the sun the Cayrel mechanism appears to raise $T(0)$ by only about 25° (Athay, 1970) if mechanical heating is ignored. The addition of a mechanical heating term to the right hand side of Equation (VII-72) gives

$$\frac{1}{H_0^c} \frac{dH_c}{d\tau_c} = \omega\alpha - \frac{T^4}{T_{\text{eff}}^4}\left(1 - \frac{1}{b_c}\right) - \frac{\nabla \cdot H_{\text{mech}}}{H_0^c}. \tag{VII-73}$$

Dissipation of mechanical energy ($\nabla \cdot H_{\text{mech}} > 0$) will raise the value of T. This, in turn, will increase both the amount of cooling due to line blanketing and the heating due to the Cayrel mechanism. It is clearly evident in the case of the sun, however, that the line cooling term will always dominate over the continuum heating term and that mechanical energy dissipation will be balanced by the cooling effect of the lines.

Model atmosphere computations for early type stars (Auer and Mihalas, 1969a, b, c; Mihalas and Auer, 1970) indicate that line blanketing and departures from LTE in the continuum each produce a net heating at small optical depths. Under these circumstances the addition of mechanical energy dissipation will raise the temperature, but it is not clear how far the temperature will rise before the line blanketing and/or the continuum effects produce sufficient cooling to reach equilibrium. It is possible that in these early type stars a small amount of mechanical energy dissipated in the external layers will produce a relatively large increase in temperature.

References

Athay, R. G.: 1970, *Astrophys. J.* **161**, 713.
Athay, R. G. and Skumanich, A.: 1969, *Astrophys. J.* **155**, 273.
Auer, L. H. and Mihalas, D.: 1969a, *Astrophys. J.* **156**, 157.
Auer, L. H. and Mihalas, D.: 1969b, *Astrophys. J.* **156**, 681.
Auer, L. H. and Mihalas, D.: 1969c, *Astrophys. J.* **158**, 641.
Cayrel, R.: 1966, *J. Quant. Spectr. Radiative Transfer* **6**, 621.
Chandrasekhar, S.: 1935, *Monthly Notices Roy. Astron. Soc.* **96**, 21.
Frisch, H.: 1966, *J. Quant. Spectr. Radiative Transfer* **6**, 629.
Gebbie, K. and Thomas, R. N.: 1970, *Astrophys. J.* **161**, 229.
Kourganoff, V.: 1963, *Basic Methods in Radiative Transfer*, Dover, New York.
Mihalas, D. and Auer, L. H.: 1970, *Astrophys. J.* **160**, 1161.
Milne, E. A.: 1930, *Thermodynamics of the Stars*, reprinted in D. H. Menzel (ed.), *Selected Papers on the Transfer of Radiation*, Dover, New York.
Münch, G.: 1946, *Astrophys. J.* **104**, 87.
Pecker, J.-C.: 1951, *Ann. Astrophys.* **14**, 152.
Thomas, R. N.: 1965, *Astrophys. J.* **141**, 333.

CHAPTER VIII

NUMERICAL METHODS

1. Introduction

A variety of techniques have been developed for solving the coupled radiative transfer and kinetic equilibrium equations. An extensive discussion of such techniques is beyond the purpose and scope of this text. Instead a few selected methods will be discussed in broad outline. The particular methods selected for discussion are illustrative of those that are more commonly used. For the detailed application of a particular method and for information about other methods the interested reader should turn to the original literature. Suitable references to such literature will be found in this chapter.

The choice of a particular method for solving the coupled equations of radiative transfer and kinetic equilibrium for simple problems is largely a matter of taste dictated by the prior experiences of those doing the solutions. However, a given method may work well for some classes of problems whereas it may be unsatisfactory for more complex problems. Similarly, some methods require minimum computer storage but are costly in computer time whereas others require large core storage with minimal computing time. Thus, the choice of methods for complex problems is ultimately dictated by the particular nature of the problem and the available computing facilities. Because so few complex problems have been solved to this point, there is not sufficient accumulated experience to warrant a compilation of the advantages and disadvantages of each method. Such experience is accumulating, however, and should be available within the next few years.

The specific methods chosen for illustration in this chapter are directly applicable to problems involving quiescent, plane parallel atmospheres. Multidimensional media and differentially moving media destroy symmetries that are used advantageously in the quiescent, plane parallel atmosphere. This adds additional dimensions to the numerical problem. These added dimensions, however, can be treated with appropriate revisions in numerical procedures and do not change the basic nature of the problem itself.

2. The Integral Flux-Divergence Equations

The equations relating S_L to B, ε^*, $\varepsilon^\dagger$, r_0 and ϕ_ν in Chapter II are formulated in terms of an integral of the flux divergence following (Athay and Skumanich, 1967). Specifically, Equation (II-47) is

$$S_L = \frac{\varepsilon^* + \delta}{\varepsilon^\dagger + \delta} B + \frac{1}{\varepsilon^\dagger + \delta} \int \frac{\phi_\nu}{\phi_\nu + r_0} \frac{dH_\nu}{d\tau_0} d\nu, \qquad \text{(VIII-1)}$$

where we have dropped the subscript 0 on S_L. The same basic parameters enter all formulations of the problem though they may appear in different form and with different groupings. We may, for example, replace $(\phi_\nu + r_0) d\tau_0$ with $d\tau_\nu$ and we may replace $dH_\nu / d\tau_\nu$ with its equivalent $J_\nu - S_\nu$. Thus, Equation (VIII-1) may be written as

$$S_L = \frac{\varepsilon^* + \delta}{\varepsilon^\dagger + \delta} B + \frac{1}{\varepsilon^\dagger + \delta} \int \Phi_\nu (J_\nu - S_\nu) d\nu. \qquad \text{(VIII-2)}$$

S_ν may be sub-divided into terms involving S_L and B by the use of Equation (I-10), which we write in the form

$$S_\nu = \frac{\phi_\nu}{\phi_\nu + r_0} S_L + \left(1 - \frac{\phi_\nu}{\phi_\nu + r_0}\right) B. \qquad \text{(VIII-3)}$$

The quantity J_ν is related to S_ν through the equation (Kourganoff, 1963)

$$J_\nu = \tfrac{1}{2} \int_0^\infty S_\nu E_1 (|t - \tau_\nu|) dt, \qquad \text{(VIII-4)}$$

or, in operator notation,

$$\tilde{J}_\nu = \Lambda_\nu \tilde{S}_\nu. \qquad \text{(VIII-5)}$$

As may be seen from Equation (VIII-5), the operator Λ_ν is a function of τ_ν as well as of ν. The same is true, of course, of J_ν and S_ν. In the following we regard J_ν and S_ν as arrays of column vectors whose components are the respective values of J_ν and S_ν at discrete optical depth points $\tau_1, \tau_2, \cdots \tau_N$. The arrays consist of the column vectors at the discrete frequencies $\nu_1, \nu_2, \cdots \nu_Q$. The operator Λ_ν is an array of Q matrices of order $N \times N$. (We adopt the notation N_n or, alternatively, τ_n to denote the nth optical depth point and Q_q or ν_q to denote the qth frequency point.) Each row, N_n, in a given matrix at frequency Q_q gives the vector operation on $\tilde{S}_\nu(\tau_\nu)$ that transforms it into J_ν at the particular optical depth point τ_n and the particular frequency ν_q. Thus, we write Equation (VIII-5) in the symbolic form

$$\begin{pmatrix} J_1 \\ J_2 \\ \cdot \\ \cdot \\ \cdot \\ J_n \end{pmatrix}_{\nu_q} = \begin{pmatrix} \Lambda_{11}, \Lambda_{12} \cdots \Lambda_n \\ \Lambda_{21} & & \cdot \\ \cdot & & \cdot \\ \cdot & & \cdot \\ \cdot & & \cdot \\ \Lambda_{n1} \cdots \cdots \Lambda_{nn} \end{pmatrix}_{\nu_q} \begin{pmatrix} S_1 \\ S_2 \\ \cdot \\ \cdot \\ \cdot \\ S_n \end{pmatrix}_{\nu_q} \qquad \text{(VIII-6)}$$

In a similar fashion, we may regard $\tilde{S}_\nu$ as

$$\tilde{S}_\nu = I \tilde{S}_\nu, \qquad \text{(VIII-7)}$$

where **I** is the unit matrix and we may regard the quantities ϕ_v, $\phi_v/(\phi_v+r_0)$, $(\varepsilon^\dagger+\delta)^{-1}$ and $(\varepsilon^*+\delta)/(\varepsilon^\dagger+\delta)$ as diagonal matrices. Thus, with the help of Equations (VIII-3), (VIII-5), and (VIII-7) we may write the vector S_L as

$$\tilde{S}_L = \frac{\varepsilon^*+\delta}{\varepsilon^\dagger+\delta}\tilde{B} + \frac{1}{\varepsilon^\dagger+\delta}\int_0^\infty \Phi_v(\Lambda_v - \mathbf{I})\left[\frac{\phi_v}{\phi_v+r_0}\tilde{S}_L + \mathbf{I} - \frac{\phi_v}{\phi_v+r_0}\tilde{B}\right]dv. \quad \text{(VIII-8)}$$

Equation (VIII-8) is of the form

$$\tilde{S}_L = \frac{\varepsilon^*+\delta}{\varepsilon^\dagger+\delta}\tilde{B} + \mathbf{X}\tilde{S}_L + \mathbf{Y}\tilde{B}, \quad \text{(VIII-9)}$$

which has as a solution

$$\tilde{S}_L = (\mathbf{I} - \mathbf{X})^{-1}\left(\frac{\varepsilon^*+\delta}{\varepsilon^\dagger+\delta} + \mathbf{Y}\right)\tilde{B}. \quad \text{(VIII-10)}$$

The separate terms in the coefficients of B are each matrices of order $n \times n$ and all indicated operation are matrix operations.

Since the quantities ε^*, $\varepsilon^\dagger$, r_0 and τ depend either directly or indirectly upon S_L, they must be solved for simultaneously with S_L. With this formulation, however, we have no provision for solving for ε^*, $\varepsilon^\dagger$, r_0 and τ independently of Equation (VIII-8). An appropriate iterative scheme that provides ε^*, $\varepsilon^\dagger$ and r_0 in terms of the escape coefficients, ϱ, is outlined in Chapter IV. Our concern at this point, therefore, is limited to the computation of the matrices **X** and **Y**.

In order to derive useful values for the matrix elements in Λ_v we expand $\tilde{S}_v$ in terms of a known set of functions $f_n(\tau_v)$ and perform the operation prescribed by Equation (VIII-4) on these known functions. Thus, we write

$$S_v(\tau_v, v) = \sum_1^N C_n(v) f_n(\tau_v). \quad \text{(VIII-11)}$$

The complete set of function $f_n(\tau_v)$ at a given v again define a matrix, which we denote by $\mathbf{F}_v$. Thus,

$$\tilde{S}_v = \mathbf{F}_v \tilde{C}_n \quad \text{(VIII-12)}$$

and

$$\tilde{C}_n = \mathbf{F}_v^{-1}\tilde{S}_v \quad \text{(VIII-13)}$$

so that

$$\tilde{S}_v = \mathbf{F}_v \mathbf{F}_v^{-1}\tilde{S}_v. \quad \text{(VIII-14)}$$

Since $\mathbf{F}^{-1}\tilde{S}_v$ is simply the symbolic representation of C_n, which is a set of constants, the Λ_v operator operates only on the matrix $\mathbf{F}_v$, i.e., on the set of functions f_n.

Our matrix equation is now of the form

$$\tilde{S}_L = \frac{\varepsilon^* + \delta}{\varepsilon^\dagger + \delta} \tilde{B} + \frac{1}{\varepsilon^\dagger + \delta} \int_0^\infty \Phi_\nu (\Lambda_\nu - I) F_\nu F_\nu^{-1}$$

$$\times \left[\frac{\phi_\nu}{\phi_\nu + r_0} \tilde{S}_L + \left(I - \frac{\phi_\nu}{\phi_\nu + r_0} \right) \tilde{B} \right] d\nu \qquad \text{(VIII-15)}$$

so we have

$$X = \frac{1}{\varepsilon^\dagger + \delta} \int_0^\infty \Phi_\nu (\Lambda_\nu - I) F_\nu F_\nu^{-1} \frac{\phi_\nu}{\phi_\nu + r_0} d\nu \qquad \text{(VIII-16)}$$

and

$$Y = \frac{1}{\varepsilon^\dagger + \delta} \int_0^\infty \Phi_\nu (\Lambda_\nu - I) F_\nu F_\nu^{-1} \left(I - \frac{\phi_\nu}{\phi_\nu + r_0} \right) d\nu. \qquad \text{(VIII-17)}$$

All that is needed to compute X and Y therefore is to specify the functions $f_n(\tau_\nu)$ and perform the indicated operations.

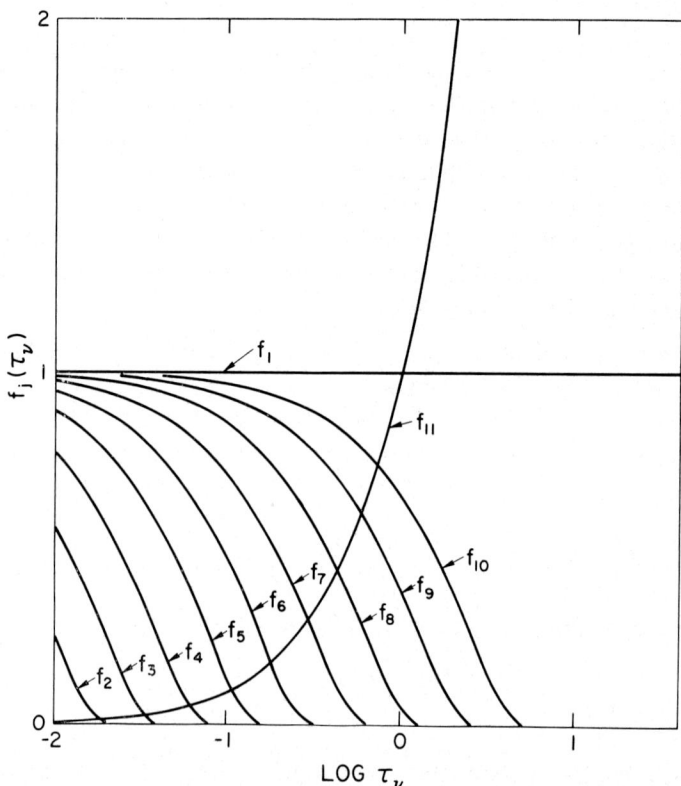

Fig. VIII-1. A plot of the f_j functions for eleven τ points equally spaced in $\log \tau_\nu$ and for $\lambda = 1$.

In principle any set of functions will suffice for $f_n(\tau_v)$ so long as the set is complete and has proper asymptotic behavior. A convenient and useful set of such functions that have the particular property of giving accurate values of J_v for semi-infinite atmospheres is given by Avrett and Loeser (1963). This set is defined by (see Figure VIII-1)

$$f_1 = 1$$
$$f_n = \left(1 - \frac{\tau_{vn}}{\tau_{vn'}}\right)\left(1 - \lambda \frac{\tau_{vn}}{\tau_{vn'}}\right), \quad \tau_{vn} < \tau_{vn'},$$
$$= 0, \quad \tau_{vn} \geq \tau_{vn'} \tag{VIII-18}$$
$$f_N = \tau_{vn}.$$

The corresponding matrix $\mathbf{F}_v$ is

$$\mathbf{F}_v = \begin{pmatrix} 1 & \left(1 - \frac{\tau_1}{\tau_2}\right)\left(1 - \lambda \frac{\tau_1}{\tau_2}\right) & \left(1 - \frac{\tau_1}{\tau_3}\right)\left(1 - \lambda \frac{\tau_1}{\tau_3}\right) & \cdots & \tau_1 \\ 1 & 0 & \left(1 - \frac{\tau_2}{\tau_3}\right)\left(1 - \lambda \frac{\tau_2}{\tau_3}\right) & \cdots & \tau_2 \\ 1 & 0 & 0 & & \cdot \\ \cdot & \cdot & \cdot & & \cdot \\ \cdot & \cdot & \cdot & & \cdot \\ 1 & 0 & 0 & & \tau_N \end{pmatrix} \tag{VIII-19}$$

The inverse of $\mathbf{F}_v$ is given by

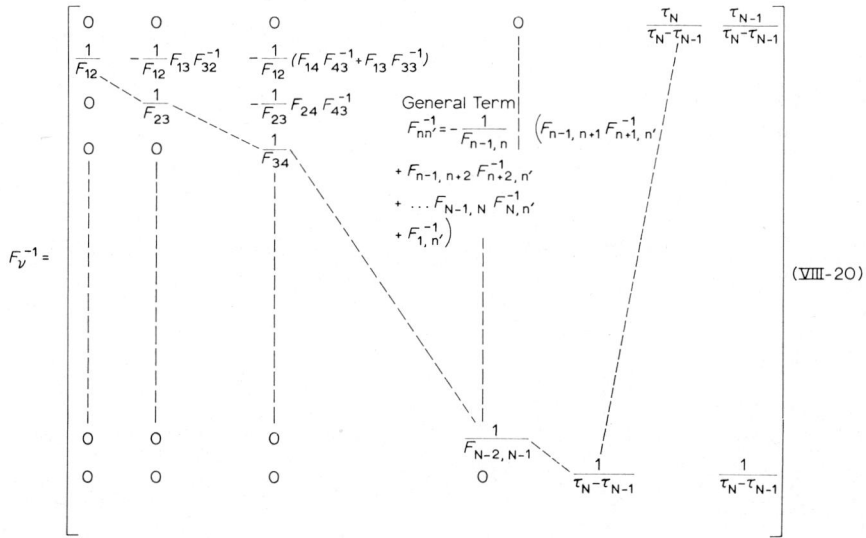

(VIII-20)

where n is the row indicator and n' is the column indicator and where $F_{n,n'}^{-1}$ means the n,n' element of $\mathbf{F}_v^{-1}$. The $\mathbf{F}_v^{-1}$ matrix is readily computed starting from the bottom row and does not require a numerical inversion of $\mathbf{F}_v$.

The matrix $\Lambda_\nu F_\nu$ is readily obtained from Equations (VIII-4) and (VIII-18). The first row of this matrix is the Λ_ν transform of unity and is equal to (Kourganoff, 1963)

$$\Lambda_\nu(1) = 1 - \tfrac{1}{2}E_2(\tau_n). \tag{VIII-21}$$

Similarly, the last row is the Λ_ν transform of τ_n and is equal to (Kourganoff, 1963)

$$\Lambda_\nu(\tau_n) = \tau_n + \tfrac{1}{2}E_3(\tau_n). \tag{VIII-22}$$

Elements in the upper triangle of $\Lambda_\nu F_\nu$, except for the first row, are given by

$$(\Lambda_\nu F_\nu)_{n,n'} = \tfrac{1}{2}\int_0^{\tau_{n'}} f_n(t) E_1(|t - \tau_n|)\, dt$$

$$= \Lambda_\nu f_n(\tau_n) - \tfrac{1}{2}\int_{\tau_{n'}}^{\infty} f_n(t) E_1(|t - \tau_n|)\, dt. \tag{VIII-23}$$

For the $f_n(\tau_n)$ defined by Equation (VIII-18)

$$\Lambda_\nu f_n(\tau_n) = 1 - \tfrac{1}{2}E_2(\tau_n) - \frac{(1+\lambda)}{\tau_{n'}}[\tau_n + \tfrac{1}{2}E_3(\tau_n)]$$

$$+ \frac{\lambda}{\tau_{n'}^2}[\tfrac{2}{3} + \tau_n^2 - E_4(\tau_n)]$$

$$= f_n(\tau_n) - \tfrac{1}{2}E_2(\tau_n) - \frac{1}{2}\frac{1+\lambda}{\tau_{n'}}E_3(\tau_n)$$

$$+ \frac{\lambda}{\tau_{n'}^2}[\tfrac{2}{3} - E_4(\tau_n)] \tag{VIII-24}$$

(Kourganoff, 1963)

To evaluate the function

$$\lambda(\tau_n) = -\tfrac{1}{2}\int_{\tau_{n'}}^{\infty} f_n(t) E_1(t - \tau_n)\, dt$$

we set

$$u = t - \tau_n \tag{VIII-25}$$

and

$$f_n(u + \tau_n) = \left(1 - \frac{\tau_n}{\tau_{n'}} - \frac{u}{\tau_{n'}}\right)\left(1 - \lambda\frac{\tau_n}{\tau_{n'}} - \lambda\frac{u}{\tau_{n'}}\right)$$

$$= \left(1 - \frac{\tau_n}{\tau_{n'}}\right)\left(1 - \lambda\frac{\tau_n}{\tau_{n'}}\right) - \left(1 - \frac{\tau_n}{\tau_{n'}}\right)\lambda\frac{u}{\tau_{n'}}$$

$$- \frac{u}{\tau_{n'}}\left(1 - \lambda\frac{\tau_n}{\tau_{n'}} - \lambda\frac{u}{\tau_{n'}}\right)$$

$$= f_n(\tau_n) - \frac{u}{\tau_{n'}}\left(1 + \lambda - 2\lambda\frac{\tau_n}{\tau_{n'}}\right) + \lambda\frac{u^2}{\tau_{n'}^2}. \tag{VIII-26}$$

This gives

$$\lambda(\tau_n) = -\tfrac{1}{2} \int_{\tau_{n'}-\tau_n}^{\infty} \left[f_n(\tau_n) - \frac{u}{\tau_{n'}}\left(1 + \lambda - 2\lambda \frac{\tau_n}{\tau_{n'}}\right) + \lambda \frac{u^2}{\tau_{n'}^2} \right] E_1(u)\, du \tag{VIII-27}$$

we then use the recurrence formula

$$E_{n-1}(u)\, du = -dE_n(u) \tag{VIII-28}$$

(Kourganoff, 1963) and integration by parts to obtain

$$\lambda_n(\tau_n) = -\tfrac{1}{2} f_n(\tau_n) E_2(\tau_{n'} - \tau_n) + \frac{1}{2}\frac{1}{\tau_{n'}}\left(1 + \lambda - 2\lambda \frac{\tau_n}{\tau_{n'}}\right)$$

$$\times [(\tau_{n'} - \tau_n) E_2(\tau_{n'} - \tau_n) + E_3(\tau_{n'} - \tau_n)] - \frac{1}{2}\frac{\lambda}{\tau_{n'}^2}$$

$$\times [(\tau_{n'} - \tau_n)^2 E_2(\tau_{n'} - \tau_n) + 2(\tau_{n'} - \tau_n) E_3(\tau_{n'} - \tau_n) + 2E_4(\tau_{n'} - \tau_n)],$$

or

$$\lambda_n(\tau_n) = \frac{(1-\lambda)}{2\tau_{n'}} E_3(\tau_{n'} - \tau_n) - \frac{\lambda}{\tau_{n'}^2} E_4(\tau_{n'} - \tau_n). \tag{VIII-29}$$

Addition of Equations (VIII-24) and (VIII-29) gives

$$(\Lambda_v F_v)_{n, n'} = f_n(\tau_n) - \tfrac{1}{2} E_2(\tau_n) - \frac{1}{2}\frac{1+\lambda}{\tau_{n'}} E_3(\tau_n) + \frac{\lambda}{\tau_{n'}^2}\left[\tfrac{2}{3} - E_4(\tau_n)\right]$$

$$+ \frac{1-\lambda}{2\tau_{n'}} E_3(\tau_{n'} - \tau_n) - \frac{\lambda}{\tau_{n'}^2} E_4(\tau_{n'} - \tau_n). \tag{VIII-30}$$

This equation is valid for $n < n'$ and $n \neq 1$ or N. For $n' = n$ it reduces to

$$(\Lambda_v F_v)_{n, n'} = -\tfrac{1}{2} E_2(\tau_n) - \frac{1}{2}\frac{1+\lambda}{\tau_n} E_3(\tau_n) - \frac{\lambda}{\tau_n^2} E_4(\tau_n) + \frac{1}{3}\frac{\lambda}{\tau_n^2} + \frac{1}{4}\frac{1-\lambda}{\tau_n}. \tag{VIII-31}$$

Elements in the lower triangle of $\Lambda_v F_v$, except for the last row, are given by

$$(\Lambda_v F_v)_{n, n'} = \tfrac{1}{2} \int_0^{\tau_{n'}} f(t) E_1(\tau_n - t)\, dt = -\tfrac{1}{2} \int_{+\tau_n}^{\tau_n - \tau_{n'}} f(\tau_n - u') E_1(u')\, du' \tag{VIII-32}$$

where

$$u' = \tau_n = t \tag{VIII-33}$$

and

$$f(\tau_n - u') = f_n(\tau_n) + \frac{u'}{\tau_{n'}}\left(1 + \lambda - 2\lambda \frac{\tau_n}{\tau_{n'}}\right) + \lambda \frac{u'^2}{\tau_{n'}^2}. \tag{VIII-34}$$

Integration of Equation (VIII-32) using Equations (VIII-28) and (VIII-34) gives

$$(\Lambda_\nu F_\nu)_{n,n'} = -\tfrac{1}{2} E_2(\tau_n) - \frac{1+\lambda}{2\tau_{n'}} E_3(\tau_n) - \frac{\lambda}{\tau_{n'}^2} E_4(\tau_n) + \frac{1-\lambda}{2\tau_{n'}}$$
$$\times E_3(\tau_n - \tau_{n'}) + \frac{\lambda}{\tau_{n'}^2} E_4(\tau_n - \tau_{n'}), \qquad n > n_1, \qquad n \neq N.$$
(VIII-35)

For $n'=n$ Equation (VIII-35) reduces to Equation (VIII-31).

Since Equations (VIII-30) and (VIII-35) specify $\Lambda_\nu F_\nu$ in terms of the discrete τ_ν points and since all of the remaining matrices in **X** and **Y** are known, the computation of **X** and **Y** is straightforward. Integration over frequency is performed numerically by computing the matrices represented in **X** and **Y** at each frequency and using some appropriate integration scheme. The method of solution described in the preceding was developed by Athay and Skumanich (1967).

The number of matrix elements in $\Lambda_\nu F_\nu$ that must be computed for each iteration on each line is $N^2 L$. Each element consists of 5 terms in $E_n(\tau_n)$ so a total of 5 $N^2 L$ values of $E_n(\tau_n)$ are needed. Most of the computing time is used for the matrix elements. Note that the number of matrix elements increases linearly with L so that a liberal use of frequency points may be used. Although the number of matrix elements increases as N^2, a notable feature of this method of solution is that the tau points need not be closely spaced. For many purposes, two or three points per decade in tau are sufficient.

The parameter λ in the f_n functions can be adjusted to help provide numerical stability. The greatest stability is achieved with $\lambda = 0$ but the solutions are more accurate with $\lambda = 1$. In problems where one of the parameters, say, B, is changing rapidly with height numerical instabilities may occur in regions of steep gradients. Reducing the value of λ dampens such instabilities quite markedly. For many problems, values of $\lambda = 0.8$ to 0.9 are found to provide sufficient stability and to maintain a high level of accuracy.

With this method of formulating and solving the combined radiative transfer and kinetic equilibrium equations any or all of the parameters ε^*, $\varepsilon^\dagger$, r_0, ϕ_ν and B may vary with depth. The numerical stability is most sensitive to changes in B and $\Delta\nu_D$ with depth, but even here relatively steep gradients can be tolerated. Nearly all of the examples given in this text have been obtained using the integral flux-divergence method or a close adaptation of it. An elegant and more detailed description of the method is contained in a paper by Avrett and Loeser (1969).

3. Required Frequency Bandwidth

Up to now we have said little about the choice of frequency points or about the required limits in frequency. Not much has been done by way of quantitative studies of the effects of different nesting of frequency points. Certain conclusions seem self evident, however.

We shall assume that ϕ_y is given by the $H(a,v)$ function (Chapter II) commonly used in astrophysics. (For a discussion of the $H(a,v)$ function, see Bohm (1960).) For convenience of discussion, we approximate $H(a,v)$ by

$$\phi_y = e^{-y^2} + \frac{a}{\pi^{1/2} y^2}, \qquad \text{(VIII-36)}$$

where it is understood that the wing term $a/\pi^{1/2} y^2$ is included only for $y > 1$, and where $y = \Delta v/\Delta v_D$.

Within the Doppler core defined by $e^{-y^2} \geqslant a/\pi^{1/2} y^2$, ϕ_y changes rapidly with y. Thus, τ_y changes rapidly with y and y points must be spaced relatively closely throughout this region. Because e^{-y^2} changes from about 10^{-3} at $y = 2.6$ to about 10^{-5} at $y = 3.4$, the Doppler core very often extends to the neighborhood of $y = 3$. Thirteen frequency points are required to cover the interval $y = 0$ to $y = 3$ in steps of $\frac{1}{4}$. Fewer points may prove sufficient.

Outside the Doppler core ϕ_y and τ_y change slowly. Three points per decade in y are usually sufficient. Thus, to go from $y = 3$ to 10^3 requires only about eight additional points in y.

Note that if ϕ_y is a function of τ_0, i.e., if Δv_D or a is a function of τ_0, it is often necessary to increase the number of y points substantially. If, for example, Δv_D increases by a factor of five through the atmosphere it may be necessary to retain close spacing in y much beyond $y = 3$, where y is defined in terms of the smallest Δv_D. The only safe procedure in such a situation is by trial and error.

Limits on the required bandwidth in y have been investigated for the two-level atom. No studies have been done by multilevel cases. However, there is little reason to suspect that the required bandwidth will depend sensitively upon the number of levels considered.

Bandwidth limits are somewhat arbitrary depending upon the acceptable limits in the accuracy of S_L. For 2% accuracy in S_L the required bandwidths (as found from trial and error Computa-tions) are

$$y_1 \approx a^{1/2}/\varepsilon, \quad \varepsilon \geqslant \delta \qquad \text{(VIII-37)}$$

and

$$y_1 \approx 10 \left(\frac{a}{r_0}\right)^{1/2} \approx 13 \frac{a}{\delta}, \quad \delta \gg \varepsilon. \qquad \text{(VIII-38)}$$

The first limit corresponds to $\tau_{y_1} \approx \varepsilon^2 \tau_0$, or to $\tau_{y_1} = 1$ at $\tau_0 = \varepsilon^{-2}$. The second corresponds to $\tau_{y_1} = 1$ at $\tau_0 = 13 a/\delta^2$. More complete results are shown in Figure VIII-2.

It is somewhat surprising that the required limits on y_1 are as large as the preceding results suggest. One might have argued that the limits on y_1 would be set reasonably well by the criterion $\tau_{y_1} = 1$ at $\tau_0 = \tau_{th}$ since this is the limit on y_1 at which the assumption of LTE breaks down. However, for the cases considered $\tau_{th} \approx a/(\varepsilon + \delta)^2$ (Chapter

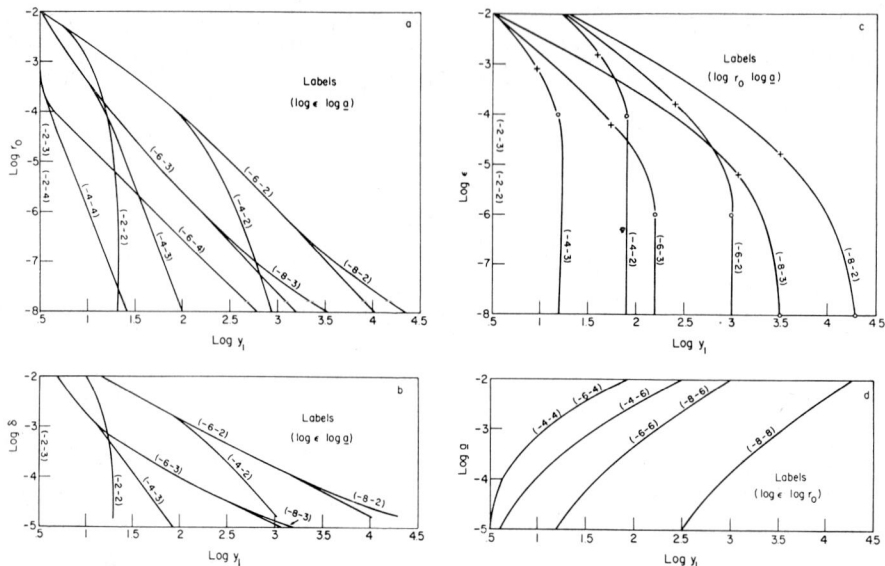

Fig. VIII-2. A plot of the values of y_1 required to attain two percent accuracy in S_L for different combinations of ε, a and r_0. – (a) Combinations of ε and a plotted with $\log r_0$ as ordinate. – (b) Same as (a) except with $\log \delta$ as ordinate. – (c) Combinations of r_0 and a with $\log \varepsilon$ as ordinate. – (d) Combinations of ε and r_0 with $\log a$ as ordinate (Athay, 1970; (*Extended Atmosphere Stars*, U.S. National Bureau of Standards, Department of Commerce).

III) which would give a y_1 limit of about $a/(\varepsilon+\delta)$. This limit is considerably smaller than those actually required. The limits established correspond more nearly to the condition $\tau_y = 1$ at $\tau_0 = (\varepsilon+\delta)^{-2}$, which is the value of τ_{th} for a pure dispersion profile.

4. Frequency Mapping

The most time consuming operation in solving for S_L is in the computation of matrix elements. In the case of strong lines where we require say 40 τ points and 25 frequency points and where we have 5 spectral lines, the number of matrix elements computed in each iteration of $\tilde{S}_L$ is $40^2 \times 25 \times 5 = 2 \times 10^5$. If 5 iterations are required the number of matrix elements increases to 10^6. For some problems this number could easily be doubled. Each matrix element involves terms in E_2, E_3, and E_4, which require computation or table look-up so the time problem in computing matrix elements is a major one.

The matrices that are computed represent the operator $(\Lambda - \mathbf{I})_v$ in Equation (VIII-15). The operator Λ_v operates on a set of functions that are uniquely defined by a set of grid points in optical depth. It follows that the matrix Λ_v is uniquely determined by this same set of grid points. All that happens in changing from one frequency to another is that the grid points are assigned different values of tau. Thus, two Λ_v matrices, say, Λ_{v1} and Λ_{v2} are related by an optical depth transformation, and, as noted by Kalkofen (1968a) it is possible to obtain each of the Λ_ν matrices by an operation on a standard matrix Λ_0 computed on a reference grid.

The standard matrix Λ_0 is defined such that

$$\tilde{J}_0 = \Lambda_0 \tilde{S}_0 \tag{VIII-39}$$

and we seek the value of Λ_ν such that Equation (VIII-5) is satisfied. Define two new matrix operators $\boldsymbol{\alpha}_\nu$ and $\boldsymbol{\beta}_\nu$ such that

$$\tilde{S}_0 = \boldsymbol{\alpha}_\nu \tilde{S}_\nu \tag{VIII-40}$$

and

$$\tilde{J}_\nu = \boldsymbol{\beta}_\nu \tilde{J}_0. \tag{VIII-41}$$

We then have, using Equations (VIII-39) and (VIII-40)

$$\tilde{J}_\nu = \boldsymbol{\beta}_\nu \Lambda_0 \tilde{S}_0 = \boldsymbol{\beta}_\nu \Lambda_0 \boldsymbol{\alpha}^\nu \tilde{S}_\nu. \tag{VIII-42}$$

It follows from Equations (VIII-5) and (VIII-42) that

$$\Lambda_\nu = \boldsymbol{\beta}_\nu \Lambda_0 \boldsymbol{\alpha}^\nu. \tag{VIII-43}$$

Since this is a matrix equation, it is equivalent to the equation

$$\Lambda_{ij\nu} = \sum_{l=1}^{N} \sum_{m=1}^{N} \beta_{il\nu} \Lambda_{lm\nu_0} \alpha^\nu_{mj} \tag{VIII-44}$$

for the individual matrix elements.

The mapping technique described here was introduced by Kalkofen (1968a) who investigated several interpolation operators $\boldsymbol{\alpha}^\nu$ and $\boldsymbol{\beta}_\nu$. A number of authors have used this or similar methods of mapping with considerable reduction in computing time (cf., Avrett, 1970; Avrett and Loeser, 1969).

Mapping techniques inevitably introduce numerical inaccuracies and care must be taken to insure that such errors are kept small. In this regard it is important to note that the standard matrix can be defined for a grid of N_0 τ points that is larger than the grid of N points on which the f_n functions are defined. Increasing the value of N_0 increases the accuracy of the interpolation but, of course, reduces the savings in computing time.

5. Free-Bound Continua

Equation (II-105) for the source function, S_1, at the head of a free-bound continuum transition is in a form closely analogous with Equation (II-46) for the line source function. Aside from the fact that we have ignored other sources of continuum opacity in Equation (II-105), the primariy difference between this equation and Equation (II-46) is that we must take into account the explicit frequency dependence of S_ν in evaluating the quantity $dH_\nu/d\tau_\nu$.

As before, we write

$$\frac{dH_\nu}{d\tau_\nu} = J_\nu - S_{\nu 1} = (\Lambda_\nu - I) S_\nu$$

where

$$S_\nu = \left(\frac{\nu}{\nu_1}\right)^3 e^{-h(\nu-\nu_1)/kT} S_1.$$ (VIII-45)

The frequency dependent coefficient of $\tilde{S}_1$ in Equation (II-46) is a known function of depth (assuming T is known). Hence, the vector form of Equation (II-45) may be written

$$\tilde{S}_\nu = N_\nu \tilde{S}_1,$$ (VIII-46)

where the matrix N_ν is defined by the coefficients in Equation (II-45).

Following the same pattern as was used for the solution of Equation (II-46) in Section 2 of this chapter, we obtain for the solution of Equation (II-105)

$$\tilde{S}_1 = (I - Z)^{-1} \frac{\varepsilon^*}{\varepsilon^\dagger} \tilde{B}_1,$$ (VIII-47)

where

$$Z = E_\nu \int_{\nu_1}^{\infty} \frac{1}{\nu^4} (\Lambda_\nu - I) F_\nu F_\nu^{-1} N_\nu \, d\nu,$$ (VIII-48)

and

$$E_\nu = \frac{\nu_1^3 e^{-h\nu_1/kT}}{E_1 \varepsilon^\dagger}.$$ (VIII-49)

Equations (VIII-30) and (VIII-35) may be used for $\Lambda_\nu F_\nu$ with F_ν given by Equation (VIII-19). Thus, Equation (VIII-47) is readily solved.

The preceding equations for S_1 have used a hydrogenic form for α_ν. An appropriate modification for other forms for α_ν is readily made. Also, other sources of continuum or line opacity are readily included in Equation (II-105) provided the source functions for the extra opacity sources are known.

An added complexity that appears in the free-bound problem is that the ionization equilibrium influences the number of free electrons. This means that the collision rates in ε^* and $\varepsilon^\dagger$ are themselves functions of S_1. In many cases, such as for low abundance elements, this dependence of the collision rates on S_1 may be ignored, but in those cases where an appreciable fraction of the free electrons come from the element in question S_1 and n_e must be evaluated simultaneously. This is readily accomplished by iteration on an initial guess at n_e, say, the LTE value.

For applications to stellar atmosphere problems it is often necessary to compute hydrostatic equilibrium atmospheric densities from a given temperature distribution and a reference density at some fixed optical or geometrical depth. To do so requires that the mean molecular weight be known as a function of depth. If the free-bound continuum under consideration appreciably influences the electron density, it will also influence the mean molecular weight. Thus, in problems of this type a double iteration may be necessary. Such an iterative scheme (Athay and Canfield, 1970) starts, for example, with an assumed vector $\tilde{S}_1$. From $\tilde{S}_1$ we obtain the mean molecular weight and, hence, the run of electron and atomic densities with depth. Using this set of atom densities to define the τ_v grid, we iterate Equation (VIII-47) with the electron density to obtain a new set of electron densities and a new set of mean molecular weights. We then recompute electron and atom densities from the new mean molecular weights and repeat the iteration of Equation (VIII-47) with the electron density. This double iterative process is followed until convergence of both iterative loops is achieved.

Although the double iterative scheme described in the preceding paragraph may seem tedious, in practice it proves to be very reasonable. Each loop usually converges in three or four iterations so that a total of about twelve solutions for S_1 are required.

For hydrogenic forms of α_v, frequency points at values of v equal to v_1, 1.25 v_1, 1.5 v_1 ... 3 v_1 are usually quite ample. Thus, a total of only about nine frequency points are required and each iteration is relatively short.

6. Simultaneous Solution of the Integral Flux-Divergence Equations

Skumanich and Domenico (1970) have extended the development of the integral flux-divergence equations for multilevel atoms to obtain simultaneous solutions for the population ratios n_j/n_i. They solve the resultant equations by a generalized Newton-Raphson method based on perturbation techniques. A similar formulation was developed earlier by Kalkofen (1968b). However, Kalkofen's method of solution was more restrictive, and therefore less useful, than the method proposed by Skumanich and Domenico. For that reason, we shall consider the latter method. Also, the complete linearization technique of Auer and Mihalas (1969) used in a differential formulation of the problem (Section 8) is based on the Newton-Raphson method and parallels closely the Skumanich and Domenico derivation.

The formalism used by Skumanich and Domenico (1970) and Kalkofen (1968b) omits continuum opacity and contributions to J_v from continuum sources. Skumanich (1971) has refined the formalism and indicated how the continuum sources may be included.

The formulation leading to Equation (II-46) represents a particular way of combining the kinetic equilibrium and radiative transfer equations such that the kinetic equilibrium equations are incorporated in ε^*, $\varepsilon^\dagger$, r_0 and S_L. An alternative approach is to incorporate the radiative transfer equations into the coefficients in the kinetic equilibrium equations. This is conveniently done via the escape coefficients.

The kinetic equilibrium equations may be written for each level j in the form

$$\left[\sum_k C_{jk} + \sum_i (A_{ji}\varrho_{ji} + C_{ji})\right] n_j - \sum_k (C_{kj} + A_{kj}\varrho_{kj}) n_k - \sum_i C_{ji}n_i = 0, \quad \text{(VIII-50)}$$

where the sums, in each case, omit $k=j$ and $i=j$ and where we adopt the convention $i<j<k$. With the equations in this form the strong cancellation in the large radiative terms is already accomplished through ϱ_{ji} and ϱ_{kj} and the demands on numerical

precision are not as severe as they would otherwise be. The and ϱ_{kj} quantities ϱ_{ji} are functionals of n_j, n_k and n_i of course.

It is convenient to divide Equation (VIII-50) by n_1 and to define $n_j/n_1 = \sigma_j$, $n_k/n_1 = \sigma_k$ and $n_i/n_1 = \sigma_i$. The equation then becomes

$$\left[\sum_k C_{jk} + \sum_i (A_{ji}\varrho_{ji} + C_{ji})\right] \sigma_j - \sum_k (C_{kj} + A_{kj}\varrho_{kj}) \sigma_k - \sum_i C_{ji}\sigma_i = 0. \tag{VIII-51}$$

Note that $\sigma_1 = 1$ so that each member of the set of equations represented by Equation (VIII-51) has a term involving C_{ij} that is independent of σ_j.

From Equation (II-16) we have

$$\varrho_{ji} S_L = S_L - \int J_\nu \Phi_\nu \, d\nu. \tag{VIII-52}$$

As before, we introduce matrix operators and write

$$\tilde{S}_L - \int \Phi_\nu \tilde{J}_\nu \, d\nu = I\tilde{S}_L - \int \Phi_\nu \Lambda_\nu \tilde{S}_\nu \, d\nu$$

$$= I\tilde{S}_L - \int \Phi_\nu \Lambda_\nu \frac{\phi_\nu}{\phi_\nu + r_0} \, d\nu \tilde{S}_L - \int \Phi_\nu \Lambda_\nu \left(I - \frac{\phi_\nu}{\phi_\nu + r_0}\right) d\nu \tilde{B}. \tag{VIII-53}$$

To simplify the notation we rewrite Equation (VIII-53) as

$$\tilde{S}_L - \int \Phi_\nu \tilde{J}_\nu \, d\nu = (I - \Lambda) \tilde{S}_L - \Lambda' B \tag{VIII-54}$$

Note that Λ and Λ' are gray operators.

We may write $\tilde{S}_L$ in the form

$$\tilde{S}_L = \frac{2h\nu^3}{c^2} \frac{\tilde{\sigma}_j}{\frac{\tilde{\omega}_j}{\tilde{\omega}_i} \tilde{\sigma}_i - \tilde{\sigma}_j} \tag{VIII-55}$$

so that Equations (VIII-52) and (VIII-54) become

$$\varrho_{ji}\tilde{\sigma}_j = \left(\frac{\tilde{\omega}_j}{\tilde{\omega}_i} \tilde{\sigma}_i - \tilde{\sigma}_j\right)(I - \Lambda)\left(\frac{\tilde{\omega}_j}{\tilde{\omega}_i} \tilde{\sigma}_i - \tilde{\sigma}_j\right)^{-1} \tilde{\sigma}_j - \frac{c^2}{2h\nu^3} \left(\frac{\tilde{\omega}_j}{\tilde{\omega}_i} \tilde{\sigma}_i - \tilde{\sigma}_j\right) \Lambda'\tilde{B}. \tag{VIII-56}$$

In principle, the set of Equations (VIII-51) can now be solved with the help of the matrix operators ϱ_{ji} defined by Equation (VIII-56). Note that these operators are similar to the operators X and Y used in Section 2. Equation (VIII-56) is not yet in its most convenient form. The Λ operator is defined in terms of the optical depths τ_ν, which, in turn, are functions of n_j. It is desirable to express τ_ν in terms of

σ_j. To accomplish this we use the particle conservative equation

$$n_T = \sum n_j = n_1 + \sum_{j=2}^{\lambda} n_j \qquad \text{(VIII-57)}$$

to obtain

$$\frac{n_T}{n_1} = 1 + \sum_{j=2}^{\lambda} \sigma_j,$$

or

$$n_1 = \frac{n_T}{1 + \sigma_2 + \sigma_3 + \cdots} = n_T \Sigma^{-1} \qquad \text{(VIII-58)}$$

and

$$n_i = n_1 \sigma_i = \sigma_i n_T \Sigma^{-1} \qquad \text{(VIII-59)}$$

(λ = total number of energy levels). The optical depth in a given transition, say, ij, is

$$\tau_{ij} = \int_0^{\tau_{ij}} (d\tau_v + d\tau_c) = \int_0^{\tau_{ij}} (\phi_v + r_0) \, d\tau_0, \qquad \text{(VIII-60)}$$

where

$$d\tau_0 = -\frac{h\nu_i}{4\pi} B_{ij} M^{-1} (\sigma_i - \sigma_j) n_T \Sigma^{-1} \, dh. \qquad \text{(VIII-61)}$$

Since τ_{ij} differs for each transition at a given depth in the atmosphere, it is convenient to transform the optical depths back to a geometrical height scale, which is common to all transitions. Thus, if we define a coordinate transform operator $\kappa_{ij\nu}$ such that

$$\tilde{\tau}_{ij\nu} = \kappa_{ij\nu} \tilde{h}, \qquad \text{(VIII-62)}$$

then

$$\tilde{h} = \kappa_{ij\nu}^{-1} \tilde{\tau}_{ij\nu} \qquad \text{(VIII-63)}$$

and Equation (VIII-56) becomes

$$\varrho_{ji} \tilde{\sigma}_j = \left(\frac{\tilde{\omega}_j}{\tilde{\omega}_i} \tilde{\sigma}_i - \tilde{\sigma}_j \right) \kappa_{ij} (\mathbf{I} - \mathbf{\Lambda}) \kappa_{ij}^{-1} \left(\frac{\tilde{\omega}_j}{\tilde{\omega}_i} \tilde{\sigma}_i - \tilde{\sigma}_j \right)^{-1} \tilde{\sigma}_j$$

$$- \frac{c^2}{2h\nu^3} \left(\frac{\tilde{\omega}_j}{\tilde{\omega}_i} \tilde{\sigma}_i - \tilde{\sigma}_j \right) \kappa_{ij} \mathbf{\Lambda}' \kappa_{ij}^{-1} \tilde{\mathbf{B}}. \qquad \text{(VIII-64)}$$

This latter equation follows from the fact that

$$\int \Phi_\nu \tilde{J}_\nu(\tau_\nu) \, d\nu = \int \Phi_\nu \Lambda_\nu(\tau_\nu) \tilde{S}_\nu(\tau_\nu) \, d\nu$$

so that

$$\int \Phi_\nu \tilde{J}_\nu(h)\,d\nu = \int \Phi_\nu \kappa_\nu [\Lambda_\nu(\tau_\nu)\,\tilde{S}_\nu(\tau_\nu)]\,d\nu$$

$$= \int \Phi_\nu \kappa_\nu \Lambda_\nu(\tau_\nu)\,\kappa_\nu^{-1} \kappa_\nu \tilde{S}_\nu(\tau_\nu)\,d\nu$$

$$= \int \Phi_\nu \kappa_\nu \Lambda_\nu(\tau_\nu)\,\kappa_\nu^{-1} \tilde{S}_\nu(h)\,d\nu. \tag{VIII-65}$$

The ij subscripts on κ_ν are dropped in Equations (VIII-65) for convenience and the ν subscript is dropped in Equation (VIII-64) to acknowledge the integration over ν. The κ_ν operators are defined by Equations (VIII-60) and (VIII-61).

Equation (VIII-64) is in a form that is convenient for the solution of the set of Equations (VIII-51). This set can be written in matrix notation as

$$\mathbf{T}_\sigma \tilde{\sigma} = \tilde{\mathbf{C}}, \tag{VIII-66}$$

whose solution is

$$\tilde{\sigma} = \mathbf{T}_\sigma^{-1} \tilde{\mathbf{C}}. \tag{VIII-67}$$

The constants $\tilde{\mathbf{C}}$ come from the C_{ij} terms in Equation (VIII-51).

Equations (VIII-66) and (VIII-67) are non-linear since $\tilde{\sigma}_j$ appears in both $\tilde{\omega}_j/\tilde{\omega}_i$ $(\tilde{\sigma}_j - \tilde{\sigma}_i)$ and in κ_ν. The non-linearity disappears if all lines terminate on the ground state so that $\sigma_j = 1$ and if stimulated emissions are negligible, i.e., $\sigma_i \ll \sigma_j$. In this case $\sigma_i - \sigma_j = 1$ and ϱ_{ji} and $\mathbf{T}_\sigma$ become independent of σ. This is the case treated by Kalkofen (1968b).

To solve the more general problem where $\mathbf{T}_\sigma$ is a function of σ Skumanich and Domenico (1970) use a Newton-Raphson perturbation technique. Let

$$\tilde{\sigma} = \tilde{\sigma}_0 + \delta \tilde{\sigma}, \tag{VIII-68}$$

where $\tilde{\sigma}_0$ is a guessed solution and $\delta \tilde{\sigma}$ is an unknown correction term. Equation (VIII-66) then becomes

$$(\mathbf{T}_{\sigma_0 + \delta \sigma})\{\tilde{\sigma}_0 + \delta \tilde{\sigma}\} = \tilde{\mathbf{C}}$$

or, equivalently,

$$(\mathbf{T}_{\sigma_0} + \delta \mathbf{T}_{\sigma_0})(\tilde{\sigma}_0 + \delta \tilde{\sigma}) = \tilde{\mathbf{C}}, \tag{VIII-69}$$

where $\delta \mathbf{T}_{\sigma_0}$ is the perturbation in $\mathbf{T}_{\sigma_0}$ induced by $\delta \tilde{\sigma}$. Since $\tilde{\mathbf{C}}$ is constant, independent of $\tilde{\sigma}$ for the line transfer problem, there is no perturbation term in $\tilde{\mathbf{C}}$. Expansion of Equation (VIII-69) to terms in first order gives

$$\mathbf{T}_{\sigma_0}\tilde{\sigma}_0 + \mathbf{T}_{\sigma_0}\delta \tilde{\sigma} + \delta \mathbf{T}_{\sigma_0}\tilde{\sigma}_0 = \tilde{\mathbf{C}}. \tag{VIII-70}$$

Let

$$\delta \mathbf{T}_{\sigma_0}\tilde{\sigma}_0 = \mathbf{T}'_{\sigma_0}\delta \tilde{\sigma} \tag{VIII-71}$$

and
$$\tilde{e}_{\sigma_0} = \tilde{C} - T_{\sigma_0}\tilde{\sigma}_0. \tag{VIII-72}$$

The prime denotes the derivative with respect to $\delta\tilde{\sigma}$ and $\tilde{e}_{\sigma_0}$ is the error term resulting from $\tilde{\sigma} = \tilde{\sigma}_0$. Equations (VIII-70), (VIII-71), and (VIII-72) give

$$(T_{\sigma_0} + T'_{\sigma_0})\delta\tilde{\sigma} = \tilde{e}_{\sigma_0}. \tag{VIII-73}$$

Thus, the correction vector $\delta\tilde{\sigma}$ is given by

$$\delta\tilde{\sigma} = (T_{\sigma_0} + T'_{\sigma_0})^{-1}\tilde{e}_{\sigma_0}. \tag{VIII-74}$$

Since T_{σ_0} and T'_{σ_0} are known and since $\tilde{e}_{\sigma_0}$ may be computed using Equation (VIII-72), the correction vectors $\delta\tilde{\sigma}$ are readily computed. After each computation of $\delta\tilde{\sigma}$, $\tilde{\sigma}$ is up-dated in accordance with Equation (VIII-70) and the computation repeated. This iterative procedure is repeated until $\tilde{e}_{\sigma_i}$ becomes smaller than some specified level, say,

$$\tilde{e}_{\sigma_i} < \varepsilon\tilde{C}, \tag{VIII-75}$$

where ε is a small number.

Solutions to multilevel problems using the technique by Skumanich and Domenico (1970) and by Domenico (1971) show convergence to an accuracy of about one percent in four iterations. There are, of course, no problems arising from inconsistencies such as are present in the iterative scheme described in Section 2.

The matrix T_σ has dimensions $\lambda N \times \lambda N$ where λ is the number of energy levels minus one and N is the number of τ points. There is one T_σ matrix for each frequency, as before. It would appear, superficially that the large matrix size is a strong disadvantage of this method. However, the T_σ matrix consists of blocks along the diagonal, each block having dimensions $N \times N$ so the total number of non-zero matrix elements is only λN^2 at each frequency and at each iteration. Thus, the time used in computing matrix elements at each iteration is comparable in the two methods. It appears that the technique described in this section converges considerably faster than the technique described in Section 2 and results in a substantial savings in computing time.

Because of the form of the T_σ matrix, it can be inverted without requiring storage of the entire matrix.

The set of functions defined by Equation (VIII-18) are useful and appropriate for evaluating the Λ transform in Equation (VIII-64). Thus, Equations (VIII-30) and (VIII-55) give the appropriate matrix elements for the Λ transform.

7. A Differential Equation Method

The coupled radiative transfer and kinetic equilibrium equations can be solved in differential form as well as in integral form. Our emphasis on the integral form is purely for illustrative purposes and is not intended to convey to the reader the impression that the integral form is necessarily best. It was chosen for illustration partly because of the author's familiarity with this particular formulation and partly because the various parameters upon which the solutions depend are clearly portrayed in a single equation.

Differential equation techniques for solving transfer problems in spectral lines are highly developed. Auer and Mihalas (1969) (see also Mihalas, 1970) have carried the development to an elegant, simultaneous solution of the kinetic equilibrium and radiative transfer equation for multilevel atoms together with the equation of radiative and hydrostatic equilibrium. Thus, they compute the model atmosphere, including the effects of line blanketing, simultaneously with the solutions of the line and continuum transfer problems. As input, they need only the effective temperature of the star, its effective gravity and its chemical composition. Clearly, such an approach would be difficult using existing integral equation techniques.

The most commonly used differential equation techniques stem from a basic method proposed by Feutrier (1964). As originally proposed, this method required substantially more computing time than the integral technique described in the preceding sections of this chapter. However, a recent modification of the method by Rybicki (1970) reduces the computing time to values fully comparable with the integral method.

The integral formulation presented in this text, for the plane-parallel atmosphere, utilizes the angular symmetry and the frequency independence of S_L to an advantage. The angular integration required for the evaluation of J_ν is treated exactly and the number of matrix elements computed increases only linearly with Q. In the differential formulation the angular variable μ is treated discretely in, say, M points, and the combined angle-frequency points increases to MQ rather than Q. The original formulation of Feutrier (1964) required evaluation and inversion of N matrices of order $MQ \times MQ$. Thus, the time and storage requirements are proportional to $N(MQ)^3$ and $N(MQ)^2$, respectively. For the integral method the corresponding proportionalities are approximately N^2Q+N^3 and $2N^2$. Since MQ may exceed or be comparable to N, the time and storage requirements of the differential method of Feutrier (1964) are increased over the integral method. The modification of Feutrier's method proposed by Rybicki (1970) consists of a change in the computational scheme to take advantage of the angular symmetry and the frequency independence of S_L.

Feutrier (1964) defines the variables

$$J_{\nu,\mu} = \tfrac{1}{2}[I(\tau, \mu, \nu) + I(\tau, -\mu, \nu)] \tag{VIII-76}$$

and

$$H_{\nu,\mu} = \tfrac{1}{2}[I(\tau, \mu, \nu) - I(\tau, -\mu, \nu)], \tag{VIII-77}$$

where, as before, the tau dependence of J and H is implicit. The transfer equation

$$\frac{\mu}{\chi_\nu}\frac{\partial I_\nu}{\partial \tau} = I_\nu - S_\nu$$

can also be written, using the above definitions of J_ν and H_ν and the notation of

Feutrier (1964) and Rybicki (1970), as

$$\frac{\mu}{\chi_v} \frac{\partial H_{v,\mu}}{\partial \tau} = J_{v,\mu} - S_v$$

and

$$\frac{\mu}{\chi_v} \frac{\partial J_{v,\mu}}{\partial \tau} = H_{v,\mu}. \tag{VIII-78}$$

Thus,

$$\frac{\mu}{\chi_v} \frac{\partial}{\partial \tau} \frac{\mu}{\chi_v} \frac{\partial J_{v,\mu}}{\partial \tau} = J_{v,\mu} - S_v. \tag{VIII-79}$$

The quantity χ_v contains the frequency dependence and scale factors to give $\chi_v\,d\tau$ as the total optical depth. Thus $\chi_v\,d\tau$ corresponds to $(\phi_v + r_0)\,d\tau_0$ in the notation used earlier in the text. In the following, however, we shall consider the case where r_0 is negligibly small. Thus, $S_v = S_L$ and $\chi_v\,d\tau = \phi_v\,d\tau_0$. We continue to use the $\chi_v\,d\tau$ notation, however.

The source function S_L contains the term $\int J_v \Phi_v\,dv$, which is related to $J_{v,\mu}$ by the equation

$$\int_0^\infty J_v \Phi_v\,dv = \tfrac{1}{2} \int_0^\infty \int_{-1}^{1} J_{v,\mu}\,d\mu \Phi_v\,dv.$$

For convenience, let $\int J_v \Phi_v\,dv = \bar{J}$ with components $\bar{J}^1, \bar{J}^2, \ldots \bar{J}^N$ at the discrete τ points. Similarly, let $J_{v,\alpha}$ at a particular set of v_i and α_i have depth components $J_i^1, J_i^2, \ldots J_i^N$. Equation (VIII-79) may then be written

$$\bar{J}^n = \sum_{i=1}^{MQ} \omega_i J_i^n \tag{VIII-80}$$

and Equation (VIII-79) becomes

$$\frac{\mu_i}{\chi_{vi}} \frac{\partial}{\partial \tau_n} \frac{\mu_i}{\chi_{vi}} \frac{\partial J_i^n}{\partial \tau_n} = J_i^n - \frac{\sum_{i=1}^{MQ} \omega_i J_i^n}{1+\varepsilon_n} - \frac{(\varepsilon B)_n}{1+\varepsilon_n}. \tag{VIII-81}$$

There are a total of NMQ such equations, one for each discrete value of n and i.

The second-order differential term in Equation (VIII-81) may be written (Auer, 1967)

$$\frac{1}{\chi_v} \frac{\partial}{\partial \tau_n} \frac{1}{\chi_{vi}} \frac{\partial J_i^n}{\partial \tau_n} = \frac{J_i^{n+1}}{\Delta_i^n \delta_i^n} - \frac{J_i^n}{\delta_i^n}\left(\frac{1}{\Delta_i^n} + \frac{1}{\nabla_i^n}\right) + \frac{J_i^{n-1}}{\nabla_i^n \delta_i^n}, \tag{VIII-82}$$

where

$$\Delta_i^n = \tfrac{1}{2}(\chi_{v_i}^{n+1} + \chi_{v_i}^n)(\tau^{n+1} - \tau^n)$$
$$\nabla_i^n = \tfrac{1}{2}(\chi_{v_i}^n + \chi_{v_i}^{n-1})(\tau^n - \tau^{n-1})$$

and

$$\delta_i^n = \tfrac{1}{2}(\Delta_i^n + \nabla_i^n).$$

It follows that Equation (VIII-81) may be written at any depth point in the form

$$-A_i^n J_i^{n-1} + B_i^n J_i^n - C_i^n J_i^{n+1} - \frac{\bar{J}^n}{1+\varepsilon_n} = \frac{(\varepsilon B)_n}{1+\varepsilon_n}, \tag{VIII-83}$$

where

$$A_i^n = \frac{\mu_i^2}{\nabla_i^n \delta_i^n}, \qquad \text{(VIII-84)}$$

$$C_i^n = \frac{\mu_i^2}{\Delta_i^n \delta_i^n}, \qquad \text{(VIII-85)}$$

and

$$B_i^n = A_i^n + C_i^n + 1. \qquad \text{(VIII-86)}$$

The expanded set of Equation (VIII-83) can be regarded as a matrix equation in which matrices operate on vectors. The transformed quantities $A_i^n J_i^{n-1}$, $B_i^n J_i^n$ and $C_c^n J_i^{n+1}$ can be grouped to define either vectors in N space or vectors in MQ space. Feutrier's (1964) formulation followed the former approach, whereas Rybicki's (1970) formulation, which we adopt, followed the latter. Rybicki writes the matrix equivalent of Equation (VIII-83) as

$$\begin{pmatrix} T_1 & & & & & U_1 \\ & T_2 & & & & U_2 \\ & & T_3 & & & U_3 \\ & & & \cdot & & \cdot \\ & & & & \cdot & \cdot \\ & & & & \cdot & \cdot \\ & & & & T_{MQ} & U_{MQ} \\ V_1 & V_2 & V_3 & \cdots & V_{MQ} & E \end{pmatrix} \begin{pmatrix} J_1 \\ J_2 \\ J_3 \\ \cdot \\ \cdot \\ \cdot \\ J_{MQ} \\ J \end{pmatrix} = \begin{pmatrix} K_1 \\ K_2 \\ K_3 \\ \cdot \\ \cdot \\ \cdot \\ K_{MQ} \\ P \end{pmatrix}. \qquad \text{(VIII-87)}$$

The expanded set of equations is given by

$$\mathbf{T}_i \mathbf{J}_i + \mathbf{U}_i \mathbf{\tilde{J}} = \mathbf{\tilde{K}}_i \qquad \text{(VIII-88)}$$

and

$$\sum_{i=1}^{MF} \mathbf{V}_i \mathbf{J}_i + \mathbf{E} \mathbf{\tilde{J}} = \mathbf{\tilde{P}}. \qquad \text{(VIII-89)}$$

Equations (VIII-83), (VIII-83), and (VIII-89) together with appropriate boundary conditions are sufficient to define the matrices $\mathbf{T}_i$, $\mathbf{U}_i$, $\mathbf{V}_i$ and $\mathbf{E}$. The latter three matrices are each diagonal with dimensions $N \times N$ and $\mathbf{T}_i$ has the tridagonal, $N \times N$ dimensional form

$$\begin{pmatrix} \times & \times & & & & & \\ \times & \times & \times & & & 0 & \\ & \times & \times & \times & & & \\ & & & \cdot & & & \\ & & & & \cdot & & \\ & & 0 & & \times & \times & \times \\ & & & & & \times & \times \end{pmatrix} \qquad \text{(VIII-90)}$$

where the ×'s represent non-zero terms.

Rybicki (1970) reduces Equation (VIII-87) by Gaussian elimination to the form

$$\begin{pmatrix} T_1 & & & & & V_1 & U_1 \\ & T_2 & & & & V_2 & U_2 \\ & & T_3 & & & & U_3 \\ & & & \cdot & & & \cdot \\ & & & & \cdot & & \cdot \\ & & & & & T_{MQ} & U_{MQ} \\ 0 & 0 & 0 & \cdots & 0 & & W \end{pmatrix} \begin{pmatrix} J_1 \\ J_2 \\ J_3 \\ \cdot \\ \cdot \\ J_{MQ} \\ J \end{pmatrix} = \begin{pmatrix} K_1 \\ K_2 \\ K_3 \\ \cdot \\ \cdot \\ K_{MQ} \\ L \end{pmatrix} \quad \text{(VIII-91)}$$

by multiplying the ith row of Equation (VIII-87) by $\mathbf{V}_i \mathbf{T}_i^{-1}$ and subtracting the result from the last row. This process is repeated for each $i = 1 \ldots MQ$ to obtain

$$\mathbf{W} = \mathbf{E} - \sum_{i=1}^{MQ} \mathbf{V}_i \mathbf{T}_i^{-1} \mathbf{U}_i \quad \text{(VIII-92)}$$

and

$$\tilde{\mathbf{L}} = \tilde{\mathbf{P}} - \sum_{i=1}^{MQ} \mathbf{V}_i \mathbf{T}_i \mathbf{K}_i. \quad \text{(VIII-93)}$$

Expansion of the last row of Equation (VIII-91) gives the result

$$\mathbf{W}\tilde{\mathbf{J}} = \tilde{\mathbf{L}}, \quad \text{(VIII-94)}$$

or

$$\tilde{\mathbf{J}} = \mathbf{W}^{-1}\tilde{\mathbf{L}}. \quad \text{(VIII-95)}$$

This value of J may be used in the equation

$$S_L = \frac{J + \varepsilon B}{1 + \varepsilon} \quad \text{(VIII-96)}$$

to give S_L. Note that W is obtained by accumulating terms in matrices of order $N \times N$ so that the full matrices in Equations (VIII-87) and (VIII-91) need not be stored. Also, the largest matrices that are inverted (T_i and W) are of order $N \times N$. This results in a marked savings in computing time and storage as compared to the alternate Feutrier solution and makes the computing time and storage requirements comparable with the integral flux divergence method.

The boundary conditions on Equation (VIII-83) are obtained from Equation (VIII-78) by writing

$$\frac{\mu}{\chi_\nu} \frac{\partial J_{\nu,\mu}}{\partial \tau} = J_{\nu,\mu} - I_{\nu,-\mu}$$

$$= I_{\nu,\mu} - J_{\nu,\mu}$$

from Equations (VIII-76) and (VIII-77). The value of $I_{\nu,-\mu}$ can be specified at $\tau = 0$ (usually $I_{\nu,-\mu} = 0$) and $I_{\nu,\mu}$ can be specified at depth. In finite, symmetric atmospheres

$I_{\nu,\mu} = J_{\nu,\mu}$ and $H_{\nu,\mu} = 0$ at the mid point. For semi-infinite atmospheres in radiative equilibrium the lower boundary condition $\bar{J} = B$ is usually sufficient, provided the computations are carried to sufficiently large values of τ. More accurate conditions on the lower boundary are given by Auer and Mihalas (1969) and Mihalas (1970).

It follows from the definition of the A_i^n, B_i^n and C_i^n coefficients entering the T_i and W matrices that the matrix elements in this method are computed from simple differences in $\chi_{\nu_i}^n$ and T^n between adjacent depth points. This eliminates the need to evaluate exponential integrals, but it increases somewhat the number of τ points that must be used.

8. Extension to the Multilevel Case and Linearization

Extension of the differential formulation to the multilevel case is accomplished via a scheme parallel to that described in Section 5 for the integral equation formulation. A common depth scale for the different lines may be defined either through some reference optical depth, τ, or through geometrical height. The kinetic equilibrium and particle conservation equations are used to provide the necessary relationship between the variables σ_j and the values of J in the different lines. It is not customary in the differential formulation to use the escape coefficients. Instead the downward rates, R_{ji} are set equal to $C_{ji} + A_{ji} + B_{ji}J_{ij}$ and the upward rates, R_{ij}, are set equal to $C_{ij} + B_{ij}J_{ij}$. Thus, the equilibrium equations define a set of simultaneous equations in the variables σ_i with coefficients involving the mean intensities J_{ij}. The latter quantities are given by Equation (VIII-95).

The set of equations defined by the kinetic equilibrium equations are, of course, exactly the same as are used in the simultaneous solution of the integral equations. All that differs in the two formulations is the method of computing J, or equivalently ϱ. The Newton-Raphson linearization technique introduced by Auer and Mihalas (1969) for the solution of the multilevel problem in differential formulation is closely parallel to that outlined in section 6 of this chapter. Since the two methods are so similar, there is little need to repeat the derivation here.

9. Core Saturation Method

After completion of the preceding manuscript the author was made aware of a new method for solving the line transfer problem that appears to offer advantages over any of the earlier methods. This latter method, due to Rybicki (1971), involves an approximation that removes 'inactive' photons from the transfer problem and considers only the photons that are 'active' in the transport of energy. The method makes use of the fact that photon energy transport is accomplished primarily by a single large step by photons for which $\tau_y \lesssim 1$ rather than by many small steps by photons for which $\tau_y \gg 1$. This is the same assumption underlying Equation (II-58) and use may be made of this to incorporate this equation directly into the transport problem.

In order to introduce Equation (II-58) into the transport problem we divide all integrals over frequency into 'core' and 'wing' components such that

$$\int_0^\infty f\,dy = \int_0^{y_1} f\,dy + \int_{y_1}^\infty f\,dy = \int_c f\,dy + \int_w f\,dy, \qquad \text{(VIII-97)}$$

where y_1 is defined by

$$\int_0^{\tau_0} \Phi_{y_1}\,d\tau_0 = \gamma. \qquad \text{(VIII-98)}$$

The value of γ remains to be chosen, but for many cases $\gamma=1$ gives good results. Rybicki (1971) notes that to the extent that $P(e)$ is zero in the core region $dI_\nu/d\tau_\nu \approx 0$ and, thus, $I_\nu = S_L$ (Equation (I-2)). Also, I_ν may be expected to be isotropic so $J_\nu = I_\nu$ and $J_\nu = S_L$. It follows that

$$\int_c \Phi_y J_y\,dy = S_L \int_c \Phi_y\,dy = [1 - P(e)]\,S_L, \qquad \text{(VIII-99)}$$

where use has been made of Equation (II-58). The effect of this approximation on the transport problem is immediately seen by substituting Equation (VIII-99) into Equations (II-31) and (II-16). This gives

$$S_L = \frac{\int_w J_y \Phi_y\,dy + \varepsilon^* B}{P(e) + \varepsilon^\dagger} \qquad \text{(VIII-100)}$$

and

$$\varrho = P(e) - \frac{\int_w J_y \Phi_y\,dy}{S_L}$$
$$= \varepsilon^* \frac{B}{S_L} - \varepsilon^\dagger \qquad \text{(VIII-101)}$$

Thus, we have reduced each of the terms in S_L and ϱ to where they are of comparable magnitude and we have largely eliminated the strong cancellation effects present in the original equations. Rybicki (1971) refers to this as 'preconditioning.'

An equivalent way of stating the assumption that $S_L = J_y$ in the core region is to write

$$\int \frac{\Phi_y}{\phi_y + r_0} \frac{dH_y}{d\tau}\,dy = \int_c \frac{\Phi_y}{\phi_y + r_0} \frac{dH_y}{d\tau_0}\,dy + \int_w \frac{\Phi_y}{\phi_y + r_0} \frac{dH_y}{d\tau_0}\,dy$$

and (cf., Equation (II-43))

$$\left(\frac{1}{\phi_y + r_0}\frac{dH_y}{d\tau_0}\right)_c = \left(J_y - \frac{\phi_y S_L}{\phi_y + r_0} - \frac{r_0}{\phi_y + r_0}B\right)_c$$

$$= \left[J_y - S_L + \frac{r_0}{\phi_y + r_0}(S_L - B)\right]_c = \left[\frac{r_0}{\phi_y + r_0}(S_L - B)\right]_c.$$

(VIII-102)

Thus, we may replace

$$\int_c \frac{\Phi_y}{\phi_y + r_0}\frac{dH_y}{d\tau_0}dy$$

in Equation (VIII-1) with $\delta_c(S_L - B)$, where

$$\delta_c = \int_c \frac{\Phi_y r_0}{\phi_y + r_0}dy$$

(VIII-103)

$$= \delta - \delta_w.$$

This gives

$$S_L = \frac{\varepsilon^* + \delta_w}{\varepsilon^\dagger + \delta_w}B + \frac{1}{\varepsilon^\dagger + \delta_w}\int_w \frac{\Phi_y}{\phi_y + r_0}\frac{dH_y}{d\tau_0}dy.$$

(VIII-104)

Thus Equation (VIII-1) retains its same form except that the flux divergence term and the continuum destruction probability term are evaluated in the wing region only.

Equation (VIII-104) can be solved by exactly the same technique as was discussed for the solution of Equation (VIII-1). Similarly, the differential equations may be solved, with appropriate redefinition of terms, without changing the form of the equations. However, Rybicki (1971) notes that to follow this approach results in only a modest savings in computer requirements. The real advantage of the method lies in the preconditioning of the equations. This makes it practical to solve Equations (VIII-100) and (VIII-101) together with the steady state equations by direct iteration. One may, for example, use the following procedure:

(a) Assume an initial set of values for ϱ_{ji}.
(b) Solve the steady state equations for τ_{ji}, ε_{ji}^* and $\varepsilon_{ji}^\dagger$.
(c) Solve Equation (VIII-100) for S_L using some appropriate expression for $J_\nu(S_\nu, \tau_\nu)$.
(d) Solve Equation (VIII-101) for a new set of ϱ_{ji} and repeat steps (a) to (d) until convergence is achieved.

Rybicki (1971) has tested a similar approach for the two level case and has found rapid convergence (10 to 20 iterations) to the correct solutions. Since the direct iterative method does not involve matrix inversion and eliminates the contributions to J_y in the line core, the use of machine core storage is minimal and each iteration proceeds much faster than when matrix inversion is used.

The core saturation method has yet to be tested for multilevel problems and for problems where gradients in Doppler width, temperature, and density are present. Nevertheless, the prospects for the method appear to be excellent.

References

Athay, R. G.: 1970, *Extended Atmosphere Stars*, NBS Special Publication No. 332, 179.
Athay, R. G. and Canfield, R. C.: 1970, *Extended Atmosphere Stars*, NBS Special Publication No. 332, 65.
Athay, R. G. and Skumanich, A.: 1967, *Ann. Astrophys.* **30**, 669.
Auer, L.: 1967, *Astrophys J. Letters* **150**, L53.
Auer, L. H. and Mihalas, D.: 1969, *Astrophys. J.* **158**, 641.
Avrett, E. H.: 1970, *Atlas Symposium No. 3*, Interdisciplinary Symposium on Applications of Transport Theory, Oxford.
Avrett, E. H. and Loeser, R.: 1963, *J. Quant. Spectr. Radiative Transfer* **3**, 201.
Avrett, E. H. and Loeser, R.: 1969, Smithsonian Astrophys. Obs. Spec. Report No. 303.
Bohm, K. H.: 1960, in J. L. Greenstein (ed.), *Stellar Atmospheres*, Univ. of Chicago, Chicago, see p. 88.
Domenico, B. A.: 1971, in press.
Feutrier, P.: 1964, *Compt. Rend. Acad. Sci. Paris* **258**, 3189.
Kalkofen, W.: 1968a, *Resonance Lines in Astrophysics*, National Center for Atmospheric Research, Boulder, Colorado.
Kalkofen, W. 1968b, *Resonance Series in Astrophysics*, National Centre for Atmosphere Research, Boulder, Colorado.
Kourganoff, V.: 1963, *Basic Methods in Radiative Transfer*, Dover, New York.
Mihalas, D.: 1970, *Stellar Atmospheres*, Freeman, San Francisco, p. 409.
Rybicki, G.: 1970, *Atlas Symposium No. 3*, Interdisciplinary Symposium on Applications of Transport Theory, Oxford.
Rybicki, G.: 1971, *Line Formation in Magnetic Fields*, National Center for Atmospheric Research, Boulder, Colorado.
Skumanich, A.: 1971, in press.
Skumanich, A. and Domenico, B. A.: 1970, *Atlas Symposium No. 3*, Interdisciplinary Symposium on Applications of Transport Theory, Oxford.

INDEX OF SUBJECTS

Absorbtivity 3, 12
Asymmetric profiles 134, 144
Atom-atom collisions 92, 96

Backwarming 203, 206, 207, 212, 228, 232, 233
Bell-shaped profiles 136, 152
Blanketing effect 51, 203, 204, 224, 225, 227, 228, 230, 232, 234, 235, 254

Cayrel mechanism 234, 235
Chromosphere 45, 46, 47, 64, 65, 89, 91, 125, 132, 133, 141, 142, 144, 145, 146, 151, 152, 181, 188, 190, 228, 230, 232, 233, 234
Coherent scattering 3, 10, 11, 58, 60, 62, 79, 145,188
Collisionally controlled lines 88, 89, 90, 91
Concavity in line wings 138, 139, 140, 144, 151
Congruent profiles 7, 142, 143, 144
Consistency checks 9, 81, 83, 84, 85, 86, 109
Continuum coupling 96, 152
Contribution function 122, 123
Crossover in line wings 140, 141
Curve-of-growth 25, 51, 176, 186, 187, 188, 189, 190, 192, 193, 210, 211, 221, 227, 228, 230, 232, 234
Cylinders 67, 69, 70, 71, 72, 73, 74, 75, 76, 77

Degradation length 27, 29, 32, 79, 80, 106, 192
Destruction probability 24, 27, 260
Dipole scattering 63
Doublets 30, 92, 94, 95, 97, 98, 99, 100, 102, 116, 151, 152, 156, 168, 169, 195, 196, 225

Eddington-Barbier relation 121, 122, 123, 124, 125, 126, 128, 129, 130, 132, 138, 141, 142, 157
Effectively thick 31
Effectively thin 31, 32, 47, 64, 65, 69, 72, 194, 195
Emission reversals 45, 178, 188
Emissivity 3, 12
Equivalent widths 25, 131, 176, 177, 186, 188, 206, 210, 214, 226, 227, 232
Escape coefficient 32, 33, 34, 62, 80, 224, 225 258
Escape probability 22, 24, 28, 29, 34

Excitation temperature 101, 103, 108, 113, 172, 179, 180, 181, 182, 183, 186, 187, 190

Fishbone effect 180
Free-bound continuum 35, 36, 81, 86, 114, 227, 247, 248
Frequency dependence 6, 7, 10, 13, 57, 125, 164
Frequency redistribution 55, 57, 58, 60

Imbedding 69, 71, 76
Interlocking 18, 27, 29, 30, 32, 33, 34, 43, 79, 80, 81, 114, 117, 119, 178, 204, 212, 213, 224, 225, 228

Kinetic equilibrium 3, 33, 144, 146, 147, 237, 244, 249, 253, 258

Limb-darkening 43, 130, 141, 142, 144, 145, 157, 158, 175, 176, 179, 181
Linearization 249, 258
Line blocking 202, 227
Line core 11, 23, 43, 44, 53, 57, 61, 76, 122, 125, 126, 131, 132, 136, 137, 142, 146, 151, 152, 153, 156, 160, 164, 165, 171, 172, 188, 259
Line shoulder 156, 172, 173, 174, 175
Line wings 11, 23, 26, 42, 43, 44, 51, 53, 61, 76, 122, 126, 136, 137, 138, 139, 140, 144, 152, 164, 165, 171, 172, 188, 259
Local thermodynamic equilibrium 2, 3, 9, 10, 11, 14, 33, 79, 119, 144, 147, 175, 177, 178, 179, 180, 181, 188, 194, 203, 204, 216, 221, 224, 225, 227, 235, 245, 248

Macroturbulence 130, 131, 132, 133
Mean-free-path 27
Microturbulence 130, 131, 133, 135, 145, 151, 181, 189, 190
Milne-Eddington Model 202, 203, 206, 216, 221, 222
Monte Carlo techniques 66
Multiplets 6, 18, 22, 92, 96, 99, 100, 152, 153, 156, 166, 187,

Net radiative bracket 33
Newton-Raphson technique 249, 252, 258
Non-coherent scattering 3, 11, 57, 58, 59, 60, 61, 62, 63, 79, 203
Number of scatterings 28, 34

Optical pumping 103, 104, 109, 111, 113, 114, 116, 117, 119

Periodic structure 67, 68, 70
Photoelectrically controlled lines 88, 89, 90, 91
Photon escape 20, 23, 31, 48, 53, 54, 65, 80, 108, 191
Photon sinks 18, 28, 30, 41, 79, 86, 88, 89
Photon sources 18, 79, 86, 88, 89
Pure absorption 2, 3, 10, 188, 203
Pure scattering 2, 3, 10, 11, 33, 203, 222

Radiative equilibrium 3, 44, 146, 222, 227, 234
Random walk 27, 28, 29, 48
Recoil 62
Roughening effect 141, 145, 159

Scaling laws 31, 46, 47, 48, 65, 69, 70, 74, 75, 82
Schuster-Schwarzschild model 50, 206, 207, 216, 221
Self-reversal 45, 91
Spectrum synthesis 6, 144, 145, 146
Stimulated emissions 12, 16, 33, 36, 62, 83, 84, 87, 88, 101, 190
Surface cooling 207, 211, 212, 222, 225, 226, 227, 228, 232, 235
Surface heating 212, 225, 226, 227, 235

Thermalization length 27, 28, 29, 30, 32, 40, 42, 43, 46, 48, 64, 65, 71, 72, 73, 79, 80, 82, 95, 98, 102, 103, 106, 116, 145, 178, 193, 195, 197
Triplets 99, 100, 140, 225

GEOPHYSICS AND ASTROPHYSICS MONOGRAPHS
AN INTERNATIONAL SERIES OF FUNDAMENTAL TEXTBOOKS

Editor:

BILLY M. MCCORMAC (Lockheed Palo Alto Research Laboratory)

Editorial Board:

R. GRANT ATHAY (High Altitude Observatory, Boulder)
P. J. COLEMAN, JR. (University of California, Los Angeles)
D. M. HUNTEN (Kitt Peak National Observatory, Tucson)
J. KLECZEK (Czechoslovak Academy of Sciences, Ondřejov)
R. LÜST (Institut für Extraterrestrische Physik, Garching-München)
R. E. MUNN (Meteorological Service of Canada, Toronto)
Z. ŠVESTKA (Fraunhofer Institute, Freiburg i. Br.)
G. WEILL (Institut d'Astrophysique, Paris)

2. J. Coulomb, *Sea Floor Spreading and Continental Drift*. 1972, X+184 pp.

Forthcoming Titles:

3. G. T. Csanady, *Turbulent Diffusion in the Environment*
4. F. E. Roach and Janet L. Gordon, *The Light of the Night Sky*
5. R. Grant Athay, *The Solar Chromosphere and Corona*
6. L. Culhane, *Introduction to X-Ray Astronomy*
7. Z. Kopal, *The Moon*
8. Z. Švestka and L. De Feiter, *Solar High Energy Photon and Particle Emission*
9. A. Vallance Jones, *The Aurora*

RAYMOND H. FOGLER LIBRARY
DATE DUE

BOOKS ARE SUBJECT TO
TER TWO WEEKS

1988